Explosive Ferroelectric Generators

From Physical Principles to Engineering

Explosive Ferroelectric Generators

From Physical Principles to Engineering

Sergey I Shkuratov

Loki Incorporated, USA

World Scientific

NEW JERSEY · LONDON · SINGAPORE · BEIJING · SHANGHAI · HONG KONG · TAIPEI · CHENNAI · TOKYO

Published by

World Scientific Publishing Co. Pte. Ltd.

5 Toh Tuck Link, Singapore 596224

USA office: 27 Warren Street, Suite 401-402, Hackensack, NJ 07601

UK office: 57 Shelton Street, Covent Garden, London WC2H 9HE

Library of Congress Control Number: 2019945885

British Library Cataloguing-in-Publication Data
A catalogue record for this book is available from the British Library.

EXPLOSIVE FERROELECTRIC GENERATORS
From Physical Principles to Engineering

ISBN 978-981-3238-93-0

For any available supplementary material, please visit
https://www.worldscientific.com/worldscibooks/10.1142/10958#t=suppl

Desk Editor: Nur Izdihar Binte Ismail

Typeset by Stallion Press
Email: enquiries@stallionpress.com

This book is dedicated to
Larry L. Altgilbers,
dynamic advocate for explosive pulsed power advancement

Preface

Poled ferroelectrics are widely used in modern electromechanical devices. Low-power ferroelectric transducers are extensively employed in marine sonar systems, ultrasound diagnostic imaging, actuators for nano-positioners, ultrasonic motors, active scalpels and numerous electro-optical devices. There is a significant interest in expanding this usage to high-power systems. High-power systems require from four to six orders of magnitude higher power than the hundreds of miliwatts associated with low-power transducers. An increase of power to the kilowatt level opens the way to the utilization of high-power transducers for therapeutic applications, specifically for high-intensity focused ultrasound therapy. A further increase of power to the megawatt level makes it possible to use poled ferroelectrics in pulsed power applications.

Explosive ferroelectric generators (FEGs) are rooted in the 1950s when the ability of shocked ferroelectric ceramics to generate high current and high voltage pulses was demonstrated for the first time at the Sandia National Laboratories. Opposite to the operation of ferroelectric materials in the low-strain mode that is piezoelectric in nature with no large-scale disorientation of the ferroelectric domains, the operation of ferroelectrics in the ultrahigh-power mode under high mechanical stress and high strain rate loading induces phase transitions and domain reorientation with possible loss of the initial remnant polarization. These transformations result in the generation of megawatt power for a brief interval of time. However, a possible change in initial polarization and piezoelectric properties limits this mode of operation in applications requiring high frequency, such as resonant devices. This raises a question: where can these systems be used? As is shown in this book, it is possible to develop really compact (even miniature), lightweight, autonomous ferroelectric generators that are capable of producing pulses of kiloamperes of current, hundreds of kilovolts of electric potential and gigawatt peak power microwave radiation. These systems can be used in applications that require portability, compactness, ultrahigh power and a limited number of events.

The research and development of ultrahigh-power ferroelectric systems has been cyclic with the most recent return starting in 1998 with a series of projects funded by the U.S. Air Force Office of Scientific Research. From 1998 to 2004 the author of this book was involved in the research of ferroelectric generators at Texas Tech University. From 2004 to the present time the author has continued this research with Loki Incorporated under Small Business Innovative Research programs. The aims of this research include both the development of practical devices and systematic studies of the behavior of ferroelectrics under high mechanical stress in order to obtain fundamental knowledge about the ultimate energy density that can be harvested from these materials and the possible limitations on operation in the ultrahigh-power mode. During the last twenty years there have been significant advances in the development of ultrahigh-power ferroelectric systems in the U.S., which are reported in this book. It reports mainly on advances made in the U.S. because these are more familiar to the author. Advances continue to be made in the U.S., China, Russia, the U.K., Germany, South Korea and other countries.

The objective of this book is to acquaint the reader with ferroelectric generators, the principles of their operation and their applications. This book will help the reader to understand the fundamentals of this type of ferroelectric system and can be used as a guide to the research and development of ultrahigh-power ferroelectric generators. This book is unique in that all experimental results presented herein were obtained with generators that had exactly the same designs and contained the same types of ferroelectric materials. To make this a self-contained book, the introduction to ferroelectric materials is presented in Chapter 1. Chapter 2 looks at the properties of lead zirconate titanate ferroelectric ceramics. Chapter 3 provides a historical review of shock studies involving ferroelectric materials. The physical principles of ferroelectric generators are discussed in detail in Chapter 4. The designs of different types of explosive ferroelectric generators are described in Chapter 5. Chapters 6 through 12 are an in-depth look at the ferroelectric generators utilizing (1) ferroelectric ceramic materials, (2) ferroelectric single crystals and (3) ferroelectric films. The applications of ferroelectric generators are presented in Chapters 13 through 17.

There are many people who made this research and this book possible. First and foremost is Larry L. Altgilbers (1945–2013), who as the Program Manager at the U.S. Army Space Missile Defense Command supported this research for many years. Larry also co-authored the chapter on the historical review of shock studies of ferroelectric materials. Others include Magne Krisitansen (1932–2017) at Texas Tech University and Robert Barker (1941–2013) at the Air Force Office of Scientific Research, who supported the basic research in explosive pulsed power, and Allen H. Stults (1957–2014) at the U.S. Army Aviation and Missile Engineering Center and Mark Rader at the U.S. Army Space Missile Defense Command, who supported this FEG research. I would like to thank Jason Baird and Vladimir G. Antipov at Loki Incorporated for their engineering, experimental and technological support, their valuable contribution to the development of ferroelectric generator designs, their ideas and their helpful discussions of our results. I would like to thank Wesley Hackenberger, Edward F. Alberta and Jun Luo at TRS Technologies, who provided ferroelectric materials for this research and valuable insights to the chapters on ferroelectrics and their properties. I would also like to thank Jay B. Chase at J.B. Chase Consulting, who provided a numerical simulation of stress distribution in Loki ferroelectric generators. In addition, I would like to acknowledge Evgueni F. Talantsev at the University of Victoria in Wellington and Christopher S. Lynch at University of California in Los Angeles for helpful discussions of the results and ideas. I would like to thank my family, Olga, Alexander and Sofia for their support.

Sergey I. Shkuratov

Contents

Chapter 2   **Lead Zirconate Titanate**
              Ferroelectric Ceramics   **25**

Chapter 6   Mechanisms of Transverse Shock
**   Depolarization of PZT 95/5**
**   and PZT 52/48   127**

Chapter 7   High-Current Generation by
**   Shock-Compressed Ferroelectric**
**   Ceramics   145**

Chapter 1

Ferroelectric Materials and Their Properties

1.1 Introduction

Explosive ferroelectric generators produce pulses of high voltage, high current and high power. A ferroelectric element of an FEG combines a few stages of a conventional pulsed power system in one, i.e. a prime power source, a high-current generator, a high-voltage generator, and a capacitive energy storage device. The properties of ferroelectric materials are essential for understanding the operation of ferroelectric generators. In this chapter, the fundamental properties of ferroelectric materials are examined. This is not an extensive review, but rather an introduction to those properties of ferroelectric materials that are important to ferroelectric generators. Since Lead Zirconate Titanate (PZT) is the material that has been extensively employed in the explosive ferroelectric generators, a detailed description of the properties of PZT ferroelectric ceramics is presented in Chapter 2. The operation of FEGs relies on the shock compression of ferroelectric materials, which has been under study since the end of the 1950s. Therefore, a brief review of the extensive literature on the shock compression of ferroelectric materials is presented in Chapter 3.

1.2 Spontaneous Polarization

Several discoveries that were made at the end of the nineteenth century looked like curiosities at that time, but now they are the basis for numerous modern engineering applications. In 1880, Pierre and Jacques Curie observed that when certain crystalline minerals are subjected to mechanical stress, an electric charge accumulates at their surface forming an electric field across the bulk of the materials. In 1881, Hankel called this phenomenon "piezoelectricity" from the Greek word *piezein*, meaning to press or squeeze [1]. Piezoelectricity is a linear reversible process. The magnitudes of piezoelectric movements, voltages produced by natural materials (tourmaline crystals, quartz) are small. However, these materials were used in the first electromechanical systems during World War I, piezoelectric ultrasonic transducers for submarine detection.

For many years natural crystals were the exclusive source of piezoelectric capabilities and many types of devices were developed with these materials. During World War II, intensive studies of piezoelectric materials were performed in the U.S., Japan and Russia [2, 3]. At the beginning of the 1940s, man-made materials, ferroelectric ceramics prepared from mixed metal oxides (TiO_2 and BaO) which formed barium titanate, $BaTiO_3$, were discovered.

Unlike piezoelectricity, ferroelectricity is the complex interaction of the dielectric and elastic properties of highly nonlinear, anisotropic, polarizable, deformable crystals [3]. The roots of ferroelectricity can be traced back to the work of Valasek in the 1920s [4–9], who at the time was investigating the piezoelectric properties of Rochelle salt (potassium sodium tartrate), which was first produced by P. Seignette in La Rochelle, France, in 1655. Valasek was the first to use the term "Curie Point" to describe the onset of polar ordering in Rochelle salts below certain temperatures. The anomalous properties of Rochelle salt, i.e., extremely high dielectric and piezoelectric responses, were for a considerable time called Seignette electricity. The term ferroelectricity was not commonly used until the early 1940s.

In the 1950s, Japanese researchers began to investigate the properties of lead titanate and lead zirconate and their mixtures [3]. This resulted in the formulation of the ferroelectric ceramic materials based on these compounds, notably lead zirconate titanate that exhibited greater sensitivity and a higher operating temperature relative to barium titanate ceramics. Ferroelectric ceramics can be hundreds of times more sensitive to mechanical or electrical input than natural crystalline materials, and the composition, shape, and dimensions of ceramics can be tailored to meet the requirements of a specific purpose. These materials enabled designers to employ the piezoelectric effect in many new applications.

Barium titanate and lead zirconate titanate have a perovskite crystal structure [10], each unit cell of which consists of a small, tetravalent metal ion, usually titanium or zirconium, in a lattice of large, divalent metal ions, usually lead or barium, and O^{2-} ions. Figure 1.1 shows schematic diagrams of a $BaTiO_3$ unit cell at temperatures above and below Curie point.

The high-temperature (above Curie point) cubic phase (Figure 1.1(a)) is easier to describe, as it consists of regular corner-sharing octahedral TiO_6 units that define a cube with O vertices and Ti-O-Ti edges. In the cubic phase, Ti^{4+} is located in the center of the cube. The unit cell is charge neutral. The cubic phase does not exhibit the ferroelectric effect.

The lower symmetry tetragonal phase is stabilized at temperatures below Curie point and involve the movement of the Ti^{4+} to an off-center position (Figure 1.1(b)). At ambient conditions, each crystal has a net dipole moment in the absence of an external electric field. This net dipole moment is due to the center of positive charge in the crystal not coinciding with the center of the negative charge due to its crystalline structure (Figure 1.1(b)). The polarization induced by a phase transformation (in this case from the cubic to the tetragonal phase) is referred to as the *spontaneous polarization, P_s.* The remarkable properties of this material arise from the cooperative behavior of the ions in the unit cell.

Ferroelectrics can be defined as polar materials that have at least two equilibrium orientations for the spontaneous polarization vector in the absence of an external electric field and that can have their spontaneous polarization switched between these two equilibrium orientations by an applied external electric field [11]. The name "ferroelectric" was given to these materials because their electrical behavior is analogous to the magnetic behavior of

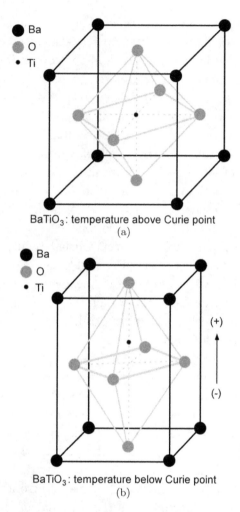

Fig. 1.1. Schematic diagrams of the crystal structure of a BaTiO$_3$ ferroelectric ceramic. (a) The high-temperature cubic phase (temperature above Curie point). Symmetric arrangement of positive and negative charges, charge neutrality. (b) The lower-temperature tetragonal phase (temperature below Curie point). The Ti^{4+} ion is off-center; the crystal has an electric dipole.

ferromagnetic materials. If an electric field is applied to a ferroelectric material and then slowly reversed and plotted against the resulting change in polarization of the material, a hysteresis loop is generated (Figure 1.2), much like for ferromagnetic materials.

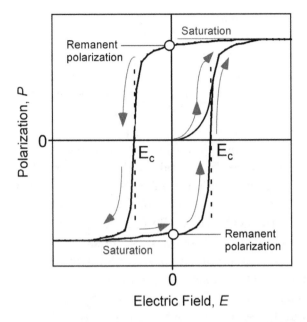

Fig. 1.2. Effect of electric field (E) on electric polarization (P) of ferroelectric material (P-E hysteresis curve or hysteresis loop). E_c is the coercive field for a ferroelectric material.

1.3 Ferroelectrics and Piezoelectrics

The piezoelectric effect is observed when a mechanical stress induces electric polarization in a material due to the distortion of the unit cells of the crystal. The direct piezoelectric effect is a linear reversible process, where the magnitude of the polarization is dependent on the magnitude of the stress and the direction of the polarity is dependent on the type of stress, i.e. tensile or compressive.

The direct piezoelectric effect is always accompanied by an inverse piezoelectric effect, where a solid is strained when placed in an external electric field. The inverse piezoelectric effect is defined to be primarily an electromechanical effect, i.e. the strain is proportional to the electric field.

A necessary condition for the piezoelectric effect to occur is that there has to be a lack of a center of symmetry in the material's crystalline structure. There are 32 classes or point groups of crystals of which 11 have a center of symmetry (centrosymmetric) and 21 do not have a center of symmetry. When there is a lack of symmetry, the net movement of positive and negative ions with respect to each other as a result of stress will produce an electric dipole. When there is a centrosymmetry, the centers of the positive and negative charges will still coincide, even after deformation of the crystal due to stress, and the unit cell will be charge neutral.

Of the 21 non-centrosymmetric crystal classes, 20 are piezoelectric. Of the 20 classes of crystals that are piezoelectric, 10 have a unique polar axis and are called polar crystals. Even when no external stress is being applied, those 10 classes of crystals with unique polar axes have a permanent electric dipole moment within their unit cells and are, thus, said to be spontaneously polarized.

The other 10 classes of piezoelectric crystals do not have polar axes. When no external stress is being applied, these 10 classes of crystals do not have a permanent electric dipole moment within their unit cells and they are not spontaneously polarized. These 10 classes of crystals possess piezoelectric effect but they are not ferroelectric.

Spontaneous polarization, P_s, is defined to be the magnitude of the dipole moment per unit volume or the magnitude of the electrical charge per unit area on the surface perpendicular to the axis of spontaneous polarization.

The magnitude of the spontaneous polarization depends on temperature. The temperature dependence of spontaneous polarization is called the *pyroelectric effect* and crystals that exhibit this property are called *pyroelectrics*. Changing the temperature of a pyroelectric changes the value of its electric dipole moment, i.e. the dipole moment is directly proportional to the temperature change.

A subgroup of the pyroelectric classes of crystals are the *ferroelectrics*. As noted earlier, ferroelectrics are polar materials that have at least two equilibrium orientations for the spontaneous polarization vector in the absence of an external electric field and that can have their spontaneous polarization switched between these two equilibrium orientations by an applied external electric field. Crystalline ferroelectric materials can be either single crystals or polycrystalline ceramics, the latter of which are currently in wide use.

A *ferroelectric crystal* is a dielectric material that has a net dipole moment even in the absence of an external electric field [12, 13]. This net dipole moment is due to the center of positive charge in the crystal not coinciding with the center of the negative charge due to its crystalline structure, which, as noted earlier, is referred to as spontaneous polarization (Figure 1.1(b)).

If an electric field is applied to a ferroelectric material and then slowly reversed and plotted against the resulting change in polarization of the material, a hysteresis loop is generated as is shown in Figure 1.2. Once a ferroelectric material has been polarized to saturation and the polarizing electric field is removed, the remaining polarization is called the *remanent polarization*, which is an important property relative to the FEG operation.

Piezoelectrics will not form a hysteresis loop and the electric moment of these materials is not affected by the application of an external electric field because of the absence of spontaneous polarization.

The difference between ferroelectric and piezoelectric materials can be summarized as follows: the ferroelectrics have spontaneous polarization and form a hysteresis loop, but piezoelectrics do not. The constitutive response of these two materials can be divided into two types. The response of a piezoelectric material to external mechanical or electrical stresses is linear, reversible, and includes the time-independent displacement of ions within a unit cell.

The constitutive response of ferroelectric materials to external mechanical stress and external electrical fields is nonlinear, irreversible, and time dependent, since the irreversible switching of the spontaneous polarization and/or movement of the domain walls occurs in an incremental fashion.

All ferroelectrics are also piezoelectric, since applied stress changes their electric polarization; similarly, an external electric field causes the material to become strained. Ferroelectric ceramics are hundreds of times more sensitive to mechanical or electrical input than natural crystalline piezoelectric materials.

However, a material may be piezoelectric without being ferroelectric. For instance, quartz is piezoelectric, but not ferroelectric, whereas barium titanate is both. All ferroelectric crystals are also pyroelectric.

The parameters of FEGs are closely related to the elastic, dielectric, and ferroelectric properties of the materials from which the ferroelectric elements of FEGs are made. Since the discovery of the ferroelectric effect, it has been found that many materials exhibit ferroelectric properties. However, only a few of these materials are suitable for practical applications. Their suitability is determined by the relationship between their mechanical stresses/deformations and their fundamental electrical parameters.

1.4 Paraelectric, Ferroelectric and Antiferroelectric States

Ferroelectricity usually disappears above a certain temperature called the *transition temperature* or *Curie temperature* (*Curie point*). When the temperature of a ferroelectric material exceeds the Curie temperature, the material is no longer ferroelectric and is said to be *paraelectric*. The high-temperature (above Curie point) cubic phase of barium titanate is shown schematically in Figure 1.1(a). The unit cell does not exhibit the ferroelectric effect.

When cooled through their transition temperature, ferroelectrics undergo a phase transformation into a *ferroelectric phase*. In the case of barium titanate the cubic phase transforms into the tetragonal phase as the material is cooled through the Curie temperature (see Figure 1.1(b)). The source of the ferroelectric effect is the shift of the central ion upward relative to the surrounding oxygen ions. Each unit cell has an electric dipole.

The unit cell is the smallest repeating unit in the crystal structure. Its behavior gives rise to electromechanical coupling in ferroelectric materials and is the source of the phenomena that occur at higher length scales. Ferroelectric material behavior at this length scale is governed by interatomic potentials, thermal energy, and the external electric field.

Like ferromagnetic materials, ferroelectric materials tend to form domains in which the unit cell dipoles that make up the domain are aligned in parallel to make the domain polar. Domains are observed at the next length scale relatively to the unit cell.

Grains are single crystal regions within a ceramic. All of the unit cells within a grain have the same alignment, yet the grains are subdivided into domains.

Dipole moments are oriented differently among different ceramic grains, or even among different regions within a single grain. Each domain carries a net dipole moment. However, because the dipoles are randomly oriented in the bulk ceramic specimen, as manufactured, the domains are randomly oriented and the specimen has no overall polarization (Figure 1.3(a)).

The ceramic specimen is polarized by exposing it to a high-DC electric field (Figure 1.3(b)), usually, but not exclusively, at a temperature slightly below the Curie point. Through this polarizing procedure, the domains most nearly aligned with the electric field expand at the expense of domains that are not aligned with the field, and the ceramic lengthens in the direction of the field.

When the electric field is removed, most of the dipoles are locked into this configuration of near alignment (Figure 1.3(c)). This gives the material a permanent polarization, called the *remanent polarization* (P_r or P_0), and a permanent deformation (elongation) that makes it anisotropic — its properties differ, according to the direction in which they are measured. Analogous to the corresponding characteristics of ferromagnetic materials, a polarized ferroelectric material exhibits hysteresis (see Figure 1.2), and dielectric constants for these materials are very high, and temperature dependent.

Crystals where polar domains are aligned anti-parallel; i.e., where half the domains are aligned with their dipoles oriented in one direction and the other half are aligned with their dipoles oriented in the opposite direction and where adjacent domains are oppositely polarized, so that there is a net center of symmetry and there is no net polarization, are called *antiferroelectrics* (Figure 1.4). This anti-polar orientation arises because the dipole-dipole interaction energy is lowered. In anti-polar crystals, where the free energy of the anti-polar state does not differ appreciably from that of the polar state, externally applied mechanical stresses or electric fields can cause the

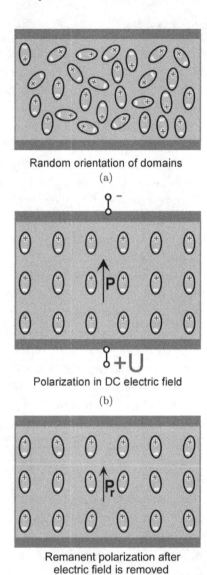

Random orientation of domains

(a)

Polarization in DC electric field

(b)

Remanent polarization after
electric field is removed

(c)

Fig. 1.3. Polarizing a ferroelectric ceramic. (a) Random orientation of domains prior to polarization. (b) Polarization in DC electric field. (Simplified for clarity. Polarization does not eliminate the random distribution of grains in the ceramic. A small percentage of grains will be constrained from aligning in the direction of polarization.) (c) Remanent polarization after electric field is removed. P_r is the remanent polarization vector.

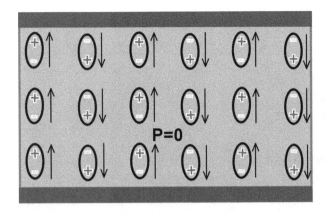

Fig. 1.4. Schematic diagram of an antiferroelectric material.

transition of the material to a parallel state (a ferroelectric state). Of course, the opposite is true as well, i.e. a stress can also cause a ferroelectric material to transition into an antiferroelectric state.

1.5 Ferroelectric and Piezoelectric Figures of Merit

There are several parameters that are used to specify the properties of ferroelectric and piezoelectric materials [12]. These parameters will be discussed in this section. The parameters of ferroelectric materials and their standard SI units are listed in Table 1.1.

1.5.1 *Remanent polarization*

When an external electric field is applied across a ferroelectric material, the dipoles within the material become increasingly aligned until no further polarization is observed. The material is then fully polarized or saturated. If the external field is reduced to zero, the dipoles become less strongly aligned; however, they do not return to their original alignment, since there are several preferred alignment directions within the ferroelectric crystal. Therefore, there is still a very high degree of alignment and the material remains polarized at some value lower than the saturation polarization (see Figures 1.2 and 1.3). The polarization at this point is called the *remanent polarization*, usually denoted by P_r or P_0. It determines the maximum amount of charge that can be coupled from the FEG into a load. The

Table 1.1.   Parameters of ferroelectric and piezoelectric materials and their standard SI units.

Parameters	Symbols	Units
Stress	T	N/m^2
Polarization	P	C/m^2
Remanent polarization	P_r or P_0	C/m^2
Curie temperature	T_C	$°C$ (degree Celsius)
Permittivity	ε	F/m
Relative permittivity or dielectric constant	ε_r or K	None
Coercive field	E_c	V/m
Breakdown field or dielectric strength	E_{break}	V/m
Acoustic impedance	Z_{mat}	$Pa \cdot s/m^3$
Elastic compliance	s	m^2/N
Young's modulus	Y	N/m^2
Piezoelectric charge constant	d	C/N
Piezoelectric voltage constant	g	$V \cdot m/N$
Electromechanical coupling factor	k	None

standard SI unit for the remanent polarization is the Coulomb per square meter (C/m^2). It is also common to see related units such as the micro-Coulomb per square centimeter ($\mu C/cm^2$).

1.5.2   *Curie temperature*

Ferroelectricity disappears above a certain temperature called the *transition temperature* or *Curie temperature*, T_C. When the temperature of a ferroelectric material exceeds the Curie temperature, the material goes into a paraelectric phase. The Curie temperature is also the temperature at which the relative permittivity (dielectric constant) of a ferroelectric material peaks. The standard unit for the Curie temperature is the degree Celsius ($°C$). The typical Curie temperature of ferroelectric materials is a few hundred degrees Celsius.

1.5.3   *Absolute permittivity and relative permittivity/ dielectric constant*

Absolute permittivity, often called *permittivity*, usually denoted by the Greek letter ε, is a physical quantity that describes how an electric field affects and is affected by a dielectric medium. It is a measure of the polarizability of a material. It is determined by measuring the ability of a material to polarize in response to an electric field, and, thereby, reduce the field inside the material. Thus, permittivity relates to a material's ability to transmit (or "permit") an electric field.

The permittivity is that property of a dielectric material that determines how much electrostatic energy can be stored per unit of volume when unit voltage is applied, and as a result it is of great importance for capacitors and capacitance calculations. For example, the permittivity determines the capacitance of the ferroelectric element, which impacts the relationship between the FEG and its load. The higher the capacitance of the FEG, the higher the capacitive load it can effectively drive. The standard SI unit for permittivity is the farad per meter (F/m).

Permittivity can be defined as a constant of proportionality that exists between the electric displacement and the electric fields. The electric displacement field, D, represents how an electric field E influences the organization of electric charges in a given dielectric medium, including charge migration and electric dipole reorientation. In the case of linear, homogeneous, isotropic materials with "instantaneous" responses to changes in the electric field, D depends linearly on the electric field:

$$D = \varepsilon \cdot E \qquad (1.1)$$

where ε is the permittivity of the substance (scalar) in Farads per meter (F/m), D is the electric flux density in Coulombs per square meter (C/m^2), and E is the electric field strength in volts per meter (V/m).

It should be mentioned that in ferroelectrics, permittivity can depend on the strength of the electric field. The value of permittivity also can vary with the position within a ferroelectric element of the FEG (whether it is in the shock-compressed zone or the uncompresseed zone).

In engineering applications, the permittivity of a dielectric medium is often expressed in relative, rather than in absolute terms, and it is represented by the ratio of its absolute permittivity to the absolute permittivity of a vacuum (also called the permittivity of free space). This dimensionless quantity is called the medium's *relative permittivity*, ε_r, also commonly referred to as the *dielectric constant*, K. It is given by:

$$\varepsilon_r = K = \varepsilon/\varepsilon_0, \qquad (1.2)$$

where ε_r is the relative permittivity (dielectric constant, K) of the substance, ε is the absolute permittivity of the substance in (F/m), and ε_0 is the absolute permittivity of a vacuum ($\varepsilon_0 = 8.85418781 \times 10^{-12}$ F/m).

1.5.4 *Breakdown field strength/dielectric strength*

The *electric breakdown field*, E_{break}, of a dielectric is the maximum electric field that it can withstand intrinsically without breaking down, i.e. without experiencing the failure of its insulating properties. It is also called the *dielectric strength*. The field at which breakdown occurs for a given case is dependent on the respective geometries of the dielectric and of the electrodes with which the electric field is applied, as well as on the rate of increase of the applied electric field. Because dielectric materials usually contain minute defects, the practical breakdown field will be a fraction of the intrinsic breakdown field seen as ideal, i.e. defect-free materials. The standard SI unit for the breakdown field is the volt per meter (V/m). It is also common to see related units such as the volt per centimeter (V/cm), the megavolt per meter (MV/m), or the kilovolt per millimeter (kV/mm).

1.5.5 *Acoustic impedance*

The *acoustic impedance* of a given material, Z_{mat}, is a measure of the opposition that a material presents to the acoustic flow resulting from an applied acoustic pressure. There is a close analogy with electrical impedance, which measures the opposition that a system presents to the electric current flow resulting from an electrical voltage applied to the system.

More specifically, acoustic impedance is a parameter that is used to evaluate the acoustic energy transfer between two materials and in a solid it is defined by the expression:

$$Z_{mat} = \rho \cdot U, \tag{1.3}$$

where ρ is the density of the material and U is the longitudinal velocity of the acoustic wave propagating in the material. The standard SI unit of acoustic impedance is the pascal by second per cubic meter ($\text{Pa} \cdot \text{s/m}^3$).

1.5.6 Coercive field

The coercive field, E_c, for a ferroelectric material is the electric field required to switch its polarization from its remanent polarization to zero polarization, $P = 0$. The value of E_c for a ferroelectric crystal is dependent on many parameters including its thermal and electrical history and environmental factors. The standard unit for the coercive field is the volt per meter (V/m). It is also common to see related units such as the volt per centimeter (V/cm), the megavolt per meter (MV/m), and the kilovolt per millimeter (kV/mm).

1.5.7 Elastic compliance

The *elastic compliance*, s, of a material is defined to be the strain produced per unit stress. It is the reciprocal of the modulus of the elasticity (Young's modulus) of the material. The elastic compliance has very different values, depending upon whether the electric field within the material is maintained at zero (short-circuited) or whether the electric displacement remains constant (open-circuited). The standard unit for the elastic compliance is the meter to the power of two per newton (m^2/N).

1.5.8 Piezoelectric charge constant/piezoelectric coefficient

The *piezoelectric charge constant*, commonly called the *piezoelectric coefficient*, d, is defined to be the electric polarization generated in a material per unit of mechanical stress:

$$P = d \cdot T, \tag{1.4}$$

where P is the electric polarization, d is the piezoelectric charge constant (piezoelectric coefficient), and T is the external mechanical stress.

Other definitions include the short circuit charge density developed per unit of applied stress and the strain developed per unit of electric field. It is a measure of the electric charge induced on the electrodes in response to a mechanical stress or the mechanical strain achieved when an electric field is applied across a ferroelectric material under constant stress. The standard unit for the piezoelectric charge constant is the coulomb per newton (C/N).

1.5.9 *Piezoelectric voltage constant*

When an external stress is applied to a piezoelectric material, the magnitude of the induced electric field is related to the applied stress by the *piezoelectric voltage constant*:

$$E = -(g \cdot T) + (D/\varepsilon), \tag{1.5}$$

where E is the induced electric field, g is the piezoelectric voltage constant, T is the external mechanical stress applied parallel to the axis of polarization, ($T < 0$: compressive stress; $T > 0$: tensile stress), D is the electric displacement, and ε is the permittivity of material at constant stress.

Other definitions of the piezoelectric voltage constant include the open circuit field developed by a given stress and the strain developed per unit of applied electric field.

The piezoelectric charge, d, and voltage, g, constants are related to each other by the following expression:

$$g = d/\varepsilon, \tag{1.6}$$

where d is the piezoelectric charge constant (see section above), and ε is the permittivity of the material. Materials with a high g constant are desirable when generating voltages by mechanical stress. The standard unit for the piezoelectric voltage charge constant is the volt by meter per newton (V $\cdot$ m/N).

1.5.10 *Electromechanical coupling factor*

The *electromechanical coupling factor*, k, is a measure of the ability of a ferroelectric material to convert mechanical energy into electrical energy and vice versa and is defined by:

$$k^2 = \text{(Stored Electrical Energy)}/\text{(Input Mechanical Energy)} \quad (1.7)$$

The values of the electromechanical coupling factor quoted in the manufacturer's specifications typically are theoretical maximum values. The electromechanical coupling factor is a dimensionless quantity.

1.6 Ferroelectric Materials

There are several classes of ferroelectric materials: ferroelectric single crystals, ferroelectric ceramics, ferroelectric polymers, ferroelectric ceramic-polymer composites, and ferroelectric films. A brief description of each class is given below.

1.6.1 *Ferroelectric single crystals*

Some of the most commonly used single-crystal piezoelectric and ferroelectric materials are quartz, lithium niobate ($LiNbO_3$), and lithium tantalite ($LiTaO_3$). Single-crystal ferroelectric materials are anisotropic and exhibit different material properties depending on the cut of the material and the direction of the bulk or surface wave propagation.

In the last several years, some kinds of ferroelectrics, so-called relaxor ferroelectrics, have received special attention because of their intriguing and extraordinary dielectric, ferroelectric and piezoelectric properties [14, 16–20]. Currently, these properties are explained by a model with polar nanoregions in a non-polar matrix. However, fundamental scientific understandings of the polar nanoregions have not been clear.

Relaxor-based ferroelectric single crystals, such as $(1\text{-}x)Pb(Mg_{1/3}Nb_{2/3})O_3\text{-}(x)PbTiO_3$ (PMN-PT) and $(1\text{-}y\text{-}x)Pb(In_{1/2}Nb_{1/2})O_3\text{-}(y)Pb(Mg_{1/3}Nb_{2/3})O_3\text{-}(x)PbTiO_3$ (PIN-PMN-PT) have triggered a revolution in electromechanical systems due to their superior piezo-electric properties. They are extensively employed for various sensors and actuator applications [14, 16–20]. These ferroelectric crystals are widely used in the development of high-power acoustic projectors and ultrasonic motors, underwater transducers for Navy sonars, and microfluidic and microbeam transducers. There is significant interest in expanding this usage to high-power and ultrahigh-power systems, such as high-intensity focused ultrasound therapy [21, 22], resonance-based transducers, and explosive ferroelectric generators.

Recently, there was a report [23, 24] on the shock depolarization of PIN-PMN-PT crystals. It was experimentally demonstrated that shock-compressed PIN-PMN-PT $[111]_C$ oriented single-domain crystals become completely depolarized under 3.9 GPa stress and release an electric charge density that has hit a record high of $0.48 \, C/m^2$ which is 50% higher than that for the PZT 95/5 ceramic widely used in explosive ferroelectric generators. It was found that the depolarization mechanism of PIN-PMN-PT crystals is distinct from those of PZT ferroelectric ceramics [24]. Another remarkable result is that the energy density generated in the shock-compressed crystals in the high-voltage mode is four times higher than that for PZT ferroelectric ceramics [23]. These results promise new applications of PIN-PMN-PT ferroelectric single crystals in ultrahigh-power ferroelectric generators and demonstrate a unique ability for precise control of their ferrroelectric properties through the sizing and crystallographic orientation of domains, which is not achievable with polycrystalline ceramics.

1.6.2   *Ferroelectric ceramics*

Ferroelectric ceramics are polycrystalline ferroelectric materials with a perovskite crystal structure that contain a large divalent metal ion, such as barium or lead, and a tetravalent metal ion, such as titanium or zirconium. Two common ferroelectric ceramics are barium titanate, $BaTiO_3$, and lead zirconate titanate, $PbZr_xTi_{1-x}O_3$. The PZT ferroelectric ceramics are widely used because of their excellent ferroelectric properties [10]. There are different formulations of PZT, depending on the ratios of Zr to Ti used and the amount and types of dopants added. For example, PZT 95/5 implies that the Zr:Ti ratio is 95:5. It has been found that when PZT 95/5 is modified with 2% niobium, the nominal state of this material is ferroelectric, but it is near an antiferroelectric state. A surface electric charge is readily released by shock compression, when the material transitions from its ferroelectric state to its antiferroelectric state. PZT ceramics are widely used in explosive ferroelectric generators.

In the last several years, lead-free ferroelectric ceramics such as $(K_{0.5}Na_{0.5})NbO_3$, $(Bi_{0.5}Na_{0.5})TiO_3$ (BNT) and others have been widely investigated [25, 26]. Among them, BNT-based ferroelectrics stand out for their excellent ferroelectric properties and relatively high Curie temperature. Forming a solution with $BiAlO_3$ (BA) was reported to further optimize their ferroelectric properties and increase remanent polarization. Recently, there was a report [27] on the shock depolarization of BNT-BA ferroelectric ceramics. It was experimentally demonstrated that shock-compressed BNT-BA ferroelectric ceramics became almost completely depolarized at shock pressure 8.2 GPa and released electric charge with density 0.38 C/m^2. These results promise a new application of BNT-BA ceramics in shock wave ferroelectric generators.

1.6.3 *Ferroelectric polymers*

Ferroelectric polymers [28–30] are a group of crystalline polar polymers that are also ferroelectric, meaning that they maintain a permanent electric polarization that can be reversed, or switched, in an external electric field.

When certain polymers, such as polyvinylidene difluoride (PVDF), are drawn and stretched in certain directions when being fabricated, they are transformed into a microscopically polar phase. Ferroelectric polymers tend to have small piezoelectric charge constants and high piezoelectric voltage constants, to be lightweight, and to have soft elasticity.

Ferroelectric polymers are used in acoustic transducers and electromechanical actuators because of their inherent piezoelectric response, and as heat sensors because of their inherent pyroelectric response [28–30].

1.6.4 *Ferroelectric ceramic-polymer composites*

Ferroelectric ceramic-polymer composites are composed of ferro-electric ceramics and a polymer phase. The advantages of these composites are their high coupling factors, low acoustic impedance, mechanical flexibility, and tailorability [31].

Ferroelectric composites can be viewed as intermediate materials between ferroelectric ceramics and ferroelectric polymers combining such attributes as flexibility and formality with a whole range of piezoelectric and pyroelectric properties which themselves depend both on the ceramic-polymer mixture and on the processing employed in its manufacturing.

1.6.5 *Ferroelectric films*

Ferroelectric films are used in the electronics industry for non-volatile ferroelectric random access memories, ferroelectric film capacitors, microelectromechanical systems and different types of ferroelectric transducers [32, 33].

1.7 Summary

- At ambient conditions, each unit cell of a ferroelectric material has a net dipole moment in the absence of an external electric field. This net dipole moment is due to the center of the positive charge in the crystal not coinciding with the center of the negative charge. It is referred to as spontaneous polarization. This remarkable property of ferroelectric materials arises from the cooperative behavior of the ions in the unit cell.

- Ferroelectric materials tend to form ferroelectric domains in which the unit cell dipoles that make up the domain are aligned in parallel to make the domain polar. Grains are single-crystal regions within a ferroelectric ceramic. All of the unit cells within a grain have the same alignment, yet the grains are subdivided into domains. Dipole moments are oriented differently among different ceramic grains. Because the dipoles are randomly oriented in the bulk ceramic specimen, as manufactured, the domains are randomly oriented and the specimen has no overall polarization.

- When an external electric field is applied across a ferroelectric material, the dipoles within the material become increasingly aligned until no further polarization is observed. The material is then fully polarized or saturated. If the external field is reduced to zero, the dipoles become less strongly aligned; however, they do not return to their original alignment. Therefore, there is still a very high degree of alignment and the material remains polarized at some value lower than the saturation polarization; a hysteresis loop is generated. The polarization at zero external electric field is called the remanent polarization, which is an important property relative to the ferroelectric generator operation. It determines the maximum amount of charge that can be coupled from the FEG into a load.

- Piezoelectrics do not form a hysteresis loop and the electric moment of these materials is not affected by the application of an external electric field because of the absence of spontaneous polarization.

- All ferroelectrics are also piezoelectric, since applied stress changes their electric polarization, similarly, an external electric field causes the material to become strained. Ferroelectric ceramics are hundreds of times more sensitive to mechanical or electrical input than natural crystalline piezoelectric materials.

- There are several parameters that are used to specify the properties of ferroelectric materials. One of these parameters is Curie temperature. When the temperature of a ferroelectric material exceeds the Curie temperature, the material is no longer ferroelectric and is said to be paraelectric. The high-temperature (above Curie point) cubic phase of ferroelectric materials is charge neutral and it does not exhibit the ferroelectric effect (spontaneous polarization). Therefore, the operation temperature of ferroelectric devices including ferroelectric generators should be below the Curie point.

- There are several classes of ferroelectric materials: single crystals, ferroelectric ceramics, ferroelectric polymers, ceramic-polymer composites, and ferroelectric films. All these materials can be used in ferroelectric generators. The results obtained with ferroelectric ceramics, ferroelectric single crystals and ferroelectric films are presented in the following chapters of this book.

Bibliography

1. J. Zelenka, *Piezoelectric Resonators and Their Applications* (Elsevier, Amsterdam, 1986).
2. C.A. Randall, R.E. Newnham and L.E. Cross, History of the First Ferroelectric Oxide, BaTiO$_3$, The Electronics Division, http://209.115.31.62/ electronics division/(26 Sept. 2004).
3. L.E. Cross and R.E. Newnham, History of Ferroelectrics, *Ceramics and Civilization, Volume III, High-Technology Ceramics — Past, Present, and Future* (The American Ceramic Society, 1987), pp. 289–301.
4. J. Fousek, Joseph Valasek and the Discovery of Ferroelectricity, *Proceedings of the Ninth IEEE International Symposium on Applications of Ferroelectrics* (University Park, PA, 1991) pp. 1–5.
5. J. Valasek, The Early History of Ferroelectricity, *Ferroelectrics* **2** (1971) pp. 239–244.
6. J. Valasek, Piezoelectric and Allied Phenomena in Rochelle Salt, *Physics Review* **15** (1920) pp. 537–538.
7. J. Valasek, Piezo-electric and Allied Phenomena in Rochelle Salt, *Physics Review* **17** (1921) pp. 475–481.
8. J. Valasek, Piezo-electric Activity of Rochelle Salt under Various Conditions, *Physics Review* **19** (1922) pp. 478–491.
9. J. Valasek, Properties of Rochelle Salt related to the Piezoelectric Effect, *Physics Review* **20** (1922) pp. 639–664.
10. B. Jaffe, W.R. Cook and H. Jaffe, *Piezoelectric Ceramics* (Academic Press, London, 1971).
11. M. Budimir, *Piezoelectric Anisotropy and Free Energy Instability in Classic Perovskites*, Thesis No. 3514, Ecole Polytechnique Federal de Lausanne (2006).
12. K. Uchino, *Ferroelectric Devices* (Marcel Dekker, Inc., New York, 2000).
13. C. Kittel, *Introduction to Solid State Physics*, 5th Edition (John Wiley & Sons, New York, 1976).
14. S.E. Park and T.R. Shrout, Ultrahigh strain and piezoelectric behavior in relaxor based ferroelectric single crystals, *J. Appl. Phys.* **82**(4) (1997) pp. 1804–1811.
15. Z.W. Yin, H.S. Luo, P.C. Wang and G.S. Xu, Growth, characterization and properties of relaxor ferroelectric PMN–PT single crystals, *Ferroelectrics* **229** (1999) pp. 207–216.
16. H. Luo, G. Xu, H. Xu, P. Wang and W. Yin, Compositional homogeneity and electrical properties of lead magnesium niobate titanate single crystals grown by a modified Bridgman technique, *Jpn. J. Appl. Phys.* **39** (2000) pp. 5581–5585.

17. Y. Guo, H. Luo, T. He and Z. Yin, Peculiar properties of a high Curie temperature Pb(In1/2Nb1/2)O3–PbTiO3 single crystal grown by the modified Bridgman technique, *Solid State Commun.* **123** (2002) pp. 417–420.

18. G. Xu, K. Chen, K.D. Yang and J. Li, Growth and electrical properties of large size Pb..In1/2Nb1/2...O3–Pb..Mg1/3Nb2/3...O3–PbTiO3 crystals prepared by the vertical Bridgman technique, *Appl. Phys. Lett.* **90** (2007) p. 032901.

19. J. Tian, P. Han, X. Huang and H. Pan, Improved stability for piezoelectric crystals grown in the lead indium niobate–lead magnesium niobate–lead titanate system, *Appl. Phys. Lett.* **91** (2007) p. 222903.

20. S. Zhang, L. Jun, W. Hackenberger and T.R. Shrout, Characterization of Pb..In1/2Nb1/2...O3–Pb..Mg1/3Nb2/3...O3–PbTiO3 ferroelectric crystal with enhanced phase transition temperatures, *J. Appl. Phys.* **104** (2008) p. 064106.

21. E. Sun and W. Cao, Relaxor-based ferroelectric single crystals: Growth, domain engineering, characterization and applications, *Progress in Materials Science* **65** (2014) pp. 124–210.

22. Q. Zhou, K.H. Lam, H. Zheng, W. Qiu and K.K. Shung, Piezoelectric single crystal ultrasonic transducers for biomedical applications. *Progress in Materials Science* **66** (2014) pp. 87–111.

23. S.I. Shkuratov, J. Baird, V.G. Antipov, E.F. Talantsev, J.B. Chase, W.S. Hackenberger, J. Luo, H.R. Jo and C.S. Lynch, Ultrahigh energy density harvested from domain-engineered relaxor ferroelectric single crystals under high strain rate loading, *Sci. Rep.* **7** (2017) p. 46758.

24. S.I. Shkuratov, J. Baird, V.G. Antipov, W.S. Hackenberger, J. Luo, S. Zhang, C.S. Lynch, J.B. Chase, H.R. Jo and C.C. Roberts, Mechanism of complete stress-induced depolarization of relaxor ferroelectric single crystals without transition through non-polar phase, *Appl. Phys. Lett.* **112** (2018) p. 122903.

25. X. Liu and X. Tan, Giant strains in non-textured (Bi1/2Na1/2)TiO3-based lead-free ceramics, *Adv. Mater.* **28** (2016) pp. 574–578.

26. M. Li, M. Pietrowski, R. De Souza, H. Zhang, I. Reaney, S. Cook, J. Kilner and D. Sinclair, A family of oxide ion conductors based on the ferroelectric perovskite Na0.5Bi0.5TiO3, *Nat. Mater.* **13** (2013) pp. 31–35.

27. P. Peng, H. Nie, G. Wang, Z. Liu, F. Cao and X. Dong, Shock-driven depolarization behavior in BNT-based lead-free ceramics, *Appl. Phys. Lett.* **113** (2018) p. 082901.

28. T. Furukawa, Ferroelectric Properties of Vinylidene Fluoride Copolymers, in *Phase Transitions* **18** (1989) pp. 143–211.

29. H. Nalwa, *Ferroelectric polymers: Chemistry, physics, and applications* (Marcel Dekker, Inc., New York, 1995).

30. A.J. Lovinger, Ferroelectric Polymers, *Science* **220**(4602) (1983) pp. 1115–1121.

31. V.Yu. Topolov and C.R. Bowen, *Electromechanical Properties in Composites Based on Ferroelectrics* (Springer, Berlin, 2009).

32. P. Tran-Huu-Hue, F. Levassort, F.V. Meulen, J. Holc, M. Kosec and M. Lethiecq, Preparation and electromechanical properties of PZT/PGO thick films on alumina substrates, *J. Eur. Ceram. Soc.* **21** (2001) pp. 1445–1449.

33. N. Setter and D. Damjanovic, Ferroelectric thin films: Review of materials, properties and applications, *J. Appl. Phys.* **100** (2006) p. 051606.

Chapter 2

Lead Zirconate Titanate Ferroelectric Ceramics

2.1 Introduction

Lead zirconate titanate, the inorganic compound with the chemical formulation $PbZr_{1-x}Ti_xO_3$, was first developed at the beginning of the 1950s at the Tokyo Institute of Technology [1]. PZT ferroelectric ceramics are hundreds of times more sensitive to mechanical or electrical input than natural crystalline materials. They exhibit greater sensitivity and a higher operating temperature relative to barium titanate. PZT is widely used in ultrasonic transducers, sensors and actuators. In lead zirconate titanate binary solid solution, the ferroelectric and piezoelectric properties are affected by the Zr/Ti ratio and can be tailored to fit a variety of practical applications. PZT ferroelectric ceramics have been extensively used in explosive ferroelectric generators. The properties of PZT ferroelectric ceramics are examined in this chapter. Since the operation of FEGs relies on the depolarization of PZT via phase transitions or domain shifting, a discussion of stress-induced phase transformations and polarization switching is also presented.

2.2 PZT Perovskite-Type ABO_3 Crystal Structure

Lead zirconate titanate is synthesized from lead, titanium and zirconium oxides, and is a member of the ferroelectric ceramic family having a perovskite crystalline structure [2]. The perovskites $PbZrO_3$ and $PbTiO_3$ form a continuous series of solid solutions $PbZr_{1-x}Ti_xO_3$,

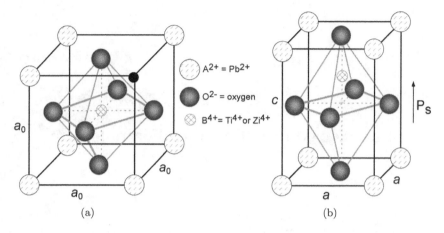

Fig. 2.1. Schematic diagrams of the lead zirconate titanate crystal structure (the ABO_3 perovskite crystal structure). (a) The high-temperature cubic phase of PZT. (b) The lower temperature tetragonal ferroelectric phase of PZT. P_S is the spontaneous polarization vector.

where x can vary from 0 to 1. These solid solutions can undergo a variety of ferroelectric (FE), paraelectric and antiferroelectric (AFE) phase transitions brought on by changes in composition, temperature, stress and/or electric field. They also exhibit large spontaneous polarization and piezoelectric properties.

The general chemical formula for perovskite molecules is ABO_3. The high-temperature cubic phase of lead zirconate titanate is shown schematically in Figure 2.1(a), with unit cell dimensions a_0. Perovskite crystals have a cubic unit cell with large cations (A) on the corners, a smaller cation (B) in the body center, and oxygen anions (O) in the centers of the faces. The A sites are occupied by Pb^{2+} ions, the B sites (the center of the unit cell) are occupied by Ti^{4+} or Zi^{4+} ions, and the face centers are occupied by O^{2-} ions.

The charge neutrality of the unit cell is readily shown by summing the contribution of the charge from each ion to the unit cell. The center ion contributes a charge of +4, each of the eight corner ions are shared with eight neighboring unit cells giving a contribution of 1/8 of their charge to the unit cell shown, so there is a net contribution of +2 from the corner ions, and each of the six face-centered oxygen ions is shared with one neighboring unit cell, so there is a net charge contribution of -6. These charge contributions sum to zero. The cubic phase (Figure 2.1(a)) of PZT is a paraelectric. It does not exhibit the ferroelectric effect.

The PZT cubic phase transforms into the tetragonal phase as the material is cooled through the Curie temperature. This is shown schematically in Figure 2.1(b). The tetragonal phase has a polar axis. The unit cell is elongated in the direction of the polar axis to length c, and contracted transverse to this direction to length a. The source of the ferroelectric effect is a shift of the central ion upward relative to the surrounding oxygen octahedron. This moves the positive charge upward relative to the negative charge resulting in dipole moment $p = Q \cdot d$, where Q is the charge that moves through a separation distance d. This dipole moment per unit cell volume is the polarization of the cell. This spontaneous polarization in the PZT unit cell, P_s, was induced by a phase transformation from the cubic into the tetragonal structure.

2.3   Intrinsic Effects

Intrinsic ferroelectricity is associated with the spontaneous polarization and spontaneous strain. The spontaneous polarization is accompanied by a spontaneous strain with principal directions aligned with the polarization (elongation) and transverse to it (contraction). The perovskite unit cell can be considered a positively charged ion (the central titanium or zirconium ion) in a negatively charged cage (the oxygen octahedron cage) (see Figure 2.1(b)).

If an electric field is applied to the perovskite unit cell, the ions move relative to one another and the crystal structure distorts. The relative motion of the ions changes the dipole moment of each unit cell. The dipole moment per unit volume is the definition of polarization. The easier it is to move the ions around, the larger is the relative permittivity (dielectric constant) of the material.

This coupling between an electric field and a mechanical field works the other way as well. If a compressive stress is applied in the polarization direction, the unit cell is shortened and tends toward the cubic configuration. In this case, the polarization decreases (the direct piezoelectric effect).

If the unit cell is "clamped" so the shape of the unit cell can't change, i.e. the ions are prevented from moving much, the dielectric constant would have a much lower value than when unclamped. A larger dielectric constant typically means flatter energy wells, which means it is easier for the ions to move. This implies a lower stiffness and larger piezoelectric constant.

The elastic compliance is the inverse of the stiffness and is typically measured at constant electric field, which means electrodes must be applied to the material and short-circuited to prevent charge buildup at the surface when the crystalline structure is deformed. If the measurements are done open circuit, an electric field is generated that tends to hold the ions in their original positions, thus stiffening the material.

2.4   Polarization Switching

When an electric field is applied in the direction of the spontaneous polarization, it pulls the positive ions in one direction and pushes the negative ions in the other direction. This elongates the unit cell, resulting in a positive strain change.

When the electric field is applied in the direction opposite to the direction of the spontaneous polarization, it tends to shrink the unit cell toward the cubic state. This results in a negative strain change.

When the compressive stress is applied in the polarization direction, the unit cell is shortened and tends toward the cubic configuration. In this case, the polarization decreases.

Polarization switching is the term that describes the process of re-orientation of the polarization of a single-unit cell. In the tetragonal structure (Figure 2.1(b)), there are six stable polarization directions toward the center of each face.

If the applied electric field exceeds the coercive field, E_c, and is applied in the direction that drives the polarization away from its current direction and toward one of the other directions, then it is a driving force for polarization re-orientation.

Polarization switching can also be driven by stress. When a compressive stress is applied in the direction that drives the unit cell toward a cubic phase, the polarization is reduced. When the applied stress reaches a critical value, the *coercive stress*, the central ion will re-orient itself to one of the other four side sites by turning 90 degrees.

2.5 Domains and Domain Walls

It was mentioned in Chapter 1 that like ferromagnetic materials, ferroelectric materials tend to form domains in which the unit cell dipoles that make up the domain are aligned in parallel to make the domain polar. Domains are observed at the next length scale relatively to the unit cell.

Grains are single crystal regions within a PZT ceramic. All of the unit cells within a grain have the same alignment, yet the grains are subdivided into domains. A typical ceramic specimen contains from 10^9 to 10^{12} grains per cubic centimeter.

Domains are regions in the PZT crystal grains with like polarization. Formation of domains reduces the energy of the crystal by relieving local stresses and local electric fields. The unit cells that form domains are from 3 to 4 angstroms in size. The size of each domain is a few hundred nanometers.

Domains are separated by domain walls. They form in a manner that minimizes energy. Figure 2.2 shows two types of compatible domain walls in the tetragonal phase of PZT: the 180° domain wall and the 90° domain wall.

When an electric field is applied, the 90° walls move, causing a large strain change due to strain differences across the walls. The domain motion tends to be an irreversible process that generates heat and results in hysteresis. When a domain wall moves, there is a large change in polarization that significantly increases the dielectric constant.

A higher dielectric constant means higher losses associated with domain motion, but it also means higher piezoelectric coefficients. Easier domain wall motion also lowers the stiffness (higher elastic compliance).

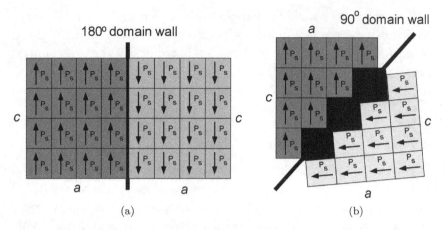

Fig. 2.2. Schematic diagrams of compatible domain walls that occur in the PZT tetragonal structure: (a) 180 degree domain wall, and (b) 90 degree domain wall. P_s is the spontaneous polarization vector.

The 180-degree domains can be reoriented by electric field only, not by stress. The 90-degree, 70.5-degree and 109.5-degree domains can be reoriented by stress and/or electric field.

Each of the crystal structures has its own set of compatible domain walls. For instance, in the rhombohedral structure, the domain types differ by 70.5-degree, 109.5-degree and 180-degree rotations of the polarization vector. This is the angle the vector must rotate through to change its orientation from pointing toward one corner of the unit cell to pointing toward the other corners.

In a ceramic, there is always a relatively large population of domain walls. These arise to accommodate intergranular stress and electric fields.

For more detailed information, it is recommended that the reader refer to the book by A.S. Sidorkin entitled Domain Structure in Ferroelectrics and Related Materials (Cambridge International Science Publishing, 2006).

2.6 Polarization/Crystal Variants

In each of the crystal structures, in the absence of external stress and external electric fields, there are stable polarization directions corresponding to equivalent energy minima. In the tetragonal structure there are six possible directions of spontaneous polarization, i.e. toward the center of each face (see Figure 2.1(b)). When the tetragonal structure is polarized in a particular one of these six directions, it will be referred to as a *polarization variant* or *crystal variant*. Stress and electric fields provide driving forces for transformation from one polarization (crystal) variant to another, i.e. domain switching. Each of the crystal structures has its own set of crystal variants.

2.7 Length Scales in Ferroelectric Ceramic Materials

A schematic representation of the phenomena that occur at multiple-length scales in ferroelectric ceramic materials is shown in Figure 2.3. The unit cell is the smallest repeating unit in the crystal structure. Material behavior at this length scale (0.3–0.4 nm) is governed by interatomic potentials, thermal energy, and external fields. The behavior of the unit cell gives rise to electromechanical coupling in ferroelectric materials and is the source of the phenomena that occur at higher length scales.

Domains are observed at the next length scale (100–500 nm). Domains are groupings of like-oriented unit cells. Domain walls separate domains. Domain walls form along crystal planes that minimize the energy of the structure.

Grains are single crystal regions (5–10 μm) within a ceramic. All of the unit cells within a grain have the same alignment, yet the grains are subdivided into domains.

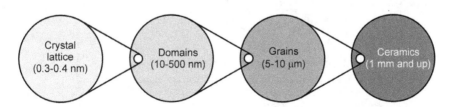

Fig. 2.3. Schematic representation of the phenomena that occur at multiple length scales in ferroelectric ceramic materials.

The ceramic is typically the working scale (>1 mm) of the material. This is then built into a sensor or actuator that is used as a part of a system.

2.8 PZT Phase Diagram

The phase diagram describes the equilibrium crystal structure at zero external stress and zero electric field. The phase diagram for PZT [2] is shown in Figure 2.4. The rhombohedral and tetragonal phases are separated by an almost temperature-independent morphotropic (composition dependent) structural boundary or morphotropic phase boundary (MPB) at which there is an abrupt change in the PZT's crystalline structure.

The MPB depends on the relative composition of lead zirconate and lead titanate. $PbZrO_3$ and $PbTiO_3$ form a continuous series of solid solutions $PbZr_{1-x}Ti_xO_3$, which exhibit a rich variety of ferroelectric and piezoelectric properties.

The PZT phase diagram displays a paraelectric cubic phase at high temperatures above Curie point, a rhombohedral ferroelectric phase region and an orthorhombic antiferroelectric phase in

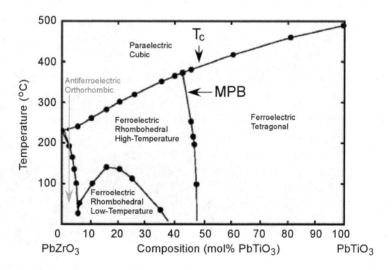

Fig. 2.4. Phase diagram for PZT.

zirconium-rich compositions, and a tetragonal ferroelectric phase in titanium-rich compositions. Each of the crystal structures can be represented as a small distortion of the high-temperature cubic phase.

Above its Curie temperature (T_C in Figure 2.4), PZT has a cubic perovskite crystal structure (see Figure 2.1(a)) and is paraelectric, i.e. no dipoles are presented. Below the Curie temperature, its cubic lattice is distorted to either a high-temperature rhombohedral, low-temperature rhombohedral, or tetragonal crystalline structure, depending on the relative compositions of $PbZrO_3$ and $PbTiO_3$, and dipoles are formed.

When the mixture is rich in Zr, the high-temperature paraelectric phase is transformed from the cubic phase structure into either a rhombohedral ferroelectric phase or orthorhombic antiferroelectric phase upon cooling below Curie point.

The rhombohedral ferroelectric region consists of two phases: a high-temperature phase of symmetry denoted by $R3m$ and a low-temperature phase of symmetry denoted by $R3c$, the latter of which has a unit cell larger than the former.

When the mixture is rich in Ti, the paraelectric phase is transformed from the cubic phase structure into a tetragonal ferroelectric phase upon cooling below Curie point.

A morphotropic phase boundary separates the tetragonal and rhombohedral phases. The piezoelectric, ferroelectric and dielectric properties of PZT peak near the MPB.

2.9 Dopants

In order to meet the specific requirements for certain applications, PZT ferroelectric ceramics can be modified by doping them with ions which have a valence different to the ions in the lattice [2]. The A and B cations in an ABO_3 perovskite structure can be replaced with other cations, such as Na^+, K^+, Bi^{3+}, and La^{3+} for the A cations, and Nb^{5+}, Fe^{3+}, Sb^{5+}, Al^{3+} and Cr^{3+} for the B cations.

These cations are classified as either being acceptor or donor dopants. If the substitute ion has a higher positive charge than the ion it is replacing, it is called a donor. If the substitute ion has a lower positive charge than the ion it is replacing, it is called an acceptor. Doped PZT ferroelectric ceramics having compositions near the MPB form so-called "soft" and "hard" PZTs.

Hard PZTs are doped with acceptor ions such as K^+, Na^+ (for the A site) and Fe^{3+}, Al^{3+}, Mn^{3+} (for the B site), creating oxygen vacancies in the lattice. Acceptor doping tends to make it harder for domain walls to move. Hard PZTs usually have lower permittivities, smaller electric losses, and lower piezoelectric coefficients. These are more difficult to pole and depole, thus making them ideal for rugged applications.

Soft PZTs are doped with donor ions such as La^{3+} (for the A sites) and Nb^{5+}, Sb^{5+} (for the B sites) leading to the creation of A and B site vacancies in the lattice. Donor doping tends to make it easier for domain walls to move (higher dielectric constants). Higher dielectric constants mean higher losses associated with domain motion, but they also mean higher piezoelectric coefficients. Easier domain wall motion also lowers the stiffness (higher elastic compliance). The soft PZTs have higher permittivities, larger losses, and higher piezoelectric coefficients, and are easier to polarize and depolarize. They can be used in applications requiring very high piezoelectric properties.

2.10 Fabrication of PZT Ceramics

Common powder technologies are used to fabricate PZT ceramics [3]. The highest ferroelectric parameters are achieved when: (1) the composition is nearly stoichiometric, (2) the fluxing agents and impurities are kept to a minimum, and (3) the density is as high as possible.

All the constituents of PZT can be precipitated from nitrate solutions or calcining citrates that contain A-site and B-site ions in a 1:1 ratio. Most compositions at present contain PbO as a major constituent which, despite its volatility at $800°C$, must be retained at a sintering temperature that could be as high as $1300°C$. Therefore, the atmosphere around the PZT, when fired, must be controlled.

Calcination (a process that causes the constituents of a mixture to interact by interdiffusion of their ions to reduce the extent of diffusion that must occur during sintering to make a homogeneous body) is carried out in alumina pots with lids. When the PZT undergoes final sintering (a process by which a coherent mass is formed by heating without melting), the mixture is surrounded by lead-rich powders, such as $PbZrO_3$. The final firing is usually carried out in batch electric kilns.

Simple shapes are formed by die-pressing, bodies of uniform shape are formed by extrusion, thin plates are formed by extrusion and rings are formed by slip casting. Electrodes are applied after machining to the desired shape or finish by applying a suitably formulated silver-bearing paint or by sputtering Ni-Cr or gold on the surface of the PZT.

The specimens are poled in transformer oil at temperatures ranging from 100 to 150°C with an applied DC electric field. The temperature and voltage are optimized to yield the maximum ferroelectric properties without allowing the leakage current to reach the levels that could cause thermal runaway and electrical breakdown [3].

2.11 Stress-Induced Phase Transformations in PZT

In the low-power applications of PZT ferroelectrics, the non-linearity and hysteresis associated with phase transformations are to be avoided. Although phase transformations involve large strain and polarization changes that could be very useful, this is typically accompanied by significant hysteresis and cyclic degradation of the material, limiting the ability to exploit these transformations in applications requiring high frequency such as in resonant devices.

There are certain applications, however, where exploitation of the domain switching and phase transformation behavior is advantageous. Domain switching and stress-induced phase transformations can produce a high current and high voltage for a short period of time. This behavior is utilized in explosive ferroelectric generators.

Stress can provide a driving force for phase transformations that involve polarization changes. Phase transformations are of particular interest in FEG design, as these transformations from a polar to a non-polar phase result in a complete release of all polarization.

2.12　PZT 95/5 Ferroelectric Ceramic

A particularly interesting member of the Zr-rich PZT family is the one having a composition of $PbZr_{0.95}Ti_{0.05}O_3$ doped with a small amount (2%) of Nb. It is commonly referred to as PZT 95/5 or PZT 95/5-2Nb. This, doped with Nb ferroelectric ceramic material, will be called PZT 95/5 hereafter in this book.

PZT 95/5 was developed specifically for explosive pulsed power applications by Sandia National Laboratories in the early 1960s. In the PZT phase diagram (see Figure 2.4) PZT 95/5 lies near an antiferroelectric orthorhombic and ferroelectric rhombohedral phase boundary. This is a composition that is ferroelectric at room temperature and can be driven into the lower volume antiferroelectric state by compressive stress. This ferroelectric material is one of a few ferroelectric materials studied in this book.

2.12.1　*Stress-induced phase transitions in PZT 95/5*

PZT 95/5 lies on the ferroelectric rhombohedral side near an antiferroelectric orthorhombic and ferroelectric rhombohedral phase boundary (Figure 2.4). In this structure, the polar axis lies in one of the eight $\langle 111 \rangle$ directions. The structure is elongated in the polar direction and contracted perpendicular to the polar direction. At ambient conditions PZT 95/5 is a composition that is ferroelectric. It can be polarized and keep remanent polarization for a long time.

The orthorhombic antiferroelectric phase has a smaller volume than that for the rhombohedral ferroelectric phase. There is also an alternating tilt of the oxygen octahedral that shifts the polarization of neighboring unit cells 180° out of phase [4]. Compositions just on the antiferroelectric side of the morphotropic boundary can be driven into the ferroelectric phase by application of a strong electric field.

A smaller volume of the AFE phase unit cell in comparison with that for the FE phase provides a possibility of an FE-to-AFE phase transition driven by compressive stress for compositions lying on the ferroelectric side near the morphotropic phase boundary.

In 1978, Fritz [5] reported on investigations of ferroelectric properties of PZT 95/5 in a hydrostatic-pressure environment and the response of PZT 95/5 to statically-applied uniaxial stress. It has been demonstrated that in hydrostatically loaded PZT 95/5 the ferroelectric rhombohedral to anti-ferroelectric orthorhombic phase transition occurs abruptly at a pressure of 0.32 GPa [5]. This pressure-induced phase transition results in a decrease of the volume of the material of about 0.1% and the complete depolarization of the material. The transformation back to the ferroelectric phase occurred at a pressure of 0.14 GPa [5].

The stress-induced transition from the ferroelectric rhombohedral phase into the non-polar antiferroelectric orthorhombic phase in PZT 95/5 can be utilized in explosive ferroelectric generators. As a result of this phase transition, PZT 95/5 is completely depolarized and surface charge can be released on the electrodes of the PZT 95/5 specimen.

2.12.2 *PZT 95/5 shock depolarization studies*

Systematic studies of the physical and electrical properties of PZT 95/5 shock-compressed by planar shock waves initiated by impacting projectiles accelerated in gas gun systems started in the 1960s at Sandia National Laboratories and continue up to the present time in laboratories all over the world.

The experimental results indicate [6–9] that PZT 95/5 ceramic specimens become completely depolarized when shock pressure is greater than 1.6 GPa. These results open the way to the development of miniature ferroelectric generators utilizing the shock-induced depolarization of PZT 95/5. The designs of these generators and the results of systematic studies of their operation are described in the following chapters of this book.

A review of early and recent studies of the physical, mechanical and electrical properties of shock-compressed PZT 95/5 is presented in Chapter 3 of this book.

2.12.3 *Fabrication of PZT 95/5 ceramics*

PZT 95/5 ferroelectric ceramic specimens were not commercially available until recent times because they are not used in sensors, actuators, and ferroelectric transducer applications due to their poor piezoelectric properties.

A new technology for the production of high-quality PZT 95/5 ferroelectric ceramics has been under development by TRS Technologies Inc. for the past few years [10]. In 2010, TRS Technologies Inc. was a recipient of the U.S. Small Business Administration's Tibbetts Award for outstanding achievements in the Small Business Innovation Research (SBIR) program. All the PZT 95/5 elements of explosive ferroelectric generators studied in this book have been supplied by TRS Technologies.

PZT 95/5 ceramic specimens are fabricated using conventional solid oxide processing, i.e. mixing oxide powders, calcining, ball milling, adding a binder, pressing, and sintering (see Section 2.10 above).

PZT 95/5 is doped with about 2% of Nb to stabilize the $R3c$ phase and to reduce dielectric losses. When cooled at a pressure of 1 bar, it transforms from the paraferroelectric phase to the $R3m$ phase and then, near room temperature, to the $R3c$ phase. As the result of this preparation procedure, PZT 95/5 lies near a boundary separating the FE rhombohedral and AFE orthorhombic phases. At ambient conditions, it is ferroelectric and can be polarized, but the application of pressure causes the $R3c$ phase to transition into the antiferroelectric phase.

PZT 95/5 ceramic specimens are formed by uniaxial die pressing. Silver electrodes are applied after machining to the desired sides of the specimens. Specimens are polarized in transformer oil at temperatures ranging from 100 to 150°C with applied DC electric fields ranging from 1 to 4 kV/mm.

2.12.4 *PZT 95/5 properties*

The mechanical, electrical, ferroelectric and piezoelectric properties of PZT 95/5 produced by TRS Technologies are presented in Table 2.1. The TRS PZT 95/5 is formulated to provide maximum discharge efficiency during shock compression.

Table 2.1. Specifications of PZT 95/5 ferroelectric ceramics produced by TRS Technologies [10].

Properties	Metric units
Density (kg/m^3)	7.7×10^3
Curie temperature, $T_C(°C)$	227
Remanent polarization (C/m^2)	0.32–0.34
Dielectric constant and Loss* — Unpoled (/ %)	350 / 2.00
Dielectric constant and Loss* — Poled (/ %)	295 / 1.97
Dielectric constant* — Depoled (−)	225
Coercive field, E_c (MV/m)	1.2
Piezoelectric charge constants, d_{33}/d_{31} ($\times 10^{-12}$ C/N)	68/−16
Piezoelectric voltage constants, g_{33}/g_{31} ($\times 10^{-3}$ V·m/N)	26.3/−6.0
Piezoelectric Coupling, $k_p/k_{31}/k_t$ (−)	0.18/0.11/0.46
Elastic compliance, $s_{11}^E/s_{12}^E/s_{33}^E$ ($\times 10^{-12}$ m^2/N)	7.68/−1.97/8.37
Acoustic Velocity/Poisson Ratio (m/s)	4200/0.26
Thickness Mode, N_t/Q_m (Hz·m)	2100/15

Quantities denoted by *measured at 25°C, 1 V, and 1 kHz.

2.13 PZT 52/48 Ferroelectric Ceramic

Another interesting member of the PZT family is one having a composition Pb(Zr$_{0.52}$Ti$_{0.48}$)O$_3$ doped with 1% of Nb$_2$O$_5$. This ferroelectric ceramic material will be called PZT 52/48 hereafter in this book.

The morphotropic phase boundary between ferroelectric tetragonal and ferroelectric rhombohedral phases in the PZT phase diagram occurs in the vicinity of Pb(Zr$_{0.535}$Ti$_{0.465}$)O$_3$ (Figure 2.4). PZT 52/48 lies near the MPB on the tetragonal side.

2.13.1 *PZT 52/48 properties*

PZT 52/48 possesses large remanent polarization, a high Curie temperature, a high dielectric constant, and high piezoelectric coefficients. It is widely used in sonar systems, actuators for nanopositioners, ultrasonic motors, active scalpels, and numerous other systems. PZT 52/48 ceramic specimens are commercially available in a variety of shapes and sizes from several manufacturers in the U.S.

Table 2.2. Specifications of PZT 52/48 (EC-64) ferroelectric ceramics produced by EDO Ceramics Corp. (ITT Inc.).

Properties	Metric units
Density (kg/m^3)	7.5×10^3
Curie temperature, T_C (°C)	320
Remanent polarization (C/m^2)	0.30–0.33
Dielectric constant (poled) measured at 1 kHz	1300
Dielectric constant (depoled) measured at 1 kHz	1140
Elastic modulus (Young's modulus) (N/m^2)	7.68/−1.97/8.37
Piezoelectric longitudinal coupling factor, k_{33}	0.71
Piezoelectric transverse coupling factor, k_{31}	0.35
Piezoelectric planar coupling factor, k_p	0.60
Piezoelectric longitudinal voltage coefficient, g_{33} ($\times 10^{-3}$ V·m/N)	25
Piezoelectric longitudinal charge coefficient, d_{33} ($\times 10^{-12}$ m/V)	295
Piezoelectric transverse voltage coefficient, g_{31} ($\times 10^{-3}$ V·m/N)	−10.7
Piezoelectric transverse charge coefficient, d_{31} ($\times 10^{-12}$ m/V)	−127
Piezoelectric Mechanical Q (thin disk)	400

All the PZT 52/48 elements of the explosive ferroelectric generators studied in this book have been supplied by EDO Ceramics Corp. (now ITT Inc. [11]). The physical, mechanical, electrical, ferroelectric and piezoelectric properties of PZT 52/48 (trade name EC-64) produced by EDO Ceramics Corp. (ITT Inc.) are presented in Table 2.2.

2.13.2 *Fabrication of PZT 52/48 ceramics*

PZT 52/48 ceramic specimens are fabricated using conventional solid oxide processing (see Section 2.10 above). PZT 52/48 is doped with 1% of Nb_2O_5. The specimens are formed by uniaxial die pressing. After the machining and deposition of silver electrodes, the specimens are polarized in transformer oil at temperatures ranging from 125 to 150°C with applied DC electric fields ranging from 1 to 1.5 kV/mm. Heating the ferroelectric elements higher than 155°C can lead to an increase in leakage currents to a level that could result in thermal runaway and electrical breakdown during the polarization procedure [3].

2.13.3   *PZT 52/48 shock depolarization studies*

Studies of the shock depolarization of PZT 52/48 started in the 1960s at Los Alamos National Laboratory. In 1962, Reynolds and Seay [12, 13] reported on the results of the detailed investigations of the depolarization of PZT 52/48 shock-compressed by planar shock waves propagated anti-parallel to the direction of polarization (longitudinal shock). The experimental results indicated that longitudinally shock-compressed PZT 52/48 was completely depolarized at shock pressures exceeding 1.1 GPa and released electric charge with density equal to its remanent polarization [13]. A review of early studies of the physical and electrical properties of shock-compressed PZT 52/48 is presented in Chapter 3 of this book.

The results of systematic studies of different types of explosive ferroelectric generators utilizing PZT 52/48 elements are presented in the following chapters of this book.

2.14   Summary

- Lead zirconate titanate, an inorganic compound with the chemical formulation $PbZr_{1-x}Ti_xO_3$, is a member of the ferroelectric ceramic family having perovskite crystalline structure. The perovskites $PbZrO_3$ and $PbTiO_3$ form a continuous series of solid solutions that can undergo a variety of ferroelectric, paraelectric and antiferroelectric transitions brought on by changes in composition and temperature.
- The crystalline structure and ferroelectric properties of PZT are affected by the Zr/Ti ratio and can be tailored to fit a variety of practical applications. In addition to this, in order to meet the specific requirements for certain applications, PZT can be modified by doping it with ions, which have a valence different to the ions in the lattice.

- PZT ferroelectric ceramics exhibit large remanent polarization and high Curie temperature, parameters that are important for FEG applications.
- PZT ferroelectric specimens can be produced by conventional ceramic manufacturing techniques and can be made in a wide range of shapes and sizes.
- A particularly interesting member of the Zr-rich PZT family is the one having a composition of $PbZr_{0.95}Ti_{0.05}O_3$ doped with a small amount of Nb. PZT 95/5 is a high energy density ferroelectric material that is important for FEG applications. The pressure-induced phase transition from the ferroelectric to antiferroelectric phase results in the complete depolarization of PZT 95/5 at relatively low pressures.
- Another interesting member of the PZT family is one having a composition $Pb(Zr_{0.52}Ti_{0.48})O_3$ doped with a small amount of Nb_2O_5. PZT 52/48 is a high-density ferroelectric that can be used for FEG applications. Experimental results indicate that PZT 52/48 can be completely depolarized under longitudinal shock compression and can release electric charge with density equal to its remanent polarization during a microsecond interval of time.

2.15 Suggested Reading on Ferroelectric Materials

- W.G. Cady, *Piezoelectricity: An Introduction to the Theory and Applications of Electromechanical Phenomena in Crystals* (Dover Publications, 1964).
- T. Mitsui, I. Tatsuzaki, and E. Nakamura, *An Introduction to the Physics of Ferroelectrics* (Gordon and Breach, New York, 1976).
- J.M. Herbert, *Ferroelectric Transducers and Sensors (Molecular Crystals and Liquid Crystals)* (Gordon and Breach, 1982).
- G. Smolenskii, *Ferroelectrics and Related Materials (Ferroelectricity and Related Phenomena)* (TH-ROUTL, 1984).
- Y. Zhiwen, *Ferroelectrics Research in the People's Republic of China: A Special Issue of the Journal Ferroelectrics* (Routledge, 1990).

- C.Z. Rosen, B.V. Hiremath, and R.E. Newnham, *Piezoelectricity* (AIP Press, 1992).
- F. Jona and G. Shirane, *Ferroelectric Crystals* (Dover Publications, 1993).
- K. Uchino, *Ferroelectric Devices* (CRC Press, 2000).
- M.E. Lines and A.M. Glass, *Principles and Applications of Ferroelectrics and Related Materials* (Oxford Classic Texts in the Physical Sciences, 2001).
- C. Pazde-Araujo, R. Ramesh, and G.W. Taylor, *Science and Technology of Integrated Ferroelectrics (Ferroelectricity and Related Phenomena)* (CRC Press, 2001).
- K. Krakauer, *Fundamental Physics of Ferroelectrics: 11th Williamsburg Ferroelectrics Workshop* (AIP Press, 2001).
- J. Yang, *Introduction to the Theory of Piezoelectricity* (Springer, 2004).
- A. Amau and A. Vives, *Piezoelectric Transducers and Applications* (Springer, 2004).
- A.S. Sidorkin, *Domain Structure in Ferroelectrics and Related Materials* (Cambridge International Science Publishing, Ltd., 2006).
- J. Yang, *Analysis of Piezoelectric Devices* (World Scientific Publishing Company, 2006).
- K. Rabe, C.H. Ahn, and J. Triscone, *Physics of Ferroelectrics: A Modern Perspective* (Springer, 2007).

Bibliography

1. L.E. Cross and R.E. Newnham, History of Ferroelectrics, from *Ceramics and Civilization, Volume III, High-Technology Ceramics — Past, Present, and Future* (The American Ceramic Society, 1987), pp. 289–301.
2. B. Jaffe, W.R. Cook and H. Jaffe, *Piezoelectric Ceramics* (Academic Press, London, 1971).
3. A.J. Moulson and J.M. Herbert, *Electroceramics: Materials, Properties, Applications.* 2nd Edition (John Wiley & Sons Ltd., West Sussex, 2003).
4. M.E. Lines and A.M. Glass, *Principles and Applications of Ferroelectrics and Related Materials* (Clarendon Press, Oxford, 1977).
5. I.J. Fritz, Uniaxial stress effects in a 95/5 lead zirconate titanate ceramic, *J. Appl. Phys.* **49**(9) (1978) pp. 4922–4928.

6. W.J. Halpin, Current from a shock-loaded short-circuited ferroelectric ceramic disk, *J. Appl. Phys.* **37**(1) (1966) pp. 153–163.

7. P.C. Lysne and C.M. Percival, Electric energy generation by shock compression of ferroelectric ceramics: normal-mode response of PZT 95/5, *J. Appl. Phys.* **46**(4) (1975) pp. 1519–1525.

8. P.C. Lysne, Dielectric properties of shock-compressed ferroelectric ceramic, *J. Appl. Phys.* **48**(3) (1977) pp. 1020–1023.

9. R.E. Setchell, Shock wave compression of the ferroelectric ceramic Pb0.99(Zr0.95Ti0.05)0.98Nb0.02O3: depoling currents, *J. Appl. Phys.* **97** (2005) p. 013507.

10. http://trstechnologies.com/

11. http://itt.com/

12. C.E. Reynolds and G.E. Seay, Multiple shock wave structures in polycrystalline ferroelectrics, *J. Appl. Phys.* **32**(7) (1961) pp. 1401–1402.

13. C.E. Reynolds and G.E. Seay, Two-wave shock structures in the ferroelectric ceramics: barium titanate and lead zirconate titanate, *J. Appl. Phys.* **33** (1962) pp. 2234–2239.

Chapter 3

Historical Perspectives of Ferroelectric Shock Depolarization Studies

3.1 Introduction

The ability of shock-compressed ferroelectric ceramics to generate high current and high voltage pulses was reported for the first time by Nielson [1] at Sandia National Laboratories in 1956 and these results were published in 1957 [2]. Since that time, systematic studies of the physical and electrical properties of ferroelectric and piezoelectric materials under shock compression have been performed by Sandia National Laboratories, Los Alamos National Laboratory and other laboratories in the U.S. and all over the world. A brief review of the extensive literature on the shock compression of ferroelectric and piezoelectric materials is presented in this chapter.

3.2 Early United States

In 1946, Gray [3] submitted a patent, which was granted in 1949, for a transducer to convert mechanical energy into electrical energy. The transducer used barium titanate ceramic in its tetragonal crystalline state.

In 1952, Kabik and Cecil [4] submitted a patent, which was granted in 1961, for an impact-responsive electric primer based on a polarized piezoelectric element composed of barium titanate ceramics to generate a voltage sufficient to initiate a high explosive charge in missiles. A plunger was abutted against one of the electrodes of the piezoelectric element. Upon impact, it was driven into the electrode, shocking the piezoelectric element and generating a voltage proportional to the rate of increase in the pressure across the element.

In 1954, Ferrara [5] submitted a patent, which was granted in 1958, for an electromechanical transducer based on the use of piezoelectric crystals or pre-polarized ceramics. The transducer used a striker to activate an explosive primer or explosive charge to stress the piezoelectric or pre-polarized ceramic material, generating a potential difference across it. Ferrara maintained that the magnitude of the potential difference induced across the transducer depended upon the manner in which the stress was applied and the degree to which it was applied. The proposed application for these transducers was as a fuse for mines and/or booby traps.

In 1956, Howe [6] submitted a patent, which was granted in 1961, for an "energy converting device" based on piezoelectric materials for converting mechanical energy into electrical energy. The device consisted of a mechanical energy source (striker or firing pin), a means for multiplying the mechanical energy (explosive primer), and a piezoelectric material for converting the mechanical energy into electrical energy. Howe pointed out that these transducers are simple, lightweight and rugged, and that they have an indefinite shelf life. It was proposed that these transducers could be used in relatively small projectiles.

In 1957, Stresau [7] submitted a patent, which was granted in 1971, for a device based on piezoelectric crystals to simultaneously initiate multiple explosive charges. The device consisted of a detonator, an explosive charge, a stack of steel or other metals having high tensile strength that were used to flatten the shock wave generated by the explosive, a piezoelectric element, and a steel block that held the piezoelectric element in place when the explosive charge was detonated. The proposed application for this device was as a simple and compact detonating system for controlling the fragmentation of bombs or projectiles, and as a fuse that could be activated mechanically, electrically, or by flame.

The first report and paper on the generation of high voltage and high current pulses by shocked ferroelectric ceramics were those by Neilson [1, 2] in 1956 and 1957, respectively. Neilson worked with two different ceramics: lead zirconate titanate (Brush Electronics "PZT-1") and barium titanate. With the PZT-1 sample, a peak current of 4 kA and total charge of 600 μC was delivered by a 0.3 μs pulse to a 0.1 Ω load and, with the barium titanate, a voltage pulse that ramped up to 60 kV in 3 μs was delivered to a 50 kΩ load.

In 1959, Berlincourt and Krueger [8] reported on the effects of mechanical and electrical stresses on tetragonal lead titanate zirconate 53% (PZT 53/47) and barium titanate. They observed that poled ferroelectric ceramics exhibit linear piezoelectric effects for relatively low mechanical stresses, but under high pressure they exhibit nonlinear effects due to domain reorientation. They measured the open circuit voltage, short-circuit charge and mechanical strains produced by ferroelectric samples that were subjected to either slowly or rapidly applied uniaxial compressive stresses ranging up to 60,000 psi, either parallel or perpendicular to the poled axis. In summary, it was found that:

- Effects in the open-circuit mode are very different to those in the short-circuit mode. In the open-circuit mode, there are no substantial nonlinearities at stresses, while stresses cause considerable nonlinearity in the short-circuit mode. This is due to charge appearing on the electrode surfaces under compression that tends to maintain a field across the ferroelectric element in the same direction as the original poling field. This field tends to prevent domain switching. If the stress is maintained, the charge will gradually bleed off and the maintaining field disappears.
- In the open-circuit mode, uniaxial compression tends to generate a field in the direction of the poling field, which tends to maintain the field across the element and prevent domain reorientation, while transverse or lateral compression tends to generate a field opposite to the direction of the poling field, which tends to encourage domain reorientation.

In 1961, Wittekindt [9] studied the shape of the current pulses generated by a shock-compressed thin ferroelectric cylinder.

In 1961 and 1962, Reynolds and Seay [10, 11] reported on the detailed experimental study of depolarization of PZT 52/48 shock-compressed by a planar shock wave propagating anti-parallel to the polarization direction (longitudinal shock). It was found that the charge released by longitudinally-shocked PZT 52/48 was a linear function of shock pressure ranging from 0.2 to 1.1 GPa [11].

The experimental results indicated that longitudinally-shocked PZT 52/48 was completely depolarized and released an electric charge density equal to its remanent polarization at shock pressures ranging from 1.1 to 1.9 GPa. Reynolds and Seay concluded that the short duration of the shock waves ruled out domain randomization and that the charge release was due to the reversible linear piezo-electric effect [11].

It was found [11] that a single impinging shock wave may generate a two-wave structure as the result of material rigidity that leads to the formation of an elastic wave from a polymorphic phase transition. It was also found that in order for the two-wave structure to form, the leading shock wave must have a lower pressure and a higher velocity than the follow-on shock wave. The first shock wave in the two-wave structure will have a pressure and a velocity characteristic of the material it is propagating through and independent of the strength of the supporting shock wave in most materials. Finally, it was found that there is a stable shock threshold, i.e. a pressure above which the two-wave structure cannot form, since the shock wave in this case is moving with a velocity greater than the velocity of the leading shock wave in the two-wave structure.

In 1964, Gilbertson from the Warhead and Terminal Ballistics Laboratory at the U.S. Navy Weapons Laboratory in Dahlgren, VA, published a report on an inexpensive compact explosively-driven single-shot high-voltage power supply for electroexplosives [12]. The device used polarized barium titanate as the ferroelectric material. It was found that the electrical output was dependent on the properties of the ceramic material, the electrical load, and the geometry of the explosive-ceramic combination.

In 1965, Graham, Nielson and Benedick [13] reported on their study of the piezoelectric current generated by shocked X-cut quartz.

They found that the amplitude which the current pulse generates and how it changes in time depend on the dielectric, piezoelectric and mechanical properties of the quartz under shock loading. One of their observations was that the electric field due to the piezoelectric field in the stressed and unstressed portions of the quartz disk was distorted by a number of mechanisms, one of which was that at the outer edge of the disk the discontinuity in the electric potential and dielectric permittivity caused electric-field fringing similar to that observed in parallel-plate capacitors. They also found that the current-time waveform was distorted when the pressure exceeded 2.5 GPa and that this distortion was due to electrical conduction (leakage current). Another observation was that the diameter-to-thickness ratio affected the distortion of the current pulse. When the disks were fully electroded, the current-time profile was

- linear when the diameter-to-thickness ratio ≥ 10;
- linear for about one-half wave-transit and then becomes nonlinear when the diameter-to-thickness ratio is equal to 5;
- nonlinear throughout the entire transit time when the diameter-to-thickness ratio < 4.

In 1966, Cutchen [14] reported on the effects of polarity on the charge released when PZT 65/35 was shocked by planar shock waves propagated parallel and anti-parallel to the direction of polarization (longitudinal shock), i.e. two polarity orientations were examined. The positive orientation was taken to be that in which the shock wave traveled opposite to the polarization direction and the negative orientation was taken to be that in which the shock wave traveled in the same direction as the polarization direction. It was found that there was sensitivity to the orientation of the polarization of the PZT sample. The data indicated that the PZT ceramic was more sensitive to small variations in the pressure profile and that the electric field degradation was more pronounced when the stress traveled in a direction opposite to that of the polarization, as opposed to when it traveled in the same direction. Since the electrical load and the stress conditions were the same for both orientations, these sensitivities must arise within the ceramic itself. These results support the earlier claims of Gerson and Jaffe [15] that PZT ceramics exhibit p-type conduction. In summary, it was concluded that conduction is more pronounced if the remnant polarization is anti-parallel to the direction of shock wave propagation.

In 1966 and 1968, Halpin [16, 17] reported on the study of the influence of the magnitude of stress on the shape of the current pulses delivered to a short-circuited load. This was accomplished by launching flat-face projectiles accelerated in the gas gun system at various velocities against ferroelectric disks coated with electrodes. The planar shock waves moved along the axis of the disk in a direction anti-parallel to the direction of polarization (longitudinal shock). Three types of ferroelectric materials were used in the experiments: normally sintered PZT 95/5, hot pressed PZT 95/5 and PSZT 68/7. The two PZT 95/5 materials were selected so that materials having the same composition, but different densities, could be compared, and he selected the PSZT so that he could compare materials having different compositions. Three observations were noted.

- The experimental results indicated that normally sintered and hot-pressed PZT 95/5 specimens were completely depolarized by longitudinal stress ranging from 1 to 2.5 GPa and released charge with density practically equal to their remanent polarization ($P_r = 0.34$ and $0.37\,\mathrm{C/m^2}$, respectively).
- The magnitude and position in time of the peak stress-induced current appeared to be stress-dependent. That is, for stresses less than 1 GPa, the magnitude of the peak current was relatively small and it appeared late in time. As the stress was increased to 1.5–2.0 GPa, the magnitude of the peak current increased and it shifted to earlier times. As the stress continued to increase beyond 2 GPa, the magnitude of the peak current decreased and it shifted to a later time.
- The total electric charge released to a short-circuited load also appeared to be stress dependent. For the low- and high-stress regimes, the total charge released is less than the remanent charge of polarization, while in the mid-stress regime, the total

charge released has a value near to that of the remanent charge polarization.

- Most of the current pulse lengths were greater than the time expected for the pressure wave to transit the ferroelectric element. Thus, it was believed that the stress wave front tilt also affected the overall shape of the current pulse.

Halpin showed that there are significant differences between the shock and static compression of ferroelectric ceramics. Under static compression, electric fields do not form in short-circuited ferroelectric ceramics whereas under shock compression large oppositely-directed electric fields are formed on either side of the shock front in short-circuited ferroelectric ceramics. If these fields exceed the coercive field of the materials, they can induce changes in the remnant polarization and large electromechanical coupling effects that must be taken into account when comparing shock and static measurements.

It was concluded in [16, 17] that the dependence of the shape of the current pulse on stress is due to the combination of stress-induced changes in remnant polarization, permittivity, and conductivity caused by rapidly applied large electric fields which accompany the process of current production.

In 1967, Houser from Picatinny Arsenal published a report [18] on the charge release due to a pressure-induced ferroelectric to antiferroelectric phase transition in the system $PBZrO_3$-$PbTiO_3$-$PbSnO_3$-Nb_2O_5.

In 1967, Linde [19] conducted a series of shock compression studies with PZT 52/48 and PZT 95/5. The specimens were shock-compressed by planar shock waves in the direction anti-parallel to the direction of polarization (longitudinal shock). The experiments were conducted with the gas gun system. The shock pressure in the ferroelectric specimens was controlled through acceleration of the impacting projectile. The shock pressure range was chosen to keep the ferroelectric specimens not mechanically destroyed. The PZT 52/48 specimens were shock-compressed at the pressure range from 0.5 to 0.9 GPa. The PZT 95/5 specimens were shock-compressed at the pressure range from 0.6 to 1.1 GPa.

The specimens of both types have been successfully recovered after shock loading. The recovered specimens were subjected to thermal heating above their Curie points in order to determine their residual polarization after shock loading.

The residual polarization of PZT 52/48 specimens recovered after shock-compression at 0.9 GPa was found to be 0.2 C/m^2, i.e. the PZT 52/48 specimens lost about 30% of their remanent polarization ($P_r = 0.34\,C/m^2$) due to the shock compression. It was concluded that the relative amount of stress-induced domain switching is significant in PZT 52/48.

The experimental results indicated that the residual polarization of PZT 95/5 specimens recovered after shock-compression at 1.1 GPa was 0.1 C/m^2. The PZT 95/5 specimens lost about 70% of their remanent polarization ($P_r = 0.35\,C/m^2$) after shock compression.

It was concluded that shock-induced depolarization of PZT 52/48 and PZT 95/5 was the result of several mechanisms, including piezoelectric reduction of dipole moments, domain switching and dynamic phase transitions (ferroelectric-to-antiferroelectric phase transitions).

In 1968, Graham and Halpin [20] reported on their study of the electrical breakdown that occurs in X-cut quartz longitudinally compressed by planar shock waves. They looked at two different orientations: −X, where the shock wave traveled in a direction opposite to the polarization, and +X, where the shock wave traveled in the same direction as the polarization. They found that breakdown occurred in the −X orientation, but not in the +X orientation. They concluded that since an electric field is present during breakdown in the −X direction, the field accelerates electrons from the shock wave front into the stressed region and initiates breakdown.

In 1968, Doran [21] reported on the study of the shock wave compression of barium titanate and PZT 95/5. He found that a two-wave structure exists in the PZT when the shock pressures exceeded 4 GPa. He also noted that the PZT compositions underwent a ferroelectric-to-antiferroelectric transition under shock compression.

In 1973, Lysne [22] extended the work of Graham and Halpin by looking at dielectric breakdown in PZT 65/35 shock-compressed by planar shock waves generated by impacting projectiles accelerated in the gas gun system. The PZT was connected across a resistive load and the shock wave traveled anti-parallel to the remanent polarization vector (longitudinal stress). In summary, it was found that:

- Dielectric breakdown is not explicitly a function of stress for stress values less than 2.3 GPa.
- Breakdown is not instantaneous.
- For electric fields greater than the breakdown threshold (≈ 5 kV/mm), the time between the introduction of the shock wave into the PZT and the onset of breakdown decreased rapidly with increasing electric field.
- Electric fields generated by PZT specimens poled to $30\,\mu C/cm^2$ and shocked by pressures of 1 GPa are sufficient to cause electrical breakdown.
- Axial release waves can travel at a velocity as much as 15% faster than the initial stress wave.

Lysne points out that one important ramification of these results is that the voltage response of a partially-poled PZT sample may be greater at higher stresses than that of a fully-poled sample shocked to the same stress values.

In a follow-on paper in 1975, Lysne [23] formulated a model to predict dielectric breakdown of longitudinally shocked PZT 65/35. He also investigated PSZT 70/30 [24] and found that polarization transformed it into a metastable ferroelectric phase. This led him to suggest that shock waves could readily transform the material into a stable antiferroelectric phase without conversion to a metastable ferroelectric phase in the absence of a strong electric field. He concluded that this could lead to an effective method for using short duration stress pulses generated by high explosives to depolarize large pieces of ceramics in explosive-driven power supplies.

In 1973, England and Gifford [25] were granted a patent for a solid state power supply activated by a pyrotechnic chain. The intended application for this two-stage power supply was as a power supply for fuses in munitions. The power supply had two piezoelectric generators. The first piezoelectric generator was stressed or shocked when the projectile was launched. It generated an electrical pulse used to trigger a sequentially-fired pyrotechnic chain of devices, each of which had a predetermined time delay and which, respectively, generated a second stress or shock wave directed against the second piezoelectric generator. The second piezoelectric generator would continue to generate power to drive other fuse circuitry until the last pyrotechnic device ceased to function.

In 1975, Lysne and Bartel [26] investigated the electromechanical response of longitudinally shock-compressed PZT 65/35 in the open-circuit mode. The PZT 65/35 specimens were shock-compressed by planar shock waves propagated in a direction anti-parallel to that of the polarization vector (longitudinal shock). They found that when the pressures were below approximately 2.0 GPa, the shock wave propagation was dispersive, while between 2.0 and 3.6 GPa, the shock wave propagation was steady. For pressures greater than 3.6 GPa, the final stress states were reached asymptotically. Also, it was found that the electrical response of a material at pressures around 6 GPa was complicated by a ferroelectric-to-paraelectric phase transformation and the kinetics by which the polarization changed. In summary, they observed a degradation of the electrical response of PZT 65/35 when the pressure exceeded 6.5 GPa. This was thought to be due to either dielectric breakdown or an increase in the electrical conductivity of the PZT material.

In 1975, Lysne and Percival [27] reported on their study of the response of PZT 95/5 compressed by planar shock waves generated by impacting projectiles accelerated in the gas gun system. The propagation direction of the shock waves was perpendicular to the direction of polarization (normal or transverse shock compression). The authors pointed out that this polarized ferroelectric ceramic has a composition that places it near an anti-ferroelectric phase boundary and that it is neither linear nor reversible in response to stresses on the order of a few tenths of a gigapascal. They also pointed out that in the past investigators had looked at loads consisting of very low (short-circuit or high-current mode) and very high (open-circuit or high-voltage mode) resistances. Therefore, one of the objectives of their investigation was to determine the response of PZT 95/5 to transverse shock waves as a function of load resistance. The PZT 95/5 specimens were shock-compressed at pressure 1.4 GPa. It was shown experimentally that when the load resistance is zero, electric fields ahead of and behind the shock wave are zero. The PZT 95/5 specimens were almost completely depolarized due to the shock compression. However, when the load resistance is not zero, electric fields exist and the dielectric properties of the shocked and unshocked material are important. The electric field ahead of the shock wave influences the mechanical properties of the unshocked material, which, in turn, affects the characteristics of the shock wave. It can be concluded that the characteristics of the shock wave are a function of the electric field, since the phase transition that the shock wave causes is dependent on the electric field.

Lysne [28] also investigated the dielectric properties of shock-compressed PZT 95/5. Planar shock waves were generated by impacting projectiles accelerated in the gas gun system. The propagation direction of the shock waves was perpendicular to the direction of polarization (transverse shock compression). On the one hand, Lysne pointed out that while axial (longitudinal) loading can be mechanically one-dimensional, the observed electrical response is complicated by electric field-induced depolarization ahead of the shock front if the specimen is connected to a low-impedance electrical load. On the other hand, transverse loading is inherently two-dimensional, but has the advantage that the electrical response is a monotonic function of time. Using transverse loading, he determined the charge released and the dielectric permittivity of PZT 95/5 when in a wide range of shock pressures. He found that when the shock pressure was equal to or greater than 1.6 GPa, the remanent polarization had vanished and the material was in a non-ferroelectric state. He also found that the permittivity of depolarized PZT 95/5 was equal to 9 nF/m and was independent of the electric field. Also, it was noted that PZT 95/5 transforms to an anti-ferroelectric state at hydrostatic pressures between 0.3 and 0.4 GPa and that complete surface charge liberation occurs between 1.3 and 1.6 GPa when operating into a short-circuit. In addition, he noted that for stress states above 1.6 GPa, the permittivity is independent of both the stress and the electric field. Lysne pointed out that knowing these parameters will be useful in the design of shock wave actuated power supplies based on phase transformations in PZT 95/5.

While attempting to determine the approximate dielectric equation of state for shock-compressed PZT 65/35, Lysne [29] discussed the possibility of the polarization of this ferroelectric ceramic with shock waves. Lysne noted that when Berlincourt and Krueger [8] measured the charge released by quasi-statically compressed PZT 52/48 and barium titanate, they found that for pressures on the order of a few tenths of a GPa that the charge released was a nonlinear function of the pressure and that it was considerably larger than that calculated using the piezoelectric coefficients.

Mazzie [30] proposed a simple model for generating energy by normally shock-compressing ferroelectric ceramics. Unlike Lysne and Percival [25], who assumed that the permittivity of the ferroelectric was different in stressed and unstressed states, he assumed that they did not differ over timescales of interest. He also assumed that the PZT 95/5 sample could be modeled as a bulk capacitor with constant capacitance connected parallel with a resistor. He found that an important design parameter was the value of the RC time constant relative to the shock transit time.

In 1978, Fritz [31] reported on investigations of the ferroelectric properties of PZT 95/5 in a hydrostatic-pressure environment and the response of PZT 95/5 to statically applied uniaxial stress. It has been demonstrated that in hydrostatically loaded PZT 95/5 the ferroelectric rhombohedral to anti-ferroelectric orthorhombic phase transition occurs abruptly at a pressure of 0.32 GPa. This pressure-induced phase transition results in a decrease of the volume of the material of about 0.1% and the complete depolarization of material. The transformation back to the ferroelectric phase occurred at a pressure of 0.14 GPa [5].

Dick and Vorthman [32] studied the effects of the electrical state of PZT 95/5 on its mechanical and electrical response to impact loading. They considered both polarized and unpolarized PZT with short-circuited and resistive loads. They found that domain reorientation occurs in polarized and unpolarized samples, but that polarization does not alter the point at which phase transition occurs nor its kinetics. They also found that domain reorientation and phase transitions cannot be separated and that domain reorientation is probably more important in unpolarized materials. This result occurs because the domains in polarized materials are already aligned normal to the applied strain direction. In addition, they found that when a high resistive load is connected to a polarized sample, there is no dramatic change in the wave front from that observed in the short-circuit case. They did see a dramatic change in the final particle velocity. It was lower in samples with high electric fields.

Brown and Chen [33] computationally investigated the electromechanical response of ferroelectric ceramics shock-loaded in the direction parallel to the direction of polarization (longitudinal shock). They found that the electric field reaches a peak value and then rapidly decays to a point where it gradually starts to rise. They concluded that the peak value of the field was due to the instantaneous piezoelectric response of the ceramic, the rapid decay was due to depolarization and the gradual rise was due to the structure of the mechanical disturbance.

In 1978, Rose and Mazzie [34] were granted a patent for a high-power, single-shot ferroelectric pulsed power source. The generator consisted of a pressure chamber, crystals (PZT or barium titanate) sandwiched between two electrodes, and a spark gap switch. Pressure exerted on one of the electrodes, which acted as a pusher plate, induced a net voltage difference across the spark gap, which discharged when the voltage across it reached its preset breakdown voltage. The pressure was created by igniting a combustible material with thermite pellets.

In 1978, Mock and Holt [35] experimentally investigated the pulse charging of a nanofarad capacitor by shock depoling PZT 56/44 and 95/5 transversely and normally. Some of their findings were:

- A decrease in charge flow from the PZT unit as the depoling stress wave passed through it. This observation was based on the decreasing amplitude of the current pulse and the concave-down shape of the voltage pulse. It was speculated that the decreasing charge flow may be due to increasing voltage on the PZT unit itself. This voltage increase may make it more difficult for the bound surface charge to be released or may cause a partial repoling of the PZT material behind the shock front. Another contributing factor may be due to stress release from the sides of the PZT unit, which was entirely encapsulated in a material having a lower shock impedance.

- An increase in charge released from identical PZT 56/44 units when the shock stress increased. This implied that the electrical conduction within the PZT unit did not increase in the stress range (4.4 to 7.8 GPa) studied.

- An increase in the load voltage as the size of the PZT unit increased. Increasing the size of the unit increased the amount of charge available for release.

In 1979, Mock and Holt [36] reported on the shock depolarization of longitudinally shock-compressed PZT 56/44 in the short-circuit mode. They looked at shock depolarization both parallel and antiparallel to the remanent polarization vector. For both orientations, they found that in the pressure range from 1.5 to 5.6 GPa, not all of the charge was released. In the pressure range 6.8 to 8.8 GPa, the charge released by the antiparallel orientation shots was less than that released by the parallel orientation shots.

Studies of shocked ferroelectric ceramics in the United States declined during the 1980s and 1990s. However, shock depolarization studies continued in other countries such as Russia, Japan and China.

3.3 Other Countries

Russia

Russia has an active FEG program which started in the 1960s. There are at least three documents that describe the early work done by Soviet researchers of ferroelectric power supplies [37–39].

They include:

- *Ferroelectricity and Ferroelectric Materials*, Report No. ATD 68-15-72-2 from the Aerospace Technology Division of the Library of Congress, Washington, D.C. (10 January 1972).
- V.A. Orlov, *Small-Sized Sources of Current*, U.S. Army Foreign Science and Technology Center Translation No. FSTC HT-23-228-72 of Malogabaritnyye Istochniki Toka, Moscow, Voyenizdat (1970).
- D.J. Butz and M.N. Golovin, *Bibliography of Soviet Literature on Shock-Compressed Piezoelectric Materials*, Battelle Report No. R-6193B (DTIC AD-B113 544) (November 1986).

Beginning in the mid-1960s, E.Z. Novitskii, V.D. Sadunov, and others [40–49] published a series of papers which reported on experimental studies and the theoretical analysis of different aspects of the operation of explosive ferroelectric generators. They investigated various ferroelectric materials for use in both longitudinal and transverse FEGs. The mentor for these authors appears to have been A.G. Ivanov, who was thought to have been affiliated with the O.Yu. Schmidt Institute of Physics of the Earth of the Russian Academy of Sciences, but more than likely was affiliated with the All-Russian Research Institute of Experimental Physics, also known as Arzamas-16, in Sarov.

V.N. Mineev may also have played a leading role in shock depolarization research. In 1976, Mineev and Ivanov [50] reported on the shock polarization and depolarization of linear and nonlinear dielectrics, semiconductors, and metals.

In 1982, V.D. Sadunov and Y.Z. Novitskii [51] received a USSR patent entitled Explosive Piezogenerator. This device was created to serve as a compact single-shot high-energy electrical power supply for lasers and X-ray sources, and to simultaneously initiate a large number of electrical blasting caps. Two variants of this generator were proposed. There were two common components in both variants; i.e., a transformer for converting explosive energy into electrical energy and a planar shock wave generator.

I.G. Tolstikov, V.D. Sadunov and E.Z. Novitskii [52] received another patent entitled Explosive Piezogenerator. This device was, in effect, an attempt to improve on their original FEG by improving its dependability and power output. An attempt was also made to prevent its initiation by unintended impacts. Thus an additional special non-polarized piezoelectrical plate assembly and a threshold switch were added to their device.

I.G. Tolstikov, V.A. Borisinok, and E.Z. Novitskii [53] were granted a patent entitled Piezoelectrical Generator. This was another attempt to improve their earlier versions without changing their basic design. Additional piezoelectric plates were added to the piezoelectric transformer in order to shorten the duration of the pulse it generated. These added plates were connected in parallel to the plates already in the transformer and distributed geometrically in a chessboard pattern. The non-polarized plate in their previous design was now polarized. The dimensions and parameters of the piezoelectric plates were chosen so that the electric charge generated by the transformer would first compensate for the depolarization of the piezoelectric plate assembly. The piezoelectric transformer and the plate assembly are both sources of electricity. These changes shortened the output electrical pulse and made the device less sensitive to impact.

I.G. Tolstikov, V.D. Sadunov, and T. V. Trishchenko [54] received a patent in 1989 entitled Piezoelectrical Generator. This patent was for changes to increase the output power and energy by optimizing the piezoelectric materials and their dimensions.

In the 2000s, Sadunov *et al.* [55] developed explosively-driven ferroelectric seed sources (prime power sources) for magnetic flux compression generators (FCGs), also called magneto-cumulative generators.

Demidov *et al.* [56–61] used a ferroelectric seed source to power a cascade of magnetic flux compression generators. The FEG provided a prime power for an FCG with a diameter of 50 mm, which in turn, was used to seed an FCG with a diameter of 100 mm through an impedance-matching transformer. They developed several different types of completely autonomous explosively-driven systems. The output energy generated by these compact autonomous systems ranged from a hundred kilojoules to the megajoule range.

Beginning in the 1980s, Prishchepenko and Tretyakov [62–65] conducted extensive testing of FEGs. They also integrated their FEGs with ferromagnetic generators to produce a very compact autonomous power supply. In 2005, Tretyakov [64] published a paper on generating high voltages by shock-loading ferroelectric elements.

There are several Russian organizations that appear to have studied shock-loaded ferroelectrics:

- The All-Russian Institute of Experimental Physics (VNIIEF), also known as Arzamas-16 (now Sarov). Some of the participants in VNIIEF's FEG program appear to have included A.G. Ivanov, E.Z. Novitskii, V.D. Sadunov, V.N. Mineyev, S.B. Kormer, L.V. Altshuler, V.A. Demidov, V.K. Chernyshev, and A.I. Pavlovskii.
- The group working with ferroelectric generators at the Chernogolovka branch of the Institute of Chemical Physics of the Russian Academy of Sciences appears to have been headed up by A.N. Dremin. Other participants included A.G. Antipenko, S.S. Nabatov, A.P. Kurto, and V.V. Yakushev.
- The group working with ferroelectric generators at the Institute of Hydrodynamics of the Russian Academy of Sciences in Novosibirsk appears to have been headed up by A.A. Deribas.
- The group working with ferroelectric generators at the Moscow State University appears to have been headed up by I.N. Polandov. This group was apparently doing very basic work on the theory and principles of the behavior of ferroelectric materials under both low and high hydrostatic pressure as well as under shock loading [66].
- Gorki State University (Nizhny Novgorod) reported on investigating the radiation emitted when a material was shock-loaded and generated an electromotive force [67].
- An unknown group published a paper on using ferroelectric materials in power supplies. Using barium titanate, they claimed to have charged a sample from 20 to 65 kV and to have produced 220 J/g [68].

France

In 1966, Besancon *et al.* [69] reported on their theoretical and experimental studies of FEGs, which they called ferroelectric transducers. They developed two different types of generators. The first was a low-impedance longitudinal FEG (the shock wave moved parallel to the polarization direction) and the second was a high-impedance transverse FEG (the shock wave moved perpendicular to the polarization direction). Two types of ferroelectrics were used in their experiments: barium titanate and lead zirconate titanate. They concluded that three phenomena accounted for the release of electric charge under shock compression: (1) the reorientation of dipoles due to the shock wave, (2) the piezoelectric effect and (3) the pyroelectric effect.

Romania

In 1987, Ludu, Nicolau and Novac [70] reported on their numerical and experimental study of the electrical pulses that could be generated by explosively-shocked PZT 51/49. Two types of shock wave attenuator were developed. The first was based on the impedance mismatch between copper and Plexiglas, and the second was based on elasto-plastic wave separation in AISI 4340 Steel. The former consisted of two pairs of copper-Plexiglas sandwiches, whose thicknesses were selected to avoid interference from reflecting shock waves while current was being generated. The highest pressure they generated in the PZT was 1.5 GPa. They found that if the pressure exceeded 5 GPa, there was a dramatic drop in current delivered to the load. They were able to reproducibly produce currents ranging from 2 to 35 A and powers ranging from 400 to 1400 W with both types of attenuator.

China

China currently has a very active FEG program [71–138]. The work in China appears to be centered at the National Key Laboratory of Shock Wave and Detonation Physics of the Southwest Institute of Fluid Physics (SWIFP) in Mianyang. According to Sun Qizhi, SWIFP began their FEG work in 2004. However, the Shanghai Institute of Ceramics, which produces PZT 95/5, appears to have begun working on FEG in the early 1980s.

Wang *et al.* [71–73] tested FEGs with PZT 95/5 with two types of dopant: Nb and Sb. They found that the Nb dopant gave the best results, since the Nb increased the electrical resistivity of the material and enhanced domain orientations, both of which increase the remnant spontaneous polarization, dielectric constant and piezoelectric constant of the PZT 95/5. They also used an FEG operating in the transverse mode to drive three types of load: a short-circuit load, a capacitive-resistive load and a capacitive-inductive load. For the short-circuit load, they typically measured currents of about 330 A. For the capacitive-resistive load, they used a 975 pF capacitive load with a 39 kΩ resistive load. They were able to generate 107 kV across the load. For the case of the capacitive-inductive load, they used PZT samples with dimensions of 0.5x7x1 cm^3 and a 3.9 nF capacitive load with a 59 μH inductive load. The maximum energy density stored in the inductor was 0.243 J/cm^3, which was significantly less than the energy density stored in the system prior to depolarization. This result indicates that a large amount of the charge was lost inside the generator, possibly due to leakage current during the depolarization process.

Wang, Dai, Sun and Chen [74] investigated possible applications for the phase transitions in PZT 95/5, one of which was explosive energy converters. They were able to deliver up to 1 kA to a short-circuit load, 15 kV to an open-circuit load and 0.5 J/cm^3 to an inductive load.

Du, Liu, Zhang, and He [75] published a paper in 2008 entitled Pulsed Power Supply of Ferroelectric Ceramics by Shock-Induced Depoling. They used PZT stacked in parallel. Operating in the normal mode, they generated a 7 kA pulse with a rise time of 500 ns.

During the last fifteen years, several Chinese research laboratories and universities have investigated different aspects of operation of explosive ferroelectric generators, conducted research on new high-energy density ferroelectric materials for FEG applications, and developed theoretical models to stimulate the operation of explosive FEGs.

In 2018, Wu *et al.* [137] reported on the results of investigations of the dielectric properties of PZT 95/5 during shock wave transit. They developed a new experimental technique for investigations of the dynamic dielectric properties of PZT 95/5 during shock compression and under high electric field. The dynamic permittivity was determined from oscillating periods in currents generated in the external oscillating circuits. This technique made it possible to measure the relative dielectric permittivity of PZT 95/5 in the stressed and unstressed zone under high electric fields. It was found that the permittivity of polarized PZT 95/5 in the unstressed zone is influenced heavily by the electric field. The permittivity increased from $\varepsilon_{r\,FE} = 198$ at zero field to $\varepsilon_{r\,FE} = 481$ at an electric field of 1.3 kV/mm. Experimental results indicated that the permittivity of shock-depolarized material in the stressed zone, on the contrary, had no correlation with the electric field. The permittivity of shock-depolarized PZT 95/5 determined from the experiments was $\varepsilon_r = 227$ under an electric field ranging from zero to 3 kV/mm.

In 2018, Peng, Nie, Wang, Liu, Cao and Dong [138] reported on the depolarization of $(Bi_{0.5}Na_{0.5})TiO_3$-$BiAlO_3$ lead-free ferroelectric ceramics under shock compression. Shocked BNT-BA ferroelectric ceramics became almost completely depolarized and released an electric charge density of $0.38\,C/m^2$. These results promise a new application of BNT-BA ceramics in shock wave ferroelectric generators.

Korea

In 2004, Seo and Ryu [139] measured the electric current and load voltages generated by explosively shocked PZT 56/44 when connected to resistive loads. They conducted tests at two shock pressure levels, 6.5 and 12 GPa, for resistive loads ranging from 10 to 1080 Ω. They concluded that the charge released by the PZT was affected by the shock pressure and load resistance. The maximum power and energy delivered to the loads used were 0.7 MW and 3.7 J, respectively, when the shock pressure was 12 GPa and the load resistances were 125 and 1080 Ω.

In 2010, Seo and Cho [140] experimentally studied the effects of impact conditions on the shock response of PZT-5 A ceramics for FEG devices. Effects of the impact on the shock strength of normally poled PZT-5 A FEG were studied within the pressure range of 2–5 GPa. Electrical breakdown was not observed inside the PZT-5 A specimen and the external circuit in the experiments. The maximum values of output voltages were almost constant in the pressure range from 2.4 to 3.8 GPa and increased in the case of 4.3 GPa pressure. The experimental results indicated that the output voltage depends on the load resistance and the load capacitance of the external circuit.

Japan

In 1986, Mashimo, Toda, and Nagayama [141] from Kumamoto University and Goto and Syono from The Research Institute for Iron, Steel, and Other Metals at Tohoku University looked at the electrical response of $BaTiO_3$ to shock compression. They looked at explosively shocking both longitudinal and transverse specimens.

In 1998, Xu, Akiyama, Nonaka, and Watanabe [142] investigated the electrical power generation characteristics of Mn-doped PZT ceramics when responding to slow mechanical stress as well as to impact stress. Although both the slow and impact stresses induce a reversible electrical response, the generation properties are distinctly different. Slow stress releases two output current peaks with opposite directions, responding to the increasing and decreasing part of the stress, respectively. However, impact stress produces a nearly one-directional signal. The output charge and energy produced by slow stress are found to be two orders of magnitude higher than those produced by impact stress. This work shows that the energy conversion efficiency of piezoelectric ceramics strongly depends on the method of stress application.

Germany

In Germany, both Diehl and Rheinmetall have investigated FEGs. In 2005, J. Bohl, J. Dommer, T. Ehlen, F. Sonnemann, and G. Staines from Diehl BGT Defense GmbH & Co. KG were granted a patent entitled High Voltage Generator, Especially for Using as a Noise Frequency Generator, U.S. Patent No. 6,969,944, on November 29, 2005 [143]. This device used stacks of ferroelectric elements to generate high voltage across a switch connected to an antenna. The mechanical pulse used to drive the two stacks was generated by a gas generator. This generator is described in detail in [144].

M. Jung, B. Schunemann, and G. Wollmann [145] from Rheinmetall Waffe Munition GMbH in Unterluss reported on a mechanically-driven compact high-voltage generator. They used two non-explosive methods for repetitively generating high-voltage pulses: a pressure spring oscillator and an accelerated piston. Both approaches were designed to operate with different ferroelectric stack sizes. They believe it is possible to build compact repetitively-pulsed high-voltage generators that are capable of generating output voltage pulses having amplitudes exceeding 100 kV.

Sweden

The Swedish Defense Research Agency (Totalforsvarets Forskningsinstitut (FOI)) began looking at FEGs in the early 2000s, but none of their work was published [146].

3.4 Recent United States

The most recent work in the U.S. has been done at Sandia National Laboratories, Loki Incorporated, Texas Tech University, Radiance Technologies, HEM Technologies, and Ktech Corporation.

In the 2000s, Setchell and others at Sandia National Laboratories revived systematic studies of the mechanical, physical and electrical properties of PZT 95/5 shock-compressed by planar shock waves initiated by impacting projectiles accelerated in a gas gun facility.

In 2003, Setchell [147] reported on detailed investigations of the mechanical properties of shock-compressed PZT 95/5. Planar-impact experiments were conducted to determine Hugoniot states and to examine constitutive mechanical properties during shock propagation. Wave profiles recorded in transmitted-wave experiments examined the effects of varying shock strength and propagation distance, poling state and orientation, and electric field strength. The collective results identified a complex material behavior governed by anomalous compressibility and incomplete phase transformation at low shock amplitudes, and a relatively slow yielding process at high shock amplitudes.

In 2005, Setchell [148] reported on the results of systematic investigations of the shock-induced depolarization of PZT 95/5. Planar-impact experiments were conducted over a wide range of conditions in order to examine the effects of varying shock strength, poling orientation, input wave shape, and electric field strength. Depolarization current waveforms were recorded in an external circuit under either short-circuit or high-field conditions, and provided a convenient means of examining the kinetics associated with the ferroelectric-to-antiferroelectric phase transition. For sufficiently strong shock waves, the measured short-circuit currents indicated that the phase transition was very rapid and essentially complete. As shock strengths were reduced, short-circuit currents showed increasing rise times and decreasing final levels at the end of shock transit. These features indicated that the transition kinetics can be characterized in terms of both a transition rate and a limiting degree of transition achieved in a given shock experiment.

In 2007, Setchell [149] looked at the effects of microstructures on PZT 95/5. Dynamic yield is the term given to the observation that an elastic precursor is followed by a plastic wave in plate-impact experiments. In the case of porosity, dynamic yield is likely associated with pore collapse and may actually initiate fracture, rather than help to avoid fracture. By adding different types and amounts of organic pore formers to PZT 95/5 prior to bisque firing and sintering, Setchell found that differences in porous microstructured materials having the same initial density have little effect on the mechanical and electrical shock properties of PZT 95/5, but that the initial density of the material has a significant impact. In a follow-on study,

Setchell found that when large pore formers are used in materials with the same initial density, the shock properties are insensitive to microstructural differences, but that the smallest pore formers did have an impact in that there was a significantly higher threshold for dynamic yielding due to increased initial density.

Increasing the yield threshold has both advantages and disadvantages. On the one hand, the shock stress required for rapid and complete depolarization is only a few tenths of a GPa below the threshold stress for yielding, so increasing the yield strength could reduce the formation of fractures and, thus, reduce the probability of electrical breakdown. On the other hand, increasing the initial density, and thus, the yield threshold, increases the bound charge and shock velocity, which may not be desirable. Setchell [149] pointed out that materials with limited microstructures would have a broader range of shock pressures for effective depolarization without altering other characteristics.

In 2007, Setchell *et al.* [150, 151] reported on the results of detailed studies of mechanical properties of shock-compressed alumina-filled epoxies. Planar-impact experiments producing states of nearly equal strain were conducted to investigate the effects of changes in the size and shape of alumina particles, and in the total volume fraction of alumina. Laser interferometry and wave timing were used to obtain transmitted wave profiles, Hugoniot states, and release wave velocities. In addition, wave profiles and velocities were obtained in "thin-pulse" experiments that examined the combined effects of compression and release properties in different compositions. Changes in the size and shape of alumina particles were found to have little effect except in the viscous spreading of wave profiles during shock compression. Increasing the volume fraction of alumina resulted in steadily increasing Hugoniot states, wave rise times, and release wave velocities. An important observation was that differences between release wave and shock wave velocities increased significantly as the alumina loading was increased. The consequences of this effect were evident in the thin-pulse experiments, which showed that increased alumina loading resulted in stronger wave attenuation.

In 2006, Setchell *et al.* [152] reported on the results of experimental investigations of the dielectric properties of PZT 95/5 during shock wave transit. To study the dielectric properties of PZT 95/5 during shock compression under high electric field the authors [152] developed new experimental methods to better isolate dielectric properties in both unstressed and stressed PZT 95/5. These methods are based on shock-driven circuits containing multiple, small PZT 95/5 elements that are displaced both parallel and perpendicular to the shock motion. For the experimental data analysis [152] a model was used that assumed that the displacement current is governed by the Debye approximation for a relaxing dielectric. The permittivity of polarized PZT 95/5 in the unstressed zone was obtained at three different levels of electric field: $\varepsilon(1.58\,kV/mm)=$ 15.8 nF/m ($\varepsilon_{r\,FE} = 1785$), $\varepsilon(3.04\,kV/mm)=$ 14.8 nF/m ($\varepsilon_{r\,FE}=$ 1672) and $\varepsilon(3.38\,\text{kV/mm}) = 13.1\,\text{nF/m}$ ($\varepsilon_{r\,FE}=$ 1480). The experimental results [152] indicated that the permittivity values of shock depolarized material in the stressed zone were significantly lower than those of polarized material in the unstressed zone.

In 2007, Setchell *et al.* [153] reported on the results of the investigations of the dielectric properties of unshocked and shocked PZT 95/5 in a wide temperature range. The authors used their method, based on shock-driven circuits with multiple PZT 95/5 elements [153], to examine dielectric properties in PZT 95/5 samples at initial temperatures from −56 to 74°C. Significant changes in permittivity with temperature were observed in both unshocked and shocked samples. Measured currents show a complex dielectric response which can only be partially predicted using a simple dielectric relaxation model.

In the early 2000s, Texas Tech University [154] looked at driving magnetic flux compression generators with FEGs. They built an FEG capable of delivering up to 1.7 J to a high inductive load. They concluded that FEGs show potential as seed sources, since they are capable of driving much larger load impedances than other methods.

Texas Tech University [155, 156] reported on their attempts to improve the performance of explosive-driven FEGs. Texas Tech investigated several FEG designs and methods to prevent surface flashover with varying results. In their FEGs they used PZT 52/48 (EC-64) disks with a thickness of 0.4 inches and a diameter of 1 inch. They tested and built several generators that generated open-circuit voltage pulses of 80 to 140 kV with a full width half maximum of 2 to 4 μs.

Radiance Technologies [157] has produced FEGs by using PZT materials fabricated in cylindrical shell geometries. This approach represented a departure from the traditional bar and disk geometries of past work, taking advantage of the symmetry and favorable scaling of cylindrical configurations. Cylindrical or conical structures offer favorable scaling since the surface area and volume of the material scale with the radius squared and height of the structure and with suitable wave shaping, allow large areas and volumes to be utilized in the FEG. With the addition of pressure wave shaping, these geometrical structures minimize the amount of explosive needed for excitation, and allow multiple pulse generation at kilohertz rates by placing additional units on the same rope like an explosive power train. In their laboratory [158], they have developed techniques to provide a reliable, simple pulse generator capable of producing large electrical pulses in a small compact geometry. The generators are inherently rugged and shock resistant. They have been used to drive resistive and capacitive loads with results consistent with the published work of others.

HEM Technologies [159] has worked extensively on power conditioning, high-G launch survivability, and the characterization of ferroelectric generators. Power conditioning is critical to ensure that the maximum small amount of energy in a generator is delivered to the load. FEGs were successfully matched to a variety of dipole and folded dipole antennas and microwave sources. HEM Technologies was the first to demonstrate the ability of an FEG to power a microwave source with an impedance-matched FEG. Power levels of > 40 kW have been achieved in these tests. Additionally, HEM Technologies was the first to demonstrate multi-pulse capabilities from a single FEG using an exploding wire array.

HEM Technologies has also tested the survivability of ferroelectrics under high-G launch conditions. Tests were conducted with FEGs launched at up to 10,000 Gs. The critical causes for energy loss in ferroelectrics associated with a high-G launch were identified and methods to mitigate these losses were also studied.

HEM Technologies [160] reported on the characterization of a variety of ferroelectric materials for FEGs. Testing was conducted to determine optimal pressures and pressure pulse shapes on the different materials. The high-voltage modifications of PZT 95/5 produced by TRS Technologies were found to have the highest energy density and voltage output.

The University of Missouri in Columbia [161–163] conducted a series of studies on kinetic-to-electrical energy piezoelectric (PZT) generators. In one series of tests, they placed the piezoelectric generators inside high-speed 30 mm projectiles. The projectiles were launched with a high-pressure helium gun to velocities of approximately 300 m/s. Upon impact, a peak output power of 25 kW was delivered to a 10 Ω load in a pulse with a length on the order of 1 μs. They also investigated those factors that affect the maximum peak power that their piezoelectric generators could produce. These factors include the piezoelectric material's thickness to area ratio, the mechanical stress force, the material's stress limit, and the external circuit's parameters. They concluded that the maximum peak power scaled with the increasing volume of piezoelectric material. The highest peak power measured in these investigations was 320 kW.

Ktech (now Raytheon Ktech) [164] conducted extensive studies of the properties of PZT 95/5 and employed them in the construction of power supplies capable of delivering 100 kV to capacitive loads and 120 kV to resistive loads.

The U.S. Army Aviation and Missile Research, Development, and Engineering Center (AMRDEC) [165] used an FEG to impulse load a sinuous antenna. They demonstrated that a single-shot FEG-sinuous antenna system can generate multiple pulses at a predictable rate and frequency. They further demonstrated that this design is flexible and mature.

AMRDEC [166] have used FEGs to initiate blasting cap arrays. A switchless FEG was designed to have suitable characteristics for initiating arrays of commercial blasting caps. This unique FEG and arrays of blasting caps were tested in many configurations including those with 36, 64, and 128 cap arrays in series. These arrays represented an inductive load for the FEG and constituted a major factor in the coupled response of the FEG-array circuit. The switchless design showed that reasonable characteristics were utilized through good explosive design of the plane-wave generator versus the more common peaking switch approach.

Loki Incorporated [167–193] has conducted systematic studies of explosive ferroelectric generators since the end of the 1990s. Loki developed and tested a variety of designs of ferroelectric generators and a wide range of ferroelectric materials. In 2006, Loki Incorporated was a recipient of the U.S. Small Business Administration's Tibbetts Award for outstanding achievements in the Small Business Innovation Research (SBIR) program. In 2008, Loki Incorporated was a recipient of the U.S. Army SBIR Achievement Award. The work done by Loki Incorporated will be presented in the remaining chapters of this book.

Bibliography

1. F.W. Neilson, Ferromagnetic and ferroelectric one-shot explosive electric transducers, *Sandia Technical Report* SCTM-230B-56(51) (1956).
2. F.W. Neilson, Effects of strong shocks in ferroelectric materials, *Bull. Am. Phys. Soc.* **11**(2) (1957) p. 302.
3. R.B. Gray, Transducer and method of the same. U.S. Patent 2,486,560. Granted November 1, 1949.
4. E. Kabik and E.L. Cecil, Impact responsive electric primer. U.S. Patent 2,972,306. Granted Feb. 21, 1961.
5. P.B. Ferrara, Energizer assembly. U.S. Patent 2,827,851. Granted March 25, 1958.
6. R.C. Howe, Energy converting device. U.S. Patent 2,970,545. Granted February 7, 1961.
7. R.H.F. Stresau, System for multiple point simultaneous initiation of explosive charges. U.S. Patent 3,589,294. Granted June 29, 1971.

8. D. Berlincourt and H.A. Krueger, Domain processes in lead titanate zirconate and barium titanate ceramics, *J. Appl. Phys.* **30**(11) (1959) pp. 1804–1810.

9. R.H. Wittekindt, Shape of the current output pulse from a thin ferroelectric cylinder under shock compression, Diamond Ordance Fuze Laboratory (DOFL) Report, TR-922 (May 1961).

10. C.E. Reynolds and G.E. Seay, Multiple shock wave structures in polycrystalline ferroelectrics, *J. Appl. Phys.* **32**(7) (1961) pp. 1401–1402.

11. C.E. Reynolds and G.E. Seay, Two-wave shock structures in the ferroelectric ceramics: barium titanate and lead zirconate titanate, *J. Appl. Phys.* **33** (1962) pp. 2234–2239.

12. W.L. Gilbertson, A practical one-shot high voltage power supply for electroexplosive devices, *U.S. Naval Weapons Laboratory Technical Memorandum* No. T-12/64 (DTIC AD No. AD813247) (July 1964).

13. R.A. Graham, F.W. Nielson and W.B. Benedick, Piezoelectric current from shock-loaded quartz — a submicrosecond stress gauge, *J. Appl. Phys.* **36**(5) (1965) pp. 1775–1783.

14. J.T. Cutchen, Polarity effects and charge liberation in lead zirconate titanate ceramics under high dynamic stress, *J. Appl. Phys.* **37**(13) (1966) pp. 4745–4750.

15. R. Gerson and H. Jaffe, Electrical conductivity in lead titanate zirconate ceramics, *Journal of Physics and Chemistry of Solids* **24**(8) (1963) pp. 979–984.

16. W.J. Halpin, Current from a shock-loaded short-circuited ferroelectric ceramic disk, *J. Appl. Phys.* **27**(1) (1966) pp. 153–163.

17. W.J. Halpin, Resistivity Estimates for Some Shocked Ferroelectrics, *J. Appl. Phys.* **39**(8) (1968) pp. 3821–3826.

18. P.E. Houser, Charge release due to a pressure enforced ferroelectric-to-antiferroelectric phase transition in the system $PBZrO_3$-$PbTiO_3$-$PbSnO_3$-Nb_2O_5", *Picatinny Technical Report 3481* (DTIC Report No. AD813182) (April 1967).

19. R.K. Linde, Depolarization of ferroelectrics at high stress rates, *J. Appl. Phys.* **38**(12) (1967) pp. 4839–4842.

20. R.A. Graham and W.J. Halpin, Dielectric breakdown and recovery of X-cut quartz under shock-wave compression, *J. Appl. Phys.* **39**(11) (1968) pp. 5077–5082.

21. D.G. Doran, Shock-wave compression of barium titanate and 95/5 lead zirconate titanate, *J. Appl. Phys.* **39**(1) (1968) pp. 40–47.

22. P.C. Lysne, Dielectric breakdown of shock-loaded PZT 95/5, *J. Appl. Phys.* **44**(2) (1973) pp. 577–582.

23. P.C. Lysne, Prediction of dielectric breakdown in shock-loaded ferroelectric ceramics, *J. Appl. Phys.* **46**(1) (1975) pp. 230–232.

24. P.C. Lysne, Kinetic effects in the electrical response of a shock-compressed ferroelectric ceramics, *J. Appl. Phys.* **46**(11) (1975) pp. 4078–4079.

25. F.E. England and J.H. Gifford, Solid state power supply activated by a pyrotechnic chain. U.S. Patent 3,756,157. Granted September 4, 1973.

26. P.C. Lysne and C.L. Bartel, Electromechanical response of PZT 65/35 subjected to axial shock loading, *J. Appl. Phys.* **46**(1) (1975) pp. 222–229.

27. P.C. Lysne and C.M. Percival, Electric energy generation by shock compression of ferroelectric ceramics: normal-mode response of PZT 95/5, *J. Appl. Phys.* **46**(4) (1975) pp. 1519–1525.

28. P.C. Lysne, Dielectric properties of shock-compressed ferroelectric ceramic, *J. Appl. Phys.* **48**(3) (1977) pp. 1020–1023.

29. P.C. Lysne, Shock-induced polarization of a ferroelectric ceramic, *J. Appl. Phys.* **48**(3) (1977) pp. 1024–1031.

30. J.A. Mazzie, Simplified model of ferroelectric energy generation by shock compression, *J. Appl. Phys.* **48**(3) (1977) pp. 1368–369.

31. I.J. Fritz, Uniaxial-stress effects in 95/5 lead zirconate titanate ceramic, *J. Appl. Phys.* **49**(9) (1978) pp. 4922–4928.

32. J.J. Dick and J.E. Vorthman, Effect of electrical state on mechanical and electrical response of a ferroelectric ceramic PZT 95/5 to impact loading, *J. Appl. Phys.* **49**(4) (1978) pp. 2494–2498.

33. W.T. Brown and P.J. Chen, On the nature of the electric field and the resulting voltage in axially loaded ferroelectric ceramics, *J. Appl. Phys.* **49**(6) (1978) pp. 3446–3450.

34. M.F. Rose and J.A. Mazzie, Ferroelectric pulsed power source. U.S. Patent 4,090,448. Submitted 29 December, 1971. Granted May 23, 1978.

35. W. Mock, Jr. and W.H. Holt, Pulse charging of nanofarad capacitors from the shock depoling of PZT 56/44 and PZT 95/5 ferroelectric ceramics, *J. Appl. Phys.* **49**(12) (1978) pp. 5846–5854.

36. W. Mock, Jr. and W.H. Holt, Axial-current-mode shock depoling of PZT 56/44 ferroelectric ceramic disks, *J. Appl. Phys.* **50**(4) (1979) pp. 2740–2748.

37. *Ferroelectricity and Ferroelectric Materials*, Report No. ATD 68-15-72-2 from the Aerospace Technology Division of the Library of Congress, Washington, D.C. (January 10, 1972).

38. V.A. Orlov, *Small-Sized Sources of Current* (U.S. Army Foreign Science and Technology Center Translation No. FSTC HT-23-228-72 of *Malogabaritnyye Istochniki Toka* (Moscow, Voyenizdat, 1970).

39. D.J. Butz and M.N. Golovin, *Bibliography of Soviet Literature on Shock-Compressed Piezoelectric Materials*, Battelle Report No. R-6193B (DTIC AD-B113 544) (November 1986).

40. A.G. Ivanov and E.Z. Novitskii, The Double Layer in an Insulator Subject to Shock Compression, *Journal of Applied Mechanics and Technical Physics* **7**(5) (1966) pp. 73–75.

41. A.G. Ivanov, V.N. Mineev, E.Z. Novitskii, Yu.N. Tyunyaev and Yu.V. Lisitsyn, Electrical effects associated with shock loading, *Combustion, Explosion, and Shock Waves* **5**(4) (1969) pp. 356–360 (translated from *Fizika Goreniya i Vzryva*).

42. E.Z. Novitskii, V.V. Kolesnikov and R.V. Vedrinskii, Shock depolarization of a piezoelectric ceramic: phenomenology, *Combustion, Explosion, and Shock Waves* **9**(6) (1973) pp. 778–783 (translated from *Fizika Goreniya i Vzryva*).

43. E.Z. Novitskii, V.A. Ogorodnikov and S.Yu. Pinchuk, Phase transition in a shock-loaded piezoelectric ceramic of the PZT system, *Combustion, Explosion, and Shock Waves* **13**(2) (1973) pp. 221–224 (translated from *Fizika Goreniya i Vzryva*).

44. E.Z. Novitsky, V.D. Sadunov and G.Ya. Karpenko, Behavior of ferro-electrics in shock waves, *Combustion, Explosion, and Shock Waves* **14**(4) (1978) pp. 505–516 (translated from *Fizika Goreniya i Vzryva*).

45. E.Z. Novitskii, V.D. Sadunov and T.V. Trishchenko, Investigation of the electrophysical properties of ferroelectrics under conditions of shock-wave loading: I. Methods of investigation, *Combustion, Explosion, and Shock Waves* **16**(1) (1980) pp. 82–92 (translated from *Fizika Goreniya i Vzryva*).

46. E.Z. Novitskii, V.D. Sadunov and T.V. Trishchenko, Electrophysical properties of ferroelectrics under shock loading conditions: II. Physical representations, *Combustion, Explosion, and Shock Waves* **16**(1) (1980) pp. 462–466 (translated from *Fizika Goreniya i Vzryva*).

47. E.Z. Novitskii, M.V. Korotchenko, M.D. Volnyanskii and V.A. Borisenok, Investigation of the dynamic piezoelectric moduli of single crystals of Bi12GeO20, Li2GeO3, and LiNbO3, *Combustion, Explosion, and Shock Waves* **16**(1) (1980) pp. 93–98 (translated from *Fizika Goreniya i Vzryva*).

48. E.Z. Novitskii and V.D. Sadunov, Development of breakdown in a shock compressed ferroelectric, *Combustion, Explosion, and Shock Waves* **20**(4) (1984) pp. 439–441 (translated from *Fizika Goreniya i Vzryva*).

49. E.Z. Novitskii and V.D. Sadunov, Energetic characteristics of a ferroelectric working body of a shock-wave energy transducer, *Combustion, Explosion, and Shock Waves* **21**(5) (1985) pp. 611–615 (translated from *Fizika Goreniya i Vzryva*).

50. V.N. Mineev and A.G. Ivanov, Electromotive force produced by shock compression of a solid, *Soviet Physics Uspekhi* **19** (1976) pp. 400–452.

51. V.D. Sadunov and E.Z. Novitskii, Explosive Piezogenerator. USSR Patent AS of SU No. 1119564A, H 01 L41/08, H 02 N 11/00, Granted July 16, 1982.

52. I.G. Tolstikov, V.D. Sadunov and E.Z. Novitskii, Explosive Piezogenerator. USSR Patent AS of SU, 08.10.87, No 1533612 A2, H 02 N2/00, H 01 L 41/08.

53. I.G. Tolstikov, V.A. Borisinok and E.Z. Novitskii, Piezoelectrical Generator, USSR Patent AS of SU, No. 1600597, H 02 N 2/00.

54. I.G. Tolstikov, V.D. Sadunov and T.V. Trishchenko, Piezoelectrical Generator, USSR Patent AS of SU, 05.09.89, 1777532 A1, H 021 N 2/00 (1989).

55. V.D. Sadunov, V.A. Demidov, A.L. Mikhaylov and T.V. Trishchenko, Ferroceramic source of EMG power supply with linear growth of current,

in book: Megagauss and Megampere Pulse Technology and Applications (Megagauss-9 / International Conference on Megagauss Magnetic Field and Related Topis), (eds. V.D. Selemir and L.N. Plyashkevich, Sarov, VNIIEF, 2004) pp. 228–231.

56. V.A. Demidov, V.D. Sadunov, S.A. Kazalov, T.V. Trishchenko, A.V. Blinov, I.K. Fetison, M.V. Kortchenko, S.N. Golosov and Ye.V. Shapovalov, Piezoceramic power supply of EMF, *in book: Megagauss and Megampere Pulse Technology and Applications* (eds. V.K. Chernyshev, V.D. Selemir and L.N. Plyashkevich, Sarov, VNIIEF, 1997) / *Proceedings of the 7th International Conference on Megagauss Magnetic Field Generation and Related Topics* (1997) pp. 336–339.

57. V.A. Demidov, V.D. Sadunov, S.A. Kazakov, L.N. Plyashkevich, T.V. Trischenko, S.N. Golosov, A.V. Blinov, I.K. Fetisov, M.V. Korotchenko and Ye.V. Shapovalov, Helical cascade FCG powered by piezogenerator, *Proceedings of the 11th IEEE International Pulsed Power Conference* (eds. G. Cooperstein and I. Vitkovitsky), Baltimore, Maryland (1997) Cat. No.97CH36127, Vol. 2, pp. 1476–1481.

58. V.A. Demidov, V.D. Sadunov, S.A. Kazakov, L.N. Plyashkevich, T.V. Trischenko, S.N. Golosov, A.V. Blinov, M.V. Korotchenko and Ye.V. Shapovalov, Helical cascade FCG powered by piezogenerator, *Proceedings of the 12th International Conference on High Power Particle Beams*, Cat. No.98EX103, Vol. 1, pp. 269–272 (1998).

59. V.A. Demidov, V.D. Sadunov, S.A. Kazakov, S.N. Golosov, A.A. Utenkov, A.S. Boriskin, M.V. Antipov, A.V. Blinov, I.V. Yurtov and L.N. Plyashkevich, Autonomous magneto-cumulative energy source, in *Proceedings of 2012 14th International Conference on Megagauss Magnetic Field Generation and Realted Topics*, Maui, HI, USA, 14–19 Oct. 2012 (eds. G.F. Kiuttu, K.W. Struve and J. Degnan) pp. 77–84 (2012). Catalogue: CFP12MEG-ART.

60. V.A. Demidov, V.D. Sadunov, S.A. Kazakov, A.S. Boriskin, S.N. Golosov, Yu. V. Vlasov, A.A. Utenkov, M.V. Antipov and A.V. Blinov, Autonomous magnetocumulative power supply, *Technical Physics* **58**(8) (2013) pp. 1219–1223 (translated from *Zhurnal Tekhnicheskoi Fiziki* Vol. 83, No. 8, (2013) pp. 135–139).

61. V.A. Demidov, V.D. Sadunov, S.A. Kazakov, S.N. Golosov, A.A. Utenkov, A.S. Boriskin, M.V. Antipov, A.V. Blinov and I.V. Yurtov, Studying autonomous magneto-cumulative energy source, *IEEE Trans. Plasma Sci.* **43**(1) (2015) pp. 339–343.

62. A.B. Prishchepenko, D.V. Tretyakov and M.V. Shchelkachev, Energy balance in frequency explosive piezoelectric generator during its operation, *in book: Megagauss and Megampere Pulse Technology and Applications* (eds. V.K. Chernyshev, V.D. Selemir and L.N. Plyashkevich, Sarov, VNIIEF, 1997) / *Proceedings of the 7th International Conference on Megagauss Magnetic Field Generation and Related Topics* (1997) pp. 310–313.

63. A.B. Prishchepenko and D.V. Tretyakov, Operation of a generator with a working body made out of ferromagnetic and ferroelectric Materials, *Electricity* **4** (2000) pp. 60–63 (in Russian).

64. D.V. Tretyakov, Generation of high-voltage pulses by shock-loading ferroelectric elements, *Instruments and Experimental Techniques* **48**(6) (2005) pp. 726–729 (translated from *Pribory i Tekhnika Eksperimenta*).

65. D.V. Tretyakov, Estimation of the parameters of the explosive voltage generator with the ferroelectric working body, *Electricity* **12** (2000) pp. 56–61 (in Russian).

66. I.N. Polandov, O.K. Gulish, B.I. Malytin and V.I. Minayev, An investigation of ferroelectrics in the system Pb(TiZr)03 with low content of Ti and doped by modifying agents, *Vestnik Moskovskogo Universiteta (Moscow State University Reprints), Seriya II (Series II), Khimiya (Chemistry)* **18**(3) (1977) pp. 344–347 (in Russian).

67. Yu.K. Bivin, V.V. Viktorov, Yu.V. Kulinich and A.S. Chursin, Electromagnetic radiation upon dynamic deformation of different materials, *Proceedings of the Fifth All-USSR Congress of Theoretical and Applied Mechanics*, Alma-Alta (1981) pp. 83–88 (in Russian).

68. V.S. Avduevskii, Yu.V. Krylov, V.P. Saltykov and I.P. Terentev, Experimental study of a ferroelectric energy storage device, *Soviet Physics Doklady* **30**(6) (1985) pp. 523–524.

69. J.E. Besancon, J. David and J. Vedel, Ferroelectric transducers, *Proceedings of the Conference on Megagauss Field Generation by Explosives and Related Experiments* (eds. H. Knoepfel and F. Herlach, Euroatom, Brussels), pp. 315–328 (1966).

70. A. Ludu, P. Nicolau and B.M. Novac, Shock wave explosive energy generator of PZT ferroelectric ceramics, *in book: Megagauss Technology and Pulsed Power Applications* (eds. C.M. Fowler, R.S. Caird and D.J. Erickson, Plenum Press, New York, 1987) pp. 369–375.

71. Y.L.Wang, W.Z. Yuen, G.R. He, S.W. Lin, Y.H. Ling, C.F. Qu and B.G. Wang, Study on shock wave explosive energy converter of PZT 95/5 ferroelectric ceramics, *Ferroelectrics* **49** (1983) pp. 169–176.

72. Wang Yong-ling, Yuan Wan-zong, He Guo-rong, Lin Sheng-wei, Ling Rong-hua and Qu Cui-feng, Ferroelectric explosion — electrical transducer experimental study, *Acta Physica Sinica* **32**(6) (1983) pp. 780–785.

73. Y.L. Wang and K.C. Kao, New ferroelectric ceramics for the generation of high energy electric pulses, *Proceeding of the 2000 IEEE International Symposium on Electrical Insulation* (2000) pp. 58–61.

74. Y.L. Wang, X. Dai, D. Sun and H. Chen, The applications of PZT 95/5 ceramics by induced phase transformation, *Proceedings of IEEE 7th International Symposium on Applications of Ferroelectrics* (1999) pp. 513–516.

75. J.M. Du, G.M. Liu, F.P. Zhang and H.L. He, Pulsed power supply of ferroelectric ceramics by shock-induced depoling, *Key Engineering materials* **368–372** (2008) pp. 18–20.

76. G. Liu, M. Jin, Y. Liu, H. Tan and H. He, Shock wave compression of PZT 95/5 ferroelectric ceramic, *Chinese Journal of High Pressure Physics* (2008).

77. G. Liu, Y. Liu, Y. Zhang, J. Du and H. He, Ferroelectric ceramic for shock driven pulsed power supply, *Materials Review* (2006).

78. J. Ju, Y. Zhang, F. Zhang, H. He and H. Wang, Large current output of PZT 95/5 ferroelectric ceramics under shock loading, *Acta Physica Sinica* (2006).

79. J. Du, G. Liu, Y. Liu, H. Wang, F. Zhang and H. He, Current output from sandwich-structured ferroelectric power supply, *Proceedings of the Conference of the American Physical Society Topical Group on Shock Compression of Condensed Matter*, (AIP Conference Proceedings, CP955, 2007) pp. 167–170 (2007).

80. Wang Yong-ling, Yuan Wan-zong, He Guo-rong, Lin Sheng-wei, Ling Rong-hua and Qu Cui-feng (Shanghai Institute of Ceramics, Academia Sinica), Study on shock wave-explosive energy conversion of ferroelectrics, *Acta Physica Sinica* **32**(6) (1983) pp. 780–785.

81. Lin Qiwen, Study of electrical response of shock wave loaded ferroelectrics, *Explosion and Shock Waves* (1984).

82. Lin Qiwen and Yuan Wanzong (Southwest Institute of Fluid Physics), Theoretical investigation of cylindrical explosive-electric transducer, *Explosion and Shock Waves* (1988).

83. Zhang Shouqi, Wang Bingcheng, Xie Mingshu, Dong Qingdong and Ma Shuyuan, Some possible applications of explosive energy, *Explosion and Shock Waves* (1984).

84. Bau Zhongxing, Gu Huicheng and Zhang Shiting, Compressibility and phase transition of PZT 95/5 ferroelectric ceramics at high pressure, *Chinese Journal of High Pressure Physics* (1987).

85. Lian Jing-yu and Wang Yong-ling (Shanghai Institute of Ceramics), The application of FE1-FE2 phase transition in Nb dope (95/5) ferroelectric ceramics to heat-electric energy conversion, *Chinese Physics Letters* **2**(6) (1985) pp. 269–270.

86. Lin Qiwen, Yuan Wanzong and Wang Weijun (Southwest Institute of Fluid Physics), Investigation on the match of ferroelectric explosive-electric transducers, *Chinese Journal of High Pressure Physics* (1988).

87. Liu Gaomin, Zhang Fuping, Du Jinmei, Tan Hua and He Hongliang, Phase transition and current properties of PZT 95/5-2Nb ferroelectric ceramic under dynamic loading, *International Journal of Modern Physics B* **22** (2008) pp. 1171–1176.

88. Wen Dianying and Lin Qiwen (Lab. for Shock Wave and Detonation Physics Research, Institute of Fluid Physics, CAEP, P. O. Box 523, Chengdu 610003), Dielectric breakdown of ferroelectric ceramics PZT-95/5 under shock compression, *Chinese Journal of High Pressure Physics* (1998).

89. He Yuan ji, Zhang Ya zhou, Li Chuan lu and Wang Hong Gang (Science College, National University of Defense Technology, Changsha 410073, China),

The numerical simulation of electric response of PZT 95/5 ferroelectric ceramics subjected to shock loading, *Chinese Journal of High Pressure Physics* (2000).

90. He Yuan ji, Zhang Ya zhou and Li Chuan lu (College of Science, National Univ. of Defense Technology, Changsha 410073, China), Study on explosive electrical exchange energy, *Journal of National University of Defense Technology* (2000).

91. Wen Dian-ying and Lin Qi-wen (Laboratory for Shock Wave and Detonation Physics Research, Institute of Fluid Physics, CAEP, Mianyang 621900,China), Electrical response of PVDF film under shock loading, *Chinese Journal of High Pressure Physics* (2000).

92. He Yan Ji, Zhang Ya Zhou and Li Chuan Zuo, Analysis of dielectric breakdown in PZT95/5 ferroelectric ceramics, *High Voltage Engineering* **27**(2) (2001).

93. He Yuan-ji, Zhang Ya-zhou and Li Chuan-lu (College of Science, National University of Defense Technology, Changsha 410073, China), Study on electrical response of PZT95/5 subjected to shock stress, *Journal of Functional Materials and Devices* (2001).

94. Du Hui, Sun Da-zhi, Zhong Ni, Qu Cui-feng, Yao Chun-hua and Jin Qi-hua (Shanghai Institute of Ceramics, Chinese Academy of Sciences, Shanghai 200050), Study on dielectric property of the modified PZT95/5 ferroelectric ceramics, *Electronic Components & Materials* (2002).

95. Liu Gao Min, Tan Hua, Yuan Wan Zong, Wang Hai Yan and Zhang Yi, Ferroelectric/antiferroelectric phase transition studies of PZT 95/5 ceramics under shock loading, *Chinese Journal of High Pressure Physics* **16**(3) (2002) pp. 231–236.

96. He Yuanji, Zhang Yazhou and Li Chuanlu (The Fourth Institute of the Second Artillery, Beijing 100085, China, College of Science, National University of Defense Technology, Changsha 410073, China), High power pulse supplier of PZT 95/5 ferroelectric ceramics for charging of nanofarad capacitors, *High Voltage Engineering* (2004).

97. Li Yingping, Lai Baitan, Chen Hejuan and Huang Xiaomao (Nanjing University of Science and Technology, Nanjing, 210094), A study on the principle of a new energy storage piezoelectric power supply, *Acta Armamentarii* (2004).

98. He Yuan-ji, Zhang Ya-zhou and Li Chuan-lu (The Fourth Institute of the Second Artillery, Beijing 100085, China; College of Science, National University of Defense Technology, Changsha 410073,China), Experimental study on electrical response of PZT 95/5 subjected to shock stress, *Journal of Functional Materials and Devices* (2004).

99. Zhang Fu-Ping, He Hong-Liang, Du, Jin-Mie, Wang Hai-Yan and Liu Gao-Min, Influence of grain size on breakdown voltage of PZT 95/5 ferroelectric ceramics under shock compression, *Journal of Inorganic Materials* **20**(4) (2005) pp. 1019–1024.

100. Chen Lang, Lu Jian-Ying, Wu Jun-Ying and Feng Chang-Gen (State Key Laboratory of Explosion Science and Technology, Beijing Institute of Technology, Beijing 100081, China), Numerical calculation of ferroelectric power supply initiating metal bridge foil, *Chinese Journal of High Pressure Physics* (2006).

101. Tang Ya-ming and Zhang Yu-qin (College of Mechanical & Electrical Engineering, Hohai University, Changzhou, China), Piezoelectric power supply of tire burst blow ignition fuse, *Journal of Hohai University Changzhou* (2006).

102. Du Jin-Mei, Zhang Fu-Ping, Zhang Yi and Wang Hai-Yan (Laboratory for Shock Wave and Detonation Physics Research, Institute of Fluid Physics, CAEP, Mianyang, China), Study of pulsed large current output of shock activated energy transfer, *Chinese Journal of High Pressure Physics* (2006).

103. Dujin Mei, Zhang Fuping, Zhang Yi and Wang Haiyan, Burst power for high pulse current output graduate, *Chinese Journal of High Pressure Physics* **20**(4) (2006) pp. 403–407.

104. Zhang Fu-Ping, Zhang Yi, DU Jin-Mei, Wang Hai-Yan and He Hong-Liang (Laboratory for Shock Wave and Detonation Physics Research, Institute of Fluid Physics, CAEP, Mianyang, China), Charge release characterization of PZT95/5 ferroelectric ceramics under tilted shock wave compression, *Chinese Journal of High Pressure Physics* (2006).

105. Liu Gaomin, Liu Yusheng, Zhang Yi, Du Jinmei and He Hongliang (Laboratory for Shock Wave and Detonation Physics Research, Institute of Fluid Physics, CAEP, Mianyang), PZT ferroelectric ceramic for shock-driven pulsed power supply, *Materials Review* (2006).

106. Du Jin-Mei, Zhang Yi, Zhang Fu-Ping, He Hong-Liang and Wang Hai-Yan (Laboratory for Shock Wave and Detonation Physics Research, Institute of Fluid Physics, China Academic of Engineering Physics, Mianyang, China), Large current output of PZT 95/5 ferroelectric ceramics under shock loading, *Acta Physica Sinica* **55**(5) (2006) pp. 2585–2585.

107. Chen Lang, Teng Xiao-Qin, Lu Jian-Ying and Feng Chang-Gen (State Key Laboratory Explosion Science and Technology, Beijing Institute of Technology, Beijing, China), Experiment and calculation of output current of ferroelectric ceramics in parallel connection loaded by shock waves, *Chinese Journal of High Pressure Physics* **21**(1) (2007) pp. 66–70.

108. Wang De-Wu, He Yuan-Ji and Tan Hui-Min (School of Aerospace Science and Engineering, Beijing Institute of Technology, Beijing 100081, China; The Second Artillery Research Academy, Beijing, China), Explosive ferroelectric power supply of flux compression generator, 北京理工大学学报 (英文版) (2007).

109. Jinmei Du, Gaomin Liu, Yusheng Liu, Haiyan Wang, Fuping Zhang and Hongliang He, Current output from sandwich-structured ferroelectric power supply, *AIP Conference Proceedings* **955** (2007) pp. 167–169 (https://doi.org/10.1063/1.2833000).

110. Zhang Fu-Ping, Du Jin-Mei, Zhang Yi, Liu Yu-Sheng and He Hong-Liang, Discharge of PZT95/5 ferroelectric ceramics under titled shock wave compression, *AIP Conference Proceedings* **955** (2007) (https://doi.org/10.1063/1.2833000).

111. Yang Shiyuan, Jin Xiaogang, Wang Junxia, Liang Xiaofeng and He Hongliang (School of Material Science and Engineering, Southwest University of Science and Technology, Mianyang 621010, Laboratory for Shock Wave and Detonation Physics Research, Institute of Fluid Physics, CAEP, Mianyang, China), Shock loading technique and the application in materials research, *Chinese Journal of Materials Research* **22**(2) (2008) pp. 120–124.

112. Liu Gao-Min, Du Jin-Mei, Liu Yu-Sheng, Tan Hua and He Hong-Liang (Laboratory for Shock Wave and Detonation Physics Research, Institute of Fluid Physics, CAEP, Mianyang, China), Shock wave compression of PZT 95/5 ferroelectric ceramic, *Chinese Journal of High Pressure Physics* (2008).

113. Y. Liu, G. Liu, F. Zhang and H. He (National Key Laboratory of Shock Wave and Detonation Physics, Institute of Fluid Physics, Mianyang, China), Experimental study of the depoling property of PZT-95/5 ferroelectric ceramic under shock wave Loading, Piezoelectrics & Acoustooptics (2008).

114. Jiang Dong-Dong, Du Jin-Mei, Gu Yan and Feng Yu-Jun (Laboratory for Shock Wave and Detonation Physics Research, Institute of Fluid Physics, China Academic of Engineering Physics, Mianyang, China; Electronic Materials Research Laboratory, Xi'an Jiaotong University, Xi'an 710049, China), Resistivity of PZT 95/5 ferroelectric ceramic under shock wave compression, *Acta Physica Sinica* (2008).

115. Chen Xuefeng, Liu Yusheng, Feng Ningbo, He Hongliang and Dong Xianlin (Shanghai Institute of Ceramics, Chinese Academy of Sciences, Shanghai, China; Institute of Fluid Physics, China Academy of Engineering Physics, Mianyang, Sichuan Province, China), Effects of shock pressure and temperature aging on the discharge behavior of PZT 95/5 ferroelectric ceramics, *Electronic Components and Materials* (2009).

116. Zhang Fu-Ping, Du Jin-Mei, Liu Yu-Sheng and He Hong-Liang (Laboratory for Shock Wave and Detonation Physics Research, Institute of Fluid Physics, Sichuan, China), Inspection of remnant polarization in the ferroelectric ceramic PZT 95/5 through pyroelectric effect, *Ceramics* **90**(8) (2007).

117. T. Zeng, X.L. Dong, H.L. He, X.F. Chen and C.H. Yao, Pressure-induced ferroelectric to antiferroelectric phase transformation in porous PZT 95/5 ceramics, *Physica Status Solidi A: Applications and Materials Science* **204**(4) (2007) pp. 1216–1220.

118. Zhang Fuping, Du Jinmei, Zhang Yi, Liu Yusheng, Liu Gaomin and He Hongliang, Discharge of PZT 95/5 ferroelectric ceramics under tilted shock wave compression, *Proceedings of 15th APS Topical Conference on Shock Compression of Condensed Matter* (2007).

119. Yang Shiyuan, Jin Xiaogang, Wang Junxia, Liang Xiaofeng and He Hongliang (School of Material Science and Engineering, Southwest University of Science and Technology, Mianyang; Laboratory of Shock Wave and Detonation Physics Research, Institute of Fluid Physics, Mianyang, China), Shock loading Technique and the Application in Materials Research, *Chinese Journal of Materials Research* **22**(2) (2008) pp. 120–124.

120. Chen Feng, Yang Shi-Yuan, Wang Jun-Xia, He Hong-Liang and Wang Guan-Cai, Effect of cylindrical shock synthesis on properties and sintering behavior of PZT 95/5 powder, *Journal of Inorganic Materials* **22**(5) (2007) pp. 827–832.

121. Jiang Dongdong, Du Jinmei, Yang Jia, Gu Yan and Feng Yujun, Microsecond current output in PbNb(Zr,Sn,Ti)O3 ceramic under shock wave compression, *Proceedings of the 17th APS Topical Conference on Shock Compression of Condensed Matter* (2011) pp. 63–67.

122. Z. Zhang, Z. Cui, J. Yan and K. Li, Phase transition study of integrated circuit chip power supplying based on ferroelectric ceramic, *Appl. Phys. Lett.* **97** (2010) p. 132906.

123. N. Feng, H. Nie, X. Chen, G. Wang, X. Dong and H. He, Depoling of porous Pb0.99(Zr0.95Ti0.05)0.98Nb0.02O3 ferroelectric ceramics under shock wave load, *Current Applied Physics* **10**(6) (2010) pp. 1387–1390.

124. D. Jiang, Y. Feng, J. Du and Y. Gu, Shock wave compression of the poled Pb0.99[(Zr0.90Sn0.10)0.96Ti0.04]0.98Nb0.02O3 ceramics: Depoling currents in axial mode and normal mode, *Chinese Science Bulletin* **57**(20) (2012) pp. 2554–2561.

125. D. Jiang, J. Du, Y. Gu and Y. Feng, Phenomenological description of depoling current in Pb0.99Nb0.02(Zr0.95Ti0.05)0.98O3 ferroelectric ceramics under shock wave compression: relaxation model, *J. Appl. Phys.* **111** (2012) p. 104102.

126. D. Jiang, J. Du, Y. Gu and Y. Feng, Shock wave compression and self-generated electric field repolarization in ferroelectric ceramics Pb0.99[(Zr0.90Sn0.10)0.96Ti0.04]0.98Nb0.02O3, *J. Phys.D: Appl. Phys.* **45**(11) (2012) p. 115401.

127. F. Zhang, H. He, G. Liu, Y. Liu, Y. Yu and Y. Wang, Failure behavior of Pb(Zr0.95Ti0.05)O3 ferroelectric ceramics under shock compression, *J. Appl. Phys.* **113** (2013) p. 183501.

128. Z. Wang, Y. Jiang, P. Zhang, X. Wang and H. He, Depolarization and electrical response of porous PZT95/5 ferroelectric ceramics under shock wave compression, *Chinese Physics Letters* **31**(7) (2014) p. 077703.

129. Y. Jiang, X. Wang, F. Zhang and H. He, Breakdown and critical field-bounds evaluation for porous PZT 95/5 ferroelectric ceramics under shock wave compression, *Smart Materials and Structures* **23**(8) (2014) p. 085020.

130. F. Zhang, Y. Liu, Q. Xie, G. Liu and H. He, Electrical response of Pb(Zr0.95Ti0.05)O3 under shock compression, *J. Appl. Phys.* **117** (2015) p. 134104.

131. J. Gao, L. Xie, H. Zhang, J. Yu, G. Liu, G. Wang, J. Bai, Y. Gu and H. He, The charge release and its mechanism for Pb(In1/2Nb1/2)–Pb(Mg1/3Nb2/3)–PbTiO3 ferroelectric crystals under one-dimensional shock wave compression, *J. Am. Ceram. Soc.* **98**(3) (2015) pp. 855–860.

132. H. Nie, Y. Yu, Y. Liu, H. He and G. Wang, Enhanced shock performance by disperse porous structure: a case study in PZT95/5 ferroelectric ceramics, *J. Am. Ceram. Soc.* **100**(12) (2017) pp. 5693–5699.

133. Y. Jiang, X. Wang, F. Zhang and H. He, Breakdown and critical field evaluation for porous PZT 95/5 ferroelectric ceramics under shock wave compression, *Smart Mater. Struct.* **23** (2014) p. 085020.

134. H. Nie, J. Yang, X. Chen, F. Zhang, Y. Yu, G. Wang, Y. Liu, H. He and X. Dong, Mechanical induced electrical failure of shock compressed PZT95/5 ferroelectric ceramics, *Current Applied Physics* **17**(4) (2017) pp. 448–453.

135. H.-C. Nie, Y.-L. Wang, H.-L. He, G.-S. Wang and X.-L. Dong, Recent Progress of Porous PZT95/5 Ferroelectric Ceramics, *Journal of Inorganic Materials* **33**(2) (2018) pp. 153–161.

136. R. Su, H. Nie, Z. Liu, P. Peng, F. Cao, X. Dong and G. Wang, The depolarization performances of 0.97PbZrO3–0.03Ba(Mg1/3Nb2/3)O3 ceramics under hydrostatic pressure, *Appl. Phys. Lett.* **112** (2018) p. 062901.

137. Y. Wu, G. Liu, Z. Gao, H. He and J. Deng, Dynamic dielectric properties of the ferroelectric ceramic Pb(Zr0.95Ti0.05)O3 in shock compression under high electrical fields, *J. Appl. Phys.* **123** (2018) p. 244102.

138. P. Peng, H. Nie, G. Wang, Z. Liu, F. Cao and X. Dong, Shock-driven depolarization behavior in BNT-based lead-free ceramics, *Appl. Phys. Lett.* **113** (2018) p. 082901.

139. M.S. Seo and J. Ryu, Explosively driven ferroelectric generator for compact pulsed power systems, *in book: Shock Compression of Condensed Matter* (eds. M.D. Furnish, Y.M. Gupta and J.W. Forbes, CP706, American Institute of Physics, New York, 2004) pp. 1313–1316.

140. C.E. Seo and K.H. Cho, Effects of impact conditions on the shock response of PZT-5A ceramics for FEG devices, *J. Korean Phys. Soc.* **57** (2010) pp. 5975–979.

141. T. Mashimo, K. Toda, K. Nagayama, T. Goto and Y. Syono, Electrical response of BaTiO3 ceramics to shock induced ferroelectric-paraelectric transition, *J. Appl. Phys.* **59**(3) (1986) pp. 748–756.

142. Chao-Nan Xu, M. Akiyama, K. Nonaka and T. Watanabe, Electrical power generation characteristics of PZT piezoelectric ceramics, *IEEE Transactions on Ultrasonics, Ferroelectrics, and Frequency Control* **45**(4) (1998) pp. 1065–1070.

143. G. Staines, J. Dommer, F. Sonnemann, J. Bohl and T. Ehlen. High Voltage Generator, Especially for Using as a Noise Frequency Generator. U.S. Patent 6,969,944. Issued November 29, 2005.

144. G. Staines, H. Hofman, J. Dommer, L.L. Altgilbers and Ya. Tkach, Compact piezo-based high voltage generator. Part II: prototype generator, *Journal of Electromagnetic Phenomenon* **4**(6) (2004) pp. 477–489.

145. M. Jung, B. Schunemann and G. Wollmann, Mechanically driven compact high voltage generator, *Proceedings of the 27th International Power Modulator Conference* (Arlington, VA, 2006) pp. 526–528.

146. Private Communication from FOI

147. R.E. Setchell, Shock wave compression of the ferroelectric ceramic Pb 0.99 (Zr 0.95 Ti 0.05) 0.98 Nb 0.02 O 3 : Hugoniot states and constitutive mechanical properties, *J. Appl. Phys.* **94**(1) (2003) pp. 573–588.

148. R.E. Setchell, Shock wave compression of the ferroelectric ceramic Pb0.99.Zr0.95Ti0.05.0.98Nb0.02O3: Depoling currents, *J. Appl. Phys.* **97** (2005) p. 013507.

149. R.E. Setchell, Shock wave compression of the ferroelectric ceramic Pb0.99,,Zr0.95Ti0.05. . .S0.98Nb0.02O3: Microstructural effects, *J. Appl. Phys.* **101** (2007) p. 053525.

150. R.E. Setchell and M.U. Anderson, Shock-compression response of an alumina-filled epoxy, *J. Appl. Phys.* **97** (2005) p. 083518.

151. R.E. Setchell, M.U. Anderson and S.T. Montgomery, Compositional effects on the shock-compression response of alumina-filled epoxy, *J. Appl. Phys.* **101** (2007) p. 083527.

152. R.E. Setchell, S.T. Montgomery, D.E. Cox and M.U. Anderson, Dielectric properties of PZT 95/5 during shock compression under high electric filed, *AIP Conf. Proc. CP845, Shock Compression of Condensed Matter — 2005*, eds. M.D. Furnish, M. Elert, T.P. Russell, and C.T. White (American Institute of Physics, 2006), pp. 278–281.

153. R.E. Setchell, S.T. Montgomery, D.E. Cox and M.U. Anderson, Initial temperature effects on the dielectric properties of PZT 95/5 during shock compression, *AIP Conf. Proceedings CP955, Shock Compression of Condensed Matter — 2006*, eds. M. Elert, M.D. Furnish, R. Cliau, N. Holmes, and J. Nguyen (American Institute of Physics, 2007), pp. 194–196.

154. N. Schoeneberg, J. Walter, A. Neuber, J. Dickens and M. Kristiansen, Ferromagnetic and ferroelectric materials as seed sources for magnetic flux compressors, *Proceedings of the 14th IEEE International Pulsed Power Conference* (2003) pp. 1069–1072.

155. S. Holt, J. Dickens, J. Walter and S. Calico, Design of explosive-driven ferroelectric pulse generators with outputs exceeding 200 kV, *Proceedings of the 15th IEEE International Pulsed Power Conference,* eds. J. Maenchen and E. Schamiloglu, Monterey, CA (2005), pp. 448–452.

156. D. Bolyard, A. Neuber, J. Krile, J. Walter, J. Dickens and M. Kristiansen, Scaling and improvement of compact explosively-driven ferroelectric generators, *Proceedings of the 2008 IEEE International Power Modulators and High Voltage Conference,* pp. 49–52 (2009).

157. Z. Roberts, S. Rendall, F. Rose, J. Sweitzer, A.H. Stults and L.L. Altgilbers, Cylindrical ferroelectric generator waveshaping techniques and performance, *Proceedings of the 18th IEEE International 18th Pulsed Power Conference,* eds. R.D. Curry and B.V. Oliver, Chicago, IL (2011), pp. 546–550.

158. Private Communication from Radiance Technologies.

159. Private Communication from HEM Technologies.

160. S.L. Holt, J.T. Krile, D.J. Hemmert, W.S. Hackenberger, E.F. Alberta, J.W. Walter, J.C. Dickens, L.L. Altgilbers and A.H. Stults, Testing of new ferroelectric elements custom engineered for explosively driven ferroelectric applications, *Proceedings of the 16th IEEE International Pulsed Power Conference*, eds. E. Schamiloglu and F. Peterkin, Albuquerque, NM (2007), pp. 1177–1180.

161. E.G. Engel, C. Keawboonchuay and W.C. Nunnally, Energy conversion and high power pulsed production using miniature piezoelectric compressors, *IEEE Transactions on Plasma Science* **28**(5) (2000) pp. 1338–1341.

162. C. Keawboonchuay and T.G. Engel, Electrical power generation characteristics of piezoelectric generator under quasi-static and dynamics stress conditions, *IEEE Transactions on Ultrasonics, Ferroelectrics, and Frequency Control* **50**(10) (2003) pp. 1377–382.

163. C. Keawboonchuay and T.G. Engel, Scaling relationships and maximum peak power generation in a piezoelectric pulse generator, *IEEE Transactions on Plasma Science* **32**(5) (2004) pp. 1879–1885.

164. Private communication from Raytheon Ktech.

165. A.H. Stults, Impulse Loading of Sinuous Antennas by Ferroelectric Generators, *Proceedings of the 2008 IEEE International Power Modulators and High Voltage Conference*, pp. 156–158 (2009).

166. A.H. Stults, Ferroelectric generator design for multiple initiation of blasting caps, *Proceedings of the 18th IEEE International 18th Pulsed Power Conference*, eds. R.D. Curry and B.V. Oliver, Chicago, IL (2011), pp. 551–554.

167. S.I. Shkuratov, M. Kristiansen, J. Dickens, A. Neuber, L.L. Altgilbers, P.T. Tracy and Ya. Tkach, Experimental study of compact explosive driven shock wave ferroelectric generators, *Proceedings of 13th International Pulsed Power Conference*, eds. R. Reinovsky and M.A. Newton, Las Vegas, Nevada, USA (2001) IEEE Catalog Number: 01CH37251, ISBN: 0-7803-7120-8, Vol. II, pp. 959–962.

168. Ya. Tkach, S.I. Shkuratov, J. Dickens, M. Kristiansen, L.L. Altgilbers and P.T. Tracy, Explosive driven ferroelectric generators, *Proceedings of 13th International Pulsed Power Conference*, eds. R. Reinovsky and M.A. Newton, Las Vegas, Nevada, USA (2001) IEEE Catalog Number: 01CH37251, ISBN: 0-7803-7120-8, Vol. II, pp. 986–989.

169. Ya. Tkach, S.I. Shkuratov, J. Dickens, M. Kristiansen, L.L. Altgilbers and P.T. Tracy, Parametric and experimental investigation of EDFEG, *Proceedings of 13th International Pulsed Power Conference*, eds. R. Reinovsky and M.A. Newton, Las Vegas, Nevada, USA (2001) IEEE Catalog Number: 01CH37251, ISBN: 0-7803-7120-8, Vol. II, pp. 990–993.

170. S.I. Shkuratov, E.F. Talantsev, L. Menon, H. Temkin, J. Baird and L.L. Altgilbers, Compact high-voltage generator of primary power based on shock wave depolarization of lead zirconate titanate piezoelectric ceramics, *Rev. Sci. Instrum.* **75**(8) (2004) pp. 2766–2769.

171. Y. Tkach, S.I. Shkuratov, E.F. Talantsev, J.C. Dickens, M. Kristiansen, L.L. Altgilbers and P.T. Tracy, Theoretical treatment of explosive-driven ferroelectric generators, *IEEE Trans. Plasma Sci.* **30**(5) (2002) pp. 1665–1673.

172. S.I. Shkuratov, E.F. Talantsev, J. Baird, H. Temkin, L.L. Altgilbers and A.H. Stults, Longitudinal shock wave depolarization of $Pb(Zr_{0.52}Ti_{0.48})O_3$ polycrystalline ferroelectrics and their utilization in explosive pulsed power, *AIP Conf. Proc. CP845, Shock Compression of Condensed Matter — 2005*, eds. M.D. Furnish, M. Elert, T.P. Russell, and C.T. White (American Institute of Physics, 2006), pp. 1169–1172.

173. S.I. Shkuratov, E.F. Talantsev, J. Baird, H. Temkin, Y. Tkach, L.L. Altgilbers and A.H. Stults, The depolarization of a $Pb(Zr_{0.52}Ti_{0.48})O_3$ polycrystalline piezoelectric energy-carrying element of compact pulsed power generator by a longitudinal shock wave, *Proceedings of the 16th IEEE International Pulsed Power Conference*, eds. E. Schamiloglu and F. Peterkin, Albuquerque, NM (2007), pp. 529–532.

174. S.I. Shkuratov, E.F. Talantsev, J. Baird, M.F. Rose, Z. Shotts, L.L. Altgilbers and A.H. Stults, Completely explosive ultracompact high-voltage nanosecond pulse-generating system, *Rev. Sci. Instrum.* **77**(4) (2006) p. 043904.

175. S.I. Shkuratov, E.F. Talantsev, J. Baird, A.V. Ponomarev, L.L. Altgilbers and A.H. Stults, Operation of the longitudinal shock wave ferroelectric generator charging a capacitor bank: experiments and digital model, *Proceedings of the 16th IEEE International Pulsed Power Conference*, eds. E. Schamiloglu and F. Peterkin, Albuquerque, NM (2007), pp. 1146–1150.

176. S.I. Shkuratov, J. Baird, E.F. Talantsev, Y. Tkach, L.L. Altgilbers, A.H. Stults and S.V. Kolossenok, Pulsed charging of capacitor bank by compact explosive-driven high-voltage primary power source based on longitudinal shock wave depolarization of ferroelectric ceramics, *Proceedings of the 16th IEEE International Pulsed Power Conference*, eds. E. Schamiloglu and F. Peterkin, Albuquerque, NM (2007), pp. 537–540.

177. S.I. Shkuratov, J. Baird, E.F. Talantsev, M.F. Rose, Z. Shotts, L.L. Altgilbers, A.H. Stults and S.V. Kolossenok, Completely explosive ultracompact high-voltage pulse generating system, *Proceedings of the 16th IEEE International Pulsed Power Conference*, eds. E. Schamiloglu and F. Peterkin, Albuquerque, NM (2007), pp. 445–448.

178. S.I. Shkuratov, J. Baird, E.F. Talantsev, A.V. Ponomarev, L.L. Altgilbers and A.H. Stults, High-voltage charging of a capacitor bank, *IEEE Trans. Plasma Sci.* **36**(1) (2008) pp. 44–51.

179. S.I. Shkuratov, J. Baird, V.G. Antipov, E.F. Talantsev, C.S. Lynch and L.L. Altgilbers, PZT 52/48 depolarization: quasi-static thermal heating versus longitudinal explosive shock, *IEEE Trans. Plasma Sci.* **38**(8) (2010) pp. 1856–1863.

180. S.I. Shkuratov, J. Baird, V.G. Antipov and E.F. Talantsev, Autonomous pulsed power generator based on transverse shock wave depolarization of ferroelectric ceramics, *Rev. Sci. Instrum.* **81**(12) (2010) p. 126102.

181. S.I. Shkuratov, J. Baird and E.F. Talantsev, Miniature 120-kV generator based on transverse shock depolarization of $Pb(Zr_{0.52}Ti_{0.48})O_3$ ferroelectrics, *Rev. Sci. Instrum.* **82**(8) (2011) p. 086107.

182. S.I. Shkuratov, J. Baird and E.F. Talantsev, The depolarization of $Pb(Zr_{0.52}Ti_{0.48})O_3$ ferroelectrics by cylindrical radially expanding shock waves and its utilization for miniature pulsed power, *Rev. Sci. Instrum.* **82**(5) (2011) p. 054701.

183. S.I. Shkuratov, E.F. Talantsev and J. Baird, Electric breakdown of longitudinally shocked $Pb(Zr_{0.52}Ti_{0.48})O_3$ ceramics, *J. Appl. Phys.* **110**(2) (2011) p. 024113.

184. S.I. Shkuratov, J. Baird and E.F. Talantsev, Effect of shock front geometry on shock depolarization of $Pb(Zr0.52Ti0.48)O3$ ferroelectric ceramics, *Rev. Sci. Instrum.* **83** (2012) p. 074702.

185. S.I. Shkuratov, J. Baird, E.F. Talantsev, E.F. Alberta, W.S. Hackenberger, A.H. Stults and L.L. Altgilbers, Miniature 100-kV explosive-driven prime power sources based on transverse shock-wave depolarization of PZT 95/5 ferroelectric ceramics, *IEEE Trans. Plasma Sci.* **40**(10) (2012) pp. 2512–2516.

186. S.I. Shkuratov, J. Baird and E.F. Talantsev, Extension of thickness-dependent dielectric breakdown law on adiabatically compressed ferroelectric materials, *Appl. Phys. Lett.* **102** (2013) p. 052906.

187. S.I. Shkuratov, J. Baird, V.G. Antipov, E.F. Talantsev, H.R. Jo, J.C. Valadez and C.S. Lynch, Depolarization mechanisms of $PbZr_{0.52}Ti_{0.48}O_3$ and $PbZr_{0.95}Ti_{0.05}O_3$ poled ferroelectrics under high strain rate loading, *Appl. Phys. Lett.* **104** (2014) p. 212901.

188. S.I. Shkuratov, J. Baird, V.G. Antipov, E.F. Talantsev, W.S. Hackenberger, A.H. Stults and L.L. Altgilbers, High voltage generation with transversely shock-compressed ferroelectrics: breakdown field on thickness dependence, *IEEE Trans. Plasma Sci.* **44**(10) (2016) pp. 1919–1927.

189. S.I. Shkuratov, J. Baird, V.G. Antipov, E.F. Talantsev, J.B. Chase, W.S. Hackenberger, J. Luo, H.R. Jo and C.S. Lynch, Ultrahigh energy density harvested from domain-engineered relaxor ferroelectric single crystals under high strain rate loading, *Sci. Rep.* **7** (2017) p. 46758.

190. S.I. Shkuratov, J. Baird, V.G. Antipov, E.F. Talantsev, J.B. Chase, W.S. Hackenberger, J. Luo, H.R. Jo and C.S. Lynch, Mechanism of complete stress-induced depolarization of relaxor ferroelectric single crystals without transition through non-polar phase, *Appl. Phys. Lett.* **112** (2018) p. 122903.

191. J. Baird and S. Shkuratov, Ferroelectric Energy Generator, System, and Method. U.S. Patent 7,560,855. Issued July 14, 2009.

192. J. Baird and S. Shkuratov, Ferroelectric Energy Generator with Voltage-Controlled Switch. U.S. Patent 7,999,445 B2. Issued August 16, 2011.

193. J. Baird and S. Shkuratov, Ferroelectric Energy Generator with Voltage-Controlled Switch. U.S. Patent 8,008,843 B2. Issued August 30, 2011.

Chapter 4

Physical Principles of Shock Wave Ferroelectric Generators

4.1 Introduction

The unique ability of ferroelectric materials to generate electric potential under mechanical stress due to the piezoelectric effect is used in a variety of modern applications. For many of these applications there are no practical alternatives. The operation of ferroelectric materials at low-strain is piezoelectric in nature and no large-scale reorientation of domains occurs. The application of low mechanical stress to the ferroelectric elements results in charge displacement, but when the stress is removed the charges flow back.

The operation of ferroelectrics under high mechanical stress and high strain rate loading is fundamentally different from that in the low-strain mode. The shock compression of ferroelectric materials induces phase transitions and domain reorientation with possible loss of their initial remanent polarization. The rapid depolarization of ferroelectrics under shock loading can result in the generation of megawatt power levels for a brief interval of time. Ultrahigh-power ferroelectric systems based on the shock compression of ferroelectrics are capable of producing pulses of kiloamperes of current and hundreds of kilovolts of electric potential. In this chapter, the physical principles of shock wave ferroelectric generators are discussed.

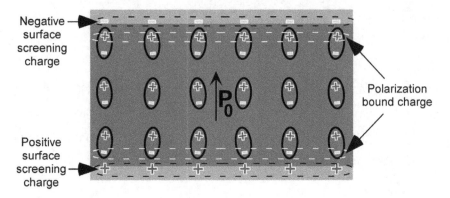

Fig. 4.1. Schematic diagram of a ferroelectric specimen illustrating the surface screening of spontaneous polarization. P_0 is the remanent polarization vector.

4.2 Surface Screening of Spontaneous Polarization in Ferroelectrics

The development of spontaneous polarization and related lattice distortion below the Curie temperature in a ferroelectric material results in the formation of regions of uniform polarization, i.e. ferroelectric domains. Figure 4.1 shows a schematic diagram of a ferroelectric specimen illustrating the surface screening of spontaneous polarization.

The ferroelectric domains are randomly oriented in the ferroelectric specimen, as manufactured, and the specimen has no overall polarization (see Figure 1.3(a) in Chapter 1). The ferroelectric specimen is polarized by exposing it to a high-DC electric field. Through the polarizing treatment, domains most nearly aligned with the applied electric field. When the electric field is removed, most of the domains are locked into this configuration near alignment (see Figures 1.3(b) and 1.3(c) in Chapter 1). This gives the ferroelectric material a permanent polarization, i.e. the remanent polarization.

Polarization discontinuities in the vicinity of the surfaces of ferroelectric specimens result in the *surface polarization charge* also called the *polarization bound charge* or *bound charge* (see Figure 4.1) that significantly affects materials properties.

The bound charge results in the domain specific adsorption of charged species from the ambience that effectively compensates the charge, i.e. the extrinsic *surface screening* processes. There are two mechanisms of the surface screening of the polarization bound charge in ferroelectrics [1–5].

The external surface screening is typically realized by free electrons in metallic electrodes, adsorbed charged species, or mobile ions and ionic species [1, 2]. The internal surface screening mechanism can be defined as that realized by electrons, or oxygen, or cationic vacancies within the ferroelectric material, with associated changes in polarization, and ionic species concentration profiles in the vicinity of the surface [3–5]. Practically, both the external and internal screening mechanisms can be realized simultaneously, with the relative contributions controlled both by the thermodynamics and kinetics of the respective screening processes [1–5].

The surface screening charges are equal and opposite to the polarization bound charges to enforce the neutrality of net charges and, correspondingly, the net electric field in a ferroelectric specimen. The screening charges are mobile, in the sense that the spatial distribution of their quasi two-dimensional density is ruled by the polarization distribution near the surface.

The surface screening charge is what one can measure in an experiment. When the polarization changes, the surface screening charge accommodates accordingly. Changes of the surface screening charge cause electric current to flow in the external circuit connected to the electrodes of a ferroelectric specimen. The generated current and voltage can be monitored by current and voltage probes.

4.3   Operation of Ferroelectrics in Piezoelectric Mode

The operation of ferroelectric materials in the low-strain mode is piezoelectric in nature. The application of low mechanical stress to the ferroelectric specimens results in charge displacement and the generation of current and voltage in the external circuit, but when the stress is removed the polarization is restored. The magnitude of compressive stress and the amplitude of electrical signals generated

Explosive Ferroelectric Generators

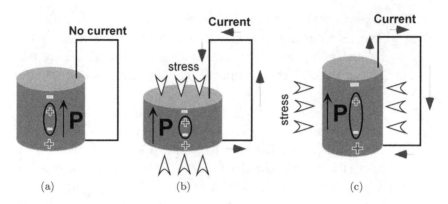

Fig. 4.2. Schematics of operation of cylindrical ferroelectric specimens in the short-circuit piezoelectric mode. (a) No stress. (b) Ferroelectric specimen is compressed along the direction of polarization (longitudinal stress). (c) Ferroelectric specimen is compressed perpendicular to the direction of polarization (transverse stress).

by applying stress to a ferroelectric specimen are linearly proportional up to a material-specific stress level that varies in a wide range from 5 MPa for $(Pb, La_{0.08})(Zr_{0.65}, Ti_{0.35})O_3$ [6] to 50 MPa for PZT 52/48 [7].

Consider the operation of ferroelectrics in the short-circuit piezoelectric mode (see schematics in Figure 4.2). Under longitudinal stress the specimen is compressed along the direction of polarization and it undergoes radial expansion due to the Poisson effect (Figure 4.2(b)). This results in the decrease in polarization due to both the reversible rotation of the dipole moment out of the polarization direction and the reduction in the magnitude of the dipole moment. The bound charge and surface screening charge become unbalanced. The charge released on the electrodes due to the applied mechanical stress is neutralized by a current flow in the external circuit. It restores the balance between polarization and screening charge in the specimen under stress. The stress-induced charge, Q_{stress}, can be determined as follows:

$$Q_{stress} = d_{33} \cdot F \tag{4.1}$$

where d_{33} is the piezoelectric charge constant in [C/N], and F is the applied force in [N].

The amplitude of the current generated in the short-circuit piezoelectric mode is directly proportional to the stress-induced charge and inversely proportional to the rise time of the mechanical stress. When the stress level is within the linear region, after the stress is removed, the polarization and screening charge return to its initial level.

Figure 4.2(c) shows a ferroelectric specimen mechanically compressed perpendicular to the direction of polarization (transverse stress). Under transverse stress the specimen is compressed in the radial direction and undergoes vertical expansion due to the Poisson effect. This results in the increase in polarization due to the reversible rotation of the dipole moment toward the polarization direction and the increase in magnitude of the dipole moment. It causes the charge unbalance in the vicinity of the surfaces of the ferroelectric specimen and as a result current is generated in the external circuit to restore the balance between the increased polarization and screening charge. The direction of the current flow is opposite to that generated under longitudinal stress (Figure 4.2(b)).

Consider the operation of ferroelectrics in the open circuit piezoelectric mode (schematics are shown in Figure 4.3). Longitudinal stress (Figure 4.3(b)) results in the decrease of polarization and

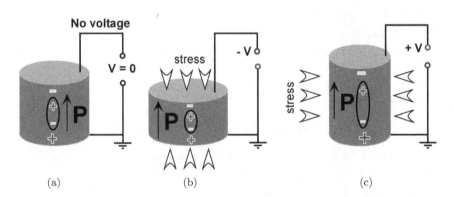

(a)   (b)   (c)

Fig. 4.3. Schematics of operation of cylindrical ferroelectric specimens in the open circuit piezoelectric mode. (a) No stress. (b) Ferroelectric specimen is compressed along the direction of polarization (longitudinal stress). (c) Ferroelectric specimen is compressed perpendicular to the direction of polarization (transverse stress).

unbalance between polarization and screening charge. The electrodes of the specimen are open in this mode of operation. A part of the screening charge is released on the electrodes and it charges the ferroelectric element itself. In the open circuit mode the longitudinal stress generates an electric field and high voltage across the specimen with polarity identical to that of the screening charge (Figure 4.3(b)). The electric field generated under stress, E_{stress}, can be determined as follows:

$$E_{stress} = g_{33} \cdot T \tag{4.2}$$

where g_{33} is the piezoelectric voltage constant in [Vm/N], and T is the applied stress in [N/m^2]. When the applied stress is removed, the polarization and screening charge are back to the initial state if the stress is within the linear region.

Figure 4.3(c) presents a schematic of the ferroelectric specimen operating in the open circuit mode under transverse stress. The expansion of the specimen along its axis results in an increase in polarization and in unbalance between the bound and screening charges. The transverse stress results in an electric field and voltage across the specimen (Figure 4.3(c)). The polarity of the voltage is opposite to that generated under longitudinal stress (Figure 4.3(b)) and, correspondingly, opposite to that of the screening charge.

Figure 4.4 shows the amplitude of the high voltage generated by the PZT 52/48 cylindrical specimen (diameter 5 mm and thickness 18 mm) as a function of compressive stress. The stress-induced voltage is practically linear proportional to the applied stress.

4.4 Shock Wave Ferroelectric Generators

One of the advantages of the piezoelectric mode is a reproducibility of generated electrical signals. However, the magnitude of piezoelectric signals is not high. An increase of the applied stress above a material-specific stress level results in an increase in generated power, but it also results in non-linear behavior of signal-stress dependence, irreversible domain rotation, and a decrease in or elimination of the piezoelectric effect.

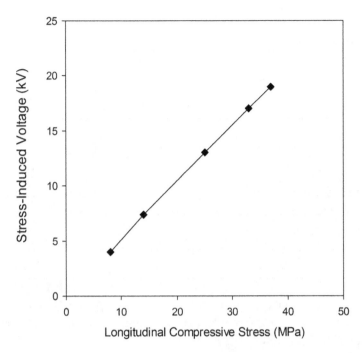

Fig. 4.4. High voltage generated by the PZT 52/48 cylindrical specimen (diameter 5 mm and thickness 18 mm) as a function of longitudinal compressive stress.

In 1957, Neilson [8] reported his pioneering results on the generation of high current and high voltage by shock-compressed lead zirconate titanate and barium titanate ferroelectric specimens. For the first time it was experimentally demonstrated that under shock loading, the ferroelectrics are capable of producing power that is six orders of magnitude higher than that in the piezoelectric mode.

The operation of ferroelectrics in the ultrahigh-power mode is fundamentally different from that in the piezoelectric mode. High stress and high strain rate loading induce phase transformations and domain reorientation with the loss of the initial remanent polarization of ferroelectric materials. This results in the generation of high current and high voltage pulses. The following sections are focused on the physical principles of the operation of shock wave ferroelectric generators and the physics of shock waves.

4.5 Shock Waves in Solids

A *shock wave* is a strong pressure wave in any elastic medium such as a solid substance, gas, or fluid, produced by strong mechanical impact, explosion, lightning, electric discharge, or other phenomena that create abrupt, discontinuous changes in pressure [9–11]. Shock waves differ from sound waves in that the shock wave front, the boundary over which compression takes place, is a region of sudden, discontinuous change in stress, density, and energy. In the shock wave front a material is stressed far beyond its elastic limit by a pressure disturbance. Because of this, shock waves propagate in a manner different from that of ordinary acoustic waves. In particular, shock waves travel faster than sound, and their speed increases as the amplitude is raised; but the intensity of a shock wave also decreases faster than does that of a sound wave. A shock wave is a discontinuous high pressure disturbance propagating through matter. The energy, e, particle velocity, u, pressure, p, and density, ρ, on both sides of a shock discontinuity propagating at a velocity U_s are as shown in Figure 4.5.

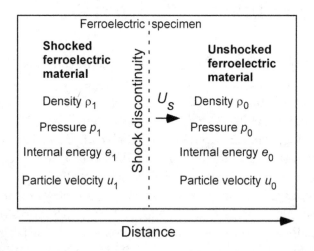

Fig. 4.5. Diagram illustrating a shock discontinuity propagating at a velocity U_S and transitioning a ferroelectric material from the initial state (ρ_0, p_0, e_0, u_0) to state (ρ_1, p_1, e_1, u_1). The subscripts 0 and 1 refer to the state of the unshocked and shocked materials, respectively.

When the boundary forces that produce a shock are removed, a decompression wave, also called a *rarefaction* or *release wave*, originates at the boundary and propagates into the interior of the material. When heat conduction is neglected, the release wave is a smooth wave, rather than the discontinuous shock wave that is observed on compression. The release wave spreads as it propagates and overtakes the shock wave. This causes attenuation, which decreases both the energy and pressure changes produced by the shock wave.

There are five variables required to describe shock waves: particle velocity, the density of the material, pressure, internal energy, and shock wave velocity. Particle velocity, u, is the velocity of a particle, atom or molecule in a medium as it transmits a shock wave. The atoms, particles or molecules of a medium are displaced from their random motion in the presence of a shock wave. The shock wave moves fast, while the particles are displaced from their original position with a relatively small particle velocity.

Five equations are required to solve for the five variables to describe shock waves. Three of the equations are based on the mass, momentum, and energy conservation laws. In the case of shock waves, these conservation equations do not depend upon the process, but merely on the initial and final states, since these variables change discontinuously across the shock front. These three conservation of mass, momentum and energy equations for the transition of unshocked material to a shocked state are known collectively as the *Rankine-Hugoniot jump conditions* or *jump equations*:

$$\frac{\rho_1}{\rho_0} = \frac{U_s - u_0}{U_s - u_1} \tag{4.3}$$

representing the conservation of mass,

$$p_1 - p_0 = \rho_0(u_1 - u_0)(U_s - u_0) \tag{4.4}$$

representing the conservation of momentum, and

$$e_1 - e_0 = \frac{p_1 u_1 - p_0 u_0}{\rho_0(U_s - u_0)} - \frac{1}{2}(u_1 + u_0) \tag{4.5}$$

representing the conservation of energy.

The subscripts 0 and 1 refer to the state of the unshocked and shocked materials, respectively. In these equations, e is the internal energy in [J/kg], p is the shock wave pressure in [Pa], ρ is the material's density in [kg/m^3], U_s is the shock velocity in [m/s], and u is the particle velocity in [m/s].

The momentun equation for shock-compressed ferroelectrics can be written as follows:

$$p = \rho_0 U_s u_p \qquad (4.6)$$

where p is the peak shock pressure in [Pa], ρ_0 is the initial density of a ferroelectric material in [kg/m^3], U_s is the shock velocity in [m/s], and $u_p = u_1$ is the particle velocity in the shocked zone in [m/s].

Usually, the shock wave pressure is reported in GPa (1 GPa $= 10^9$ Pa $= 10^9$ N/m$^2 = 10^4$ bars $= 14.5 \cdot 10^4$ psi) and the shock velocity in km/s. Conveniently, 1 km/s $= 1$ mm/μs, where the latter units are comparable to the dimensions and time intervals arising in the analysis of explosive ferroelectric generators. The peak shock pressure and shock wave velocity in the ferroelectric elements of ferroelectric generators range from 1.5 to 5 GPa and from 2.5 to 4.5 mm/μs, respectively.

If the state of the material into which a shock is propagating is known, the Rankine-Hugoniot jump equations provide three relationships with the five unknown quantities ρ_1, p_1, e_1, and $u_p = u_1$. One of these quantities, i.e. the pressure or the particle velocity in the material behind the shock, is a measure of the strength of the shock and must be specified as a boundary condition. The additional relationship needed to completely characterize the shock wave depends on the behavior of the material.

This additional equation is an experimentally obtained relationship between any two variables called *the Hugoniot curve* or simply a *Hugoniot*. Experimental measurements indicate that the response of materials in a given initial state to shock compression can often be represented by the Hugoniot relation [9]:

$$U_s = c_0 + s u_p \qquad (4.7)$$

where the parameters c_0 and s are determined from experimental fits to the data. A Hugoniot curve can relate any two of the variables in question (most commonly the shock velocity versus particle velocity or shock pressure versus particle velocity) and it can be transformed into a relationship between any other two of the five variables.

Setchell [12] performed detailed experiments with PZT 95/5 to determine Hugoniot states for this ferroelectric material. Pressure-volume Hugoniot curves for PZT 52/48 ferroelectrics were measured by Reynolds and Seay [13].

4.6 Physical Principles of Shock Wave Ferroelectric Generators

The direction of shock propagation in the ferroelectric element of an FEG can be parallel (axial) or perpendicular (normal) to the direction of the remanent polarization in the element. The generator where the shock wave travels parallel or anti-parallel to the polarization direction is referred to as *the longitudinal shock wave FEG* or *longitudinal FEG*. The generator where the shock wave travels perpendicular to the polarization direction is referred to as *the transverse shock wave FEG* or *transverse FEG*.

4.6.1 *Operation of longitudinal shock wave ferroelectric generator*

A schematic diagram of longitudinally shock-compressed ferroelectric elements of the FEG operating in the short-circuit mode is shown in Figure 4.6. The stress wave moves through the element along the direction of polarization and depolarizes the material. The element is divided into two zones, an unstressed zone and a stressed zone. The stressed zone is depolarized. The polarization in the unstressed zone is equal to the remanent polarization, P_0.

Due to the shock depolarization, the negative surface screening charge on the left electrode is separated from the positive polarization bound charge in the left face of the unstressed zone. This results in an electric field E_1 in the stressed zone. For voltage balance in the short-circuit mode the presence of E_1 requires the presence of a counter field E_0 across the unstressed zone. This counter field results in the reconfiguration of the positive and negative polarization

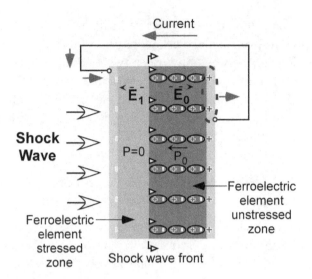

Fig. 4.6. Schematic diagram of a longitudinally shock-compressed ferroelectric element operating in the short-circuit mode.

bound charges in the unstressed zone. This in turn releases some surface screening charges on both electrodes, which are neutralized by a current flow in the external circuit. The direction of shock-induced current is identical to that of the current generated in the short-circuit piezoelectric mode under longitudinal stress (see Figure 4.2(b)). However, the origins of these two currents are different.

Opposing electric fields that are generated in stressed and unstressed zones during shock wave transit can cause complex behavior in longitudinally shocked ferroelectric elements and distortions in the stress-induced current waveforms. This complex behavior is discussed in Chapters 7 and 10 of this book.

A longitudinally shock-compressed ferroelectric element operating in the open circuit mode is shown schematically in Figure 4.7. The electrodes of the element are open in this mode of operation. An electric field and high voltage are generated across the element due to the separation of negative surface screening charge (left electrode) and positive polarization bound charge in the unstressed zone. Some repolarization will take place in the stressed zone because it is now a linear dielectric. The polarity of shock-induced voltage is identical

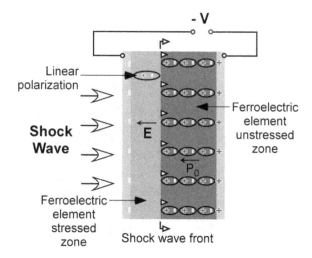

Fig. 4.7. Schematic diagram of a longitudinally shock-compressed ferroelectric element operating in the open circuit mode.

to the polarity of the surface screening charge (Figure 4.7) similar to that in the open circuit piezoelectric mode (see Figure 4.3(c)).

The stress-induced voltage across the ferroelectric element is the sum of two electric potentials. One of them is the potential across the stressed zone and another one is the potential across the unstressed zone. These potentials depend on the volume of each zone and its dielectric properties. The dielectric properties of the compressed zone are under change during shock wave transit. The division of voltage and, correspondingly, electric field between two zones can cause early electric breakdown within the ferroelectric element. The complex behavior of the longitudinally shocked ferroelectric element in the open circuit mode is discussed in Chapter 9 of this book.

4.6.2 Operation of transverse shock wave ferroelectric generator

A schematic diagram of a transversely shock-compressed ferroelectric element operating in the short-circuit mode is shown in Figure 4.8. The shock wave travels through the element perpendicular to the direction of polarization. The stressed zone of the element is

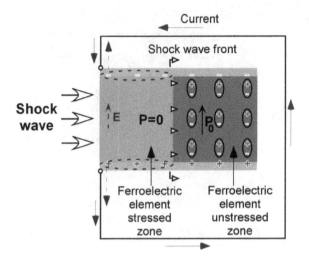

Fig. 4.8. Schematic diagram of a transversely shock-compressed ferroelectric element operating in the short-circuit mode.

depolarized due to the shock compression. This results in unbalance between the surface screening charge and polarization in the element. Some surface screening charge is released on both electrodes. It causes a current flow in the external circuit.

The direction of current generated under transverse shock compression (Figure 4.8) is opposite to the direction of current generated under transverse stress in the piezoelectric mode (see Figure 4.2(c)). The piezoelectric current is caused by the increase in polarization of the ferroelectric element due to the application of transverse stress, while the shock-induced current is caused by the depolarization of a part of the ferroelectric element due to the shock compression.

The duration of a shock-induced current pulse is equal to the transit time of the shock wave front through the element:

$$\tau_p = l/U_s \qquad (4.8)$$

where τ_p is the current pulse duration in [μs], l is the length of the ferroelectric element in [mm], and U_s is the velocity of a shock wave in the element in [mm/μs]. The current pulse duration for the PZT 95/5 element with length 50 mm would be about 12 μs. The amplitude of the depolarization current is directly proportional to

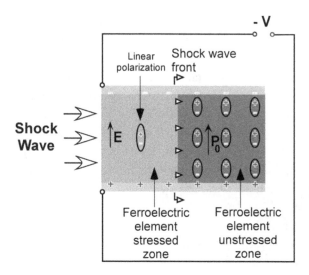

Fig. 4.9. Schematic diagram of a transversely shock-compressed ferroelectric element operating in the open circuit mode.

the remanent polarization of the PZT 95/5 element and the area of its electrodes.

Consider the operation of a transversely shocked ferroelectric element in the open circuit mode. A schematic diagram is shown in Figure 4.9. The surface screening charge released due to the shock depolarization of the stressed zone is not neutralized by a current flow in the external circuit because the electrodes are open in this mode of operation. This charge results in a high electric field and high voltage across the element.

The polarity of the voltage generated under transverse shock compression is identical to the polarity of the screening charge that balanced the polarization (Figure 4.9) but it is opposite to the polarity of the voltage produced by ferroelectrics under transverse stress in the open circuit piezoelectric mode (see Figure 4.3(c)).

4.7 Summary

- The surface screening charges are equal and opposite to the polarization bound charges to enforce the neutrality of net charges and, correspondingly, net electric field in a ferroelectric specimen.

- The surface screening charges are mobile, in the sense that the spatial distribution of their quasi two-dimensional density is ruled by the polarization distribution near the surface.
- The surface screening charge is what one can measure in an experiment. When the polarization changes, the surface screening charge accommodates accordingly. Changes in the surface screening charge cause electric current to flow in the external circuit connected to the electrodes of a ferroelectric specimen. The surface screening charges play an important part in the electrical operation of shock wave ferroelectric generators and piezoelectric generators.
- The physical principles of shock wave ferroelectric generators are fundamentally different from those of low-strain piezoelectric systems.
- The application of low mechanical stress to the ferroelectric elements results in charge displacement, but when the stress is removed the polarization is restored because no large-scale reorientation of domains occurs.
- The shock compression of ferroelectric materials induces phase transformations and domain reorientation with the loss of their initial remanent polarization. Rapid depolarization under shock loading results in the generation of megawatt power levels during a brief interval of time.
- The characteristics of shock waves, such as the shock pressure and geometry of the shock wave front are important factors of the shock-induced depolarization of ferroelectric materials and the electrical operation of shock wave ferroelectric generators.

Bibliography

1. S.V. Kalinin, D.A. Bonnell, T. Alvarez, X. Lei, Z. Hu, J. Ferris, Q. Zhang and S. Dunn, Atomic polarization and local reactivity on ferroelectric surfaces: A new route toward complex nanostructures, *Nano Letters*, **2**(6) (2002) pp. 589–594.
2. S. Jesse, A.P. Baddorf and S.V. Kalinin, Ferroelastic domain wall dynamics in ferroelectric bilayers, *Appl Phys Lett.* **88** (2006) p. 062908.
3. J. Bardeen, Surface states and rectification at a metal semi-conductor contact, *Phys. Rev.* **71** (1947) pp. 717–727.

4. L. Pintilie, V. Stancu, L. Trupina and I. Pintilie, Ferroelectric Schottky diode behavior from a SrRuO3-Pb(Zr0.2Ti0.8)O3-Ta structure, *Phys. Rev. B* **82** (2010) p. 085319.

5. A.K. Tagantsev, L.E. Cross and J. Fousek, *Domains in Ferroic Crystals and Thin Film* (Springer, New York, 2010).

6. C.S. Lynch, The effect of uniaxial stress on the electro-mechanical response of 8/65/35 PLZT, *Acta Materialia* **44**(10) (1996) pp. 4137–4148.

7. ITT Corp. data sheet for EC-64 ferroelectic material.

8. F.W. Neilson, Effects of strong shocks in ferroelectric materials, *Bull. Am. Phys. Soc.* **11**(2) (1957) p. 302.

9. M.B. Boslough and J.R. Assay, *High-Pressure Shock Compression of Solids*, eds. J.R. Assay and M. Shahinpoor (Springer-Verlag, New York, 1993), pp. 7–42.

10. D.S. Drumheller, *Introduction to Wave Propagation in Nonlinear Fluids and Solids* (Cambridge University Press, Cambridge, 1998).

11. R. Cheret, *Detonation of Condensed Explosives* (Springer-Verlag, New York, 1993).

12. R.E. Setchell, Shock wave compression of the ferroelectric ceramics Pb0.99. Zr0.95Ti0.05.0.98Nb0.02O3: Hugoniot states and constitutive mechanical properties, *J. Appl. Phys.* **94**(1) (2003) pp. 573–588.

13. C.E. Reynolds and G.E. Seay, Two-wave shock structures in the ferroelectric ceramics: Barium titanate and lead zirconate titanate, *J. Appl. Phys.* **33**(7) (1962) pp. 2234–2241.

Chapter 5

Design of Miniature Explosive Ferroelectric Generators

5.1 Introduction

It is shown in the previous chapter that the shock depolarization of ferroelectric elements within FEGs is a key factor in the operation of ferroelectric generators. Shock waves in ferroelectric elements can be initiated either by an impacting projectile that has been accelerated to high speed, or by a detonation wave from a high-explosive charge. The initiation of shock waves by impacting projectiles accelerated using gas gun facilities provides precise control of shock front geometry and shock pressure [1–5]. However, these gas gun facilities can only be used for academic studies, not for engineering applications, due to their large size (from 10 to 30 meters), their large mass (more than 1000 kg) and their complicated control systems.

One of the goals of this research was to develop compact or even miniature completely autonomous pulsed power devices utilizing the energy stored in ferroelectric materials. To miniaturize the ferroelectric generators high explosives were used to initiate shock waves in the ferroelectric elements in these devices. A small (from 1 to 10 grams) high-explosive charge is capable of initiating the high-amplitude shock wave in a ferroelectric element of an FEG. This approach made it possible to reduce the size of ferroelectric generators to a few centimeters.

The energy stored in the ferroelectric elements of FEGs is released by explosive-generated shock waves. The properties of detonation waves and high-explosive materials are examined in this chapter.

107

There are many explosive materials, although only a few have sufficient power and detonate with the reproducibility required for use in high-performance ferroelectric generators. Only a few of these materials satisfy the additional requirements of safety and stability.

Three different designs of miniature explosively-driven ferroelectric generators are described in detail in this chapter. The FEGs based on these designs have proven to be operationally reliable prime power sources. Construction materials for explosive ferroelectric generators are discussed.

5.2 Detonation Waves

There are two possible processes by which the reaction zone can propagate into the unreacted material when explosive materials undergo combustion: deflagration and detonation. *Deflagration* is when the reaction zone passes relatively slowly ($\sim$1 mm/μs) across the unreacted energetic material and the gaseous combustion products flow in a direction opposite to that of the reaction zone. *Detonation* is when the reaction zone passes at supersonic speeds ($\sim$2–10 mm/μs) across the unreacted energetic material and the gaseous combustion products flow in the same direction as that of the reaction zone. The detonation process is strongly directive, which is responsible for an essential property that is not obvious; that is, the detonation products just behind the shock wave flow in the same direction as the motion of the detonation wave. It is this condition that is responsible for sustaining the detonation front.

A detonation wave is essentially a shock wave that is supported by a trailing exothermic chemical reaction. It involves a wave travelling through a highly combustible or chemically unstable medium, such as a high-explosive or an oxygen-methane mixture. The chemical reaction of the medium occurs following the shock wave, and the chemical energy of the reaction drives the wave forward maintaining its strength.

A detonation wave follows slightly different rules from an ordinary shock wave (see the description of shock waves in Chapter 4 of this book) since it is driven by the chemical reaction occurring behind the shock wave front. When the chemical energy liberated at a shock front exceeds that required to sustain the shock wave, the detonation wave will grow in strength. When the chemical energy liberated at a shock front is too little, the detonation wave will be attenuated by an overtaking decompression wave, as in the case of a nonreactive shock.

In the case of a detonation, there exists an equilibrium amplitude at which the chemical energy produced is just sufficient to maintain a steady shock. In this case, the shock propagates at a constant velocity called the detonation velocity and has a constant pressure and particle velocity called the Chapman-Jouguet pressure and velocity, respectively.

5.3 Explosives

In ferroelectric generators, the source of energy is that stored in their ferroelectric elements, which is released by explosive-generated shock waves. There are many explosive materials, although only a few have sufficient power and detonate with the reproducibility required for use in high-performance ferroelectric generators. Only a few of these materials satisfy the additional requirements of safety and stability that may be associated with specific applications.

The explosives used in explosively driven ferroelectric generators are almost entirely those in the class called secondary high explosives, a category comprised of materials with a sensitivity between that of the easily detonable primary explosives and that of the very insensitive tertiary explosives. Primary explosives will detonate when subjected to heat or shock, while secondary explosives will only detonate when subjected to shocks produced by primary explosives.

Primary explosives are used in blasting caps and other initiating devices, but are unsuitable for use as main-charges. Tertiary explosives are the safest materials to use, but often prove impracticably difficult to initiate, do not function well in small amounts, and are less powerful than the best available secondary explosives.

Table 5.1. Properties of secondary high explosives [6].

Explosive	Chemical formula	Density (kg/m^3)	Detonation velocity (km/s)	Chapman-Jouguet pressure (km/s)
TNT	$C_7H_5N_3O$	1650	6.93	21
RDX	$C_3H_6N_6O_6$	1810	8.7	34
HMX	$C_4H_8N_8O_8$	1900	9.11	39

Almost all secondary explosives are organic compounds formed from the elements C, H, N, and O and are called CHNO explosives. Their molecular structures are usually quite complicated. Explosive materials are often mixtures of explosive compounds and include polymeric binders that affect their mechanical properties, safety, etc., but are not intended to alter their detonation performance. Explosives of this class liberate their detonation energy by molecular decomposition and reformation into detonation products such as CO_2, N_2, and H_2. The detonation process for many of these explosives does not require atmospheric oxygen, since they contain both the fuel and the oxidizer.

A traditional secondary high explosive is trinitrotoluene (TNT), which is easily melted and cast into the shapes required for explosive pulsed power sources and other applications. More powerful explosives in the same class are hexahydrotrinitrotriazine (also called RDX, cylonite, or hexogen) and cyclotetramethylenetetranitamine (HMX). Desensitized RDX high explosives were used in all the experiments described in this book. The properties of these secondary high explosives are listed in Table 5.1.

5.4   Shock Impedance

When an explosive is in contact with another material, the detonation wave, which is itself a shock wave, interacts with the surface of the material and generates a shock wave within it. The degree of this interaction depends on the shock impedance of the explosive and of the material. The shock impedance of a material, Z_{mat}, is defined to be the product of its initial density, ρ_0, and the shock velocity, U_s:

$$Z_{mat} = \rho_0 U_s \tag{5.1}$$

The shock impedance of the explosive products, Z_{explos}, is defined to be the initial density of the unreacted explosive, ρ, and the detonation velocity, U_{det}:

$$Z_{explos} = \rho U_{det} \tag{5.2}$$

From the momentum equation (see Eq. (4.4) in Chapter 4 of this book)

$$p = \rho_0 U_s u_p = Z_{mat} u_p \tag{5.3}$$

it can be seen that the shock velocity increases as the pressure increases. However, although the impedance also increases, it does so rather slowly, which implies that it can be treated as a constant within reasonable ranges. The impedance is sufficiently constant to allow a differentiation between low-impedance and high-impedance materials. When a shock wave passes from a low-impedance material into a high-impedance material, the shock pressure increases and vice versa.

Therefore, when the explosive products of detonation interact with the surface of a material, two possible cases must be considered: when the impedance of the material is greater than that of the explosive and when the impedance of the explosive is greater than that of the material.

When $Z_{mat} > Z_{explos}$, the detonation produces a shock pressure at the interface that is higher than the Chapman-Jouguet pressure that propagates into the material and a shock wave that propagates into the explosive products. When $Z_{mat} < Z_{explos}$, the detonation produces a shock pressure at the interface that is lower than the Chapman-Jouguet pressure and a partial rarefaction wave is rejected back into the compressed reaction product gases.

A brief description of detonation waves and high explosives is presented in previous sections. For more detailed information, it is recommended that the reader refer to the book by P.W. Cooper entitled *Explosives Engineering* [6].

5.5　Design of Transverse Planar Shock Wave FEG

Planar shock waves initiated in solids with impacting projectiles accelerated in gas gun systems have certain advantages over non-planar shock waves, i.e. stable shock wave front geometry, constant pressure along the front, and constant velocity of the shock wave front. Each of these advantages makes it possible to simplify theoretical models used for the analysis of shock wave experiments.

The shock profile must be considered in FEG design, since the wrong profiles can lead to fracturing and/or early electric breakdown within the ferroelectric elements of the FEG. The profile of the shock wave can affect the amplitude of the output voltage and current pulses produced by FEGs.

To initiate a planar shock wave in the ferroelectric element of an explosive FEG we used a metallic impactor, a *flyer plate*, like it is arranged in gas gun systems [7]. A schematic of the transverse planar shock wave FEG is shown in Figure 5.1. It contains an explosive chamber, a flyer plate and a ferroelectric element incorporated in a plastic body. The direction of shock wave propagation in the

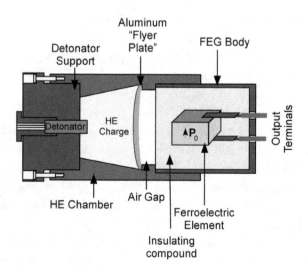

Fig. 5.1.　Schematic of the transverse planar shock wave FEG. P_0 is the remanent polarization vector.

Fig. 5.2. Transverse planar shock wave FEG loaded with HE charge and prepared for explosive and electrical operation (courtesy of Loki Incorporated).

ferroelectric element is perpendicular to the direction of the remanent polarization (transverse shock compression).

The body of the FEG is made of electrically insulating material (epoxy or urethane compounds) to avoid electric discharge during the operation of the generator. A photograph of a transverse planar shock wave FEG is shown in Figure 5.2.

The explosive part of the FEG contains a high-explosive (HE) charge, a detonator support made of polycarbonate or high-density polyethylene, with an RP-501 exploding bridgewire detonator (supplied by RISI). Desensitized RDX high explosives and RP-501 detonators were used in all the experiments described in this book (see Table 5.1 for the properties).

The flyer plate is responsible for the initiation of the shock wave in the ferroelectric element. The plate is made of 6061 aluminium alloy. It has a thickness of 3 mm. To provide planar impact on the ferroelectric element, the shape of the flyer plate is curved. The plate is deformed into a quasi-flat structure under blast loading. To derive the correct curvature and thickness of the flyer plate, the arrival times and locations of the explosive shock are estimated by applying the principles of geometric optics and impedance matching to detonation waves. Then, the shock transit time through the flyer plate material was estimated in order to calculate the arrival time of the shock wave along concentric circular loci on the target side of the flyer plate. The combination of shock arrival times and locations on the plate surface enables calculation of the plate curvature needed to produce an essentially flat flyer plate upon impact with the ferroelectric target. The air gap between the flyer plate and the top of the FEG plastic body (acceleration path) is 5 mm. There is a thin shoulder on the perimeter of the plate. This shoulder holds the plate in place until the pressure of the gases reaches the critical level behind the flyer plate. We performed experiments with different thicknesses of the shoulder. The optimum thickness was found to be 0.7 mm.

The electrical contacts cannot be soldered to the electrodes of the ferroelectric elements because thermal heating decreases their polarization and, correspondingly, the energy produced by the FEGs. The electrical connection to the ferroelectric elements was provided by gluing copper foil strips to the electrodes with silver-based conductive epoxy adhesive (Chemtronic CW2400) that cures at room temperature.

The operation of the generator (Figures 5.1 and 5.2) is as follows. After detonation of the HE charge, a shock wave is generated and high-pressure gases are produced within the explosive chamber of the FEG. The high-pressure gases are used to accelerate a flyer plate. The collision of the flyer plate with the plastic body of the FEG initiates a planar shock wave in the body. The shock wave front propagates through a potting material and into the ferroelectric element. The shock propagation direction is perpendicular to the polarization vector (transverse shock). As a result of the transverse shock compression, the ferroelectric element generates voltage and current in the load circuit. The experimental results obtained with ferroelectric generators based on this design are presented in the following chapters of this book.

5.6 Design of Transverse FEG Based on Spherically Expanding Planar Shock Waves

In the transverse planar shock wave FEG described in the previous section, a flyer plate does not initiate the shock wave directly in the ferroelectric element but in the encapsulating material.

The encapsulation of the ferroelectric element in transverse FEGs is necessary to suppress electric discharge between electrodes of the element caused by the flyer plate and detonation plasma. The shock wave travels through the potting compound and enters the ferroelectric element. The difference in shock impedance of polyurethane or epoxy compounds and ferroelectrics could result in the distortion of the shock wave front entering the ferroelectric element. This shock impedance difference could make the shock front not perfectly planar.

In addition to this, the increase in the FEG output energy achieved through scaling up the cross-section of the ferroelectric element leads to a significant increase in the size and mass of the flyer plate, and the amount of HE (correspondingly, its destructive power), mass, and weight of the generator.

Based on everything mentioned above we made a decision to develop the design of the transverse shock wave FEG without a metallic impactor of any kind. This generator utilizes not planar, but spherically expanding shock waves [8–13]. A schematic diagram of the transverse FEG based on this design is shown in Figure 5.3. The FEG contains two parts: a detonation chamber and a ferroelectric element incorporated in a plastic body. The ferroelectric element

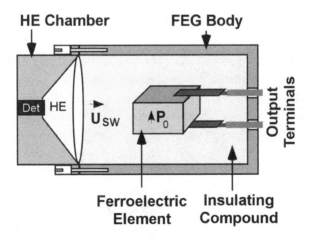

Fig. 5.3.   Schematic of the transverse FEG based on spherically expanding shock waves. P_0 is the remanent polarization vector.

Fig. 5.4. A transverse FEG based on spherically expanding shock waves loaded with high explosives and prepared for explosive and electrical operation (courtesy of Loki Incorporated). The diameter of the FEG was 38 mm.

is encapsulated with urethane or epoxy compound as the electrical insulating material. The shock waves in the ferroelectric element of this FEG are initiated from the HE detonation that is in direct contact with the top of the plastic body of the generator.

A photograph of the transverse FEG based on spherically expanding shock waves is shown in Figure 5.4. The diameter of this FEG was 38 mm. The distance between the top of the urethane potting materials and the top of the PZT 95/5 ferroelectric element was 25 mm. The mass of the HE charge was 11 grams of RDX.

The operation of the FEG (Figures 5.3 and 5.4) is as follows. After the initiation of the detonator, the detonation wave propagates through high explosives. The HE charge is in direct contact with the top of the plastic body of the FEG. The HE detonation shock propagates through a urethane potting material and into the ferroelectric element. The shock wave travels across the polarization vector of the ferroelectric element.

The advantages of this design are simplicity, fewer components in comparison with the planar shock wave FEG (Figure 5.1), the smaller mass of HE, and the smaller size and weight of the FEG. The experimental results obtained with ferroelectric generators based on this design are presented in the following chapters of this book.

The shock waves generated in transverse FEGs of both types (Figures 5.1 and 5.3) are not perfectly planar. This could result in a more complex electrical response by the shock-compressed ferroelectric elements in miniature FEGs in comparison with that obtained in those experiments in which planar shock waves are generated by gas gun facilities. The experimental results obtained with FEGs of both designs are presented in the following chapter of this book.

5.7   Design of Longitudinal Planar Shock Wave FEG

A schematic of the longitudinal planar shock wave FEG developed by Loki Inc. [14–22] is shown in Figure 5.5. The explosive part of this FEG is identical to that for the transverse planar shock wave FEG (see Figure 5.1). However, the ferroelectric element holder and positioning of the ferroelectric element are different from those for the transverse planar shock wave FEG. The direction of shock wave propagation in the ferroelectric element is parallel or anti-parallel to the direction of the remanent polarization (longitudinal shock compression).

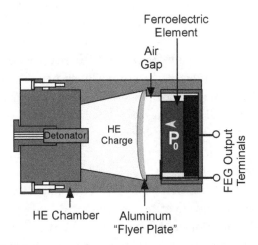

Fig. 5.5.   Schematic of the longitudinal planar shock wave FEG.

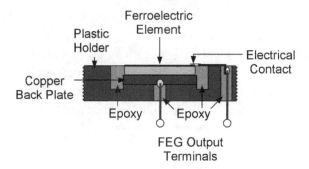

Fig. 5.6.   Schematic of the ferroelectric element holder of the longitudinal shock wave FEG.

A schematic of the ferroelectric element holder of longitudinal FEG is shown in Figure 5.6. The ferroelectric element is bonded to a copper back plate with thickness 5 mm. The copper back plate provides mechanical impedance matching in order to minimize the reflection of the stress wave when it reaches the back electrode of the ferroelectric element. A silver-loaded epoxy (Chemtronic CW2400) was used to bond the ferroelectric element to the back plate. The silver-loaded epoxy serves as the electrical contact.

The ferroelectric element, along with the copper back plate, is centered in the plastic holder. The two electrical output terminals of the FEG are connected to the electrodes of the ferroelectric element. One terminal is bonded to the front electrode of the element (which was subjected to flyer plate impact) with silver-loaded epoxy. The other terminal is soldered to the copper back plate. The gap between the ferroelectric element ceramic sidewall and the inside wall of the plastic holder is filled with epoxy to avoid surface flashover, electric discharge initiated by detonation-generated plasma, and electric breakdown due to the impact of the aluminum flyer plate.

The body of the FEG and ferroelectric element holder were made of electrically insulating materials. We conducted a series of experiments to determine the effect of the generator body and its thickness on the explosive and electrical operation of the FEG. Three types of plastics were used in these experiments: high-density polyethylene, polyvinylchloride (PVC), and polycarbonate. The experimental results show that all of these materials work well. There was no significant difference in the generation of high voltage and current pulses by FEGs having body-wall thicknesses ranging from 0.5 mm to 30 mm. From this, it can be concluded that the generator body can be a lightweight thin plastic shell that holds together parts of the FEG and does not disintegrate until the explosive charge has completed detonation.

The operation of the generator (Figure 5.5) is as follows. After detonation of the HE charge, a shock wave is generated and high-pressure gases are produced within the explosive chamber of the FEG. The high-pressure gases accelerate a flyer plate. The collision of the flyer plate with the front electrode of the ferroelectric element initiates a planar shock wave in the element without any distortions as it occurs in the transverse planar shock wave FEGs (see Figure 5.1). The planar shock wave propagates through the element. The shock propagation direction is parallel or anti-parallel to the polarization vector (longitudinal shock). As the result of the longitudinal shock compression, the ferroelectric element generates voltage and current in the load circuit. The experimental results obtained with ferroelectric generators based on this design are presented in the following chapters of this book.

5.8 Encapsulating Materials

The ferroelectric generators work under extreme conditions, i.e. high explosive detonation, high pressure gases, shock compression, and high electric field. To hold ferroelectric elements in place, to control the properties of the shock waves, and to avoid electric break-down within the generators, the ferroelectric elements of transverse and longitudinal FEGs were encapsulated with electrical insulating materials.

We tested a wide range of commercially available epoxy and urethane compounds as encapsulating materials for the FEGs. We found that the Smooth-On Crystal Clear 204 compound facilitates the reliable and reproducible operation of the generators, and is technologically compatible. Crystal Clear 204 is a general-purpose

Table 5.2. Properties of encapsulating materials used in the explosive ferro-electric generators.

Property	Crystal clear 204	Stycact 2651-40
Base chemistry	Urethane	Epoxy
Liquid type	Clear liquid	Black liquid
Density (10^3 kg/m^3)	1.04	1.5
Cure time (hrs)	48	24
Viscosity	600	2200
Work life/Pot life (min)	120	60
Dielectric Strength (kV/mm)	10.4	17.7
Tensile Strength	2500	6300
Volume Resistivity (Ohm-cm)	$1.4 \cdot 10^{15}$	$1.0 \cdot 10^{14}$
Durometer/Hardness	72 D	85 D

urethane compound. The parameters of Crystal Clear 204 are presented in Table 5.2. A photograph of the transverse FEG prepared with this compound is shown in Figure 5.4. The bodies of generators prepared with Crystal Clear 204 are absolutely transparent. This makes it possible to observe and control the position of the ferroelectric elements and other parts within FEGs.

Another encapsulating material we used in the FEGs was Emerson and Cuming Stycast 2651-40 (with Catalyst 23 LV). The parameters of Stycast 2651-40 are presented in Table 5.2. Stycast 2651-40 has almost double the electric breakdown field in comparison with that of Crystal Clear 204. In addition to this, Stycast 2651-40 has significantly higher density and hardness than Crystal Clear 204. It can provide a better match of the shock impedance of ceramic ferroelectric elements to the encapsulating material and, correspondingly, this results in lower distortion of the shock wave profiles within the FEGs.

A photograph of the body of a transverse FEG prepared with Stycast 2651-40 potting material is shown in Figure 5.7. The results obtained with the generators prepared with the two encapsulating materials are presented in the following chapters of this book.

Setchell *et al.* [22, 23] have investigated the shock properties of materials used for the encapsulation of ferroelectric specimens

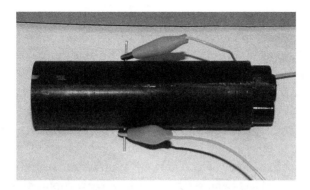

Fig. 5.7. Body of a transverse shock wave FEG prepared with Stycast 2651-40 potting material (courtesy of Loki Incorporated).

in gas gun shock depolarization studies. The purpose of these investigations was to find a way to match the shock properties of encapsulating materials to those of the ferroelectrics. The effects of different compositions of alumina (polycrystalline Al_2O_3) additive on the shock properties of epoxy were investigated. It was found that increasing the volume fraction of alumina resulted in steadily increasing Hugoniot states, shock wave rise times, and release wave velocities.

It was observed that the difference between the release wave and shock wave velocities increased significantly as the concentration of alumina increased, resulting in stronger shock wave attenuation. Changes in the size and shape of the alumina particles had little effect with the exception that they tended to cause viscous spreading of the shock wave profiles during shock compression. Based on the obtained results the conclusion was reached that the shock properties of the epoxy-based encapsulating material can be controlled through adding a small fraction of alumina to epoxy [22, 23].

5.9 Summary

- In explosive ferroelectric generators, the source of energy is that stored in their ferroelectric elements, which is released by explosive-initiated shock waves. The properties of detonation waves and high-explosive materials are examined in this chapter.

It is shown which high explosives have sufficient power, detonate with the reproducibility required for use in high-performance ferroelectric generators, and satisfy the additional requirements of safety and stability.

- Explosive ferroelectric generators based on three different designs are described in detail in this chapter. The utilization of high explosives makes it possible to reduce the size of shock wave ferroelectric generators to a few cubic centimeters.

- In two of the explosive FEG designs, planar transverse shock waves (transverse FEGs where the shock wave propagation direction is perpendicular to the direction of polarization) and planar longitudinal shock waves (longitudinal FEGs where the shock wave propagation direction is parallel or anti-parallel to the direction of polarization) are initiated in ferroelectric elements with explosively driven metallic impactors (flyer plates).

- Another FEG design does not utilize metallic impactors of any kind. This generator is based on spherically expanding shock waves. The shock waves in the ferroelectric element of this FEG are initiated from the HE detonation that is in direct contact with the plastic body of the generator. This design can be used for both the transverse and longitudinal FEGs.

- The FEGs based on the designs described in this chapter have proven to be operationally reliable prime power sources. Each of these designs can be modified in order to provide optimum output parameters for specific tasks. The results obtained with generators based on these designs are presented in the following chapters of this book.

- Due to the small size of explosive FEGs, the initiated shock waves can be not perfectly planar. This results in a more complex electrical response by the shock-compressed ferroelectric elements in miniature FEGs in comparison with that obtained in those experiments in which planar shock waves are generated by gas-gun facilities.

- The shock profile must be considered in FEG design, since the wrong profiles can lead to fracturing and/or early electric break-down. The profile of the shock wave can affect the amplitude of the output voltage and current pulses produced by both longitudinal and transverse shock wave FEGs.

- Construction materials for ferroelectric generators should provide reliable explosive and electrical operation of the FEGs. The construction materials discussed in this chapter proved their reliability in a number of experiments performed with explosive ferroelectric generators of different designs.

Bibliography

1. W.J. Halpin, Current from a shock-loaded short-circuited ferroelectric ceramic disk, *J. Appl. Phys.* **27**(1) (1966) 153–163. R. K. Linde, Depolarization of ferroelectrics at high stress rates, *J. Appl. Phys.* **38**(12) (1967) pp. 4839–4842.

2. R.E. Setchell, Shock wave compression of the ferroelectric ceramics Pb0.99.Zr0.95Ti0.05.0.98Nb0.02O3: Hugoniot states and constitutive mechanical properties, *J. Appl. Phys.* **94**(1) (2003) pp. 573–588.

3. R.E. Setchell, Shock wave compression of the ferroelectric ceramic Pb0.99(Zr0.95Ti0.05)0.98Nb0.02O3: depoling currents, *J. Appl. Phys.* **97** (2005) p. 013507.

4. R.E. Setchell, Shock wave compression of the ferroelectric ceramic $Pb_{0.99}(Zr_{0.95}Ti_{0.05})_{0.98}Nb_{0.02}O_3$: microstructural effect, *J. Appl. Phys.* **101** (2007) p. 053525.

5. P.W. Cooper, *Explosives Engineering* (Wiley-VCH, New York, 1996).

6. S.I. Shkuratov, J. Baird and E.F. Talantsev, Autonomous pulsed power generator based on transverse shock wave depolarization of ferroelectric ceramics, *Rev. Sci. Instrum.* **81** (2010) p. 126102.

7. S.I. Shkuratov, J. Baird and E.F. Talantsev, Miniature 120-kV generator based on transverse shock depolarization of $Pb(Zr_{0.52}Ti_{0.48})O_3$ ferroelectrics, *Rev. Sci. Instrum.* **82** (2011) p. 086107.

8. S.I. Shkuratov, J. Baird and E.F. Talantsev, Effect of shock front geometry on shock depolarization of Pb(Zr0.52Ti0.48)O3 ferroelectric ceramics, *Rev. Sci. Instrum.* **83** (2012) p. 074702.

9. S.I. Shkuratov, J. Baird, E.F. Talantsev, E.F. Alberta, W.S. Hackenberger, A.H. Stults and L.L. Altgilbers, Miniature 100-kV explosive-driven prime power sources based on transverse shock-wave depolarization of PZT 95/5 ferroelectric ceramics, *IEEE Trans. Plasma Sci.* **40**(10) (2012) pp. 2512–2516.

10. S.I. Shkuratov, J. Baird and E.F. Talantsev, Extension of thickness-ependent dielectric breakdown law on adiabatically compressed ferroelectric materials, *Appl. Phys. Lett.* **102** (2013) p. 052906.

11. S.I. Shkuratov, J. Baird, V.G. Antipov, E.F. Talantsev, H.R. Jo, J.C. Valadez and C.S. Lynch, Depolarization mechanisms of PbZr0.52Ti0.48O3

and PbZr0.95Ti0.05O3 poled ferroelectrics under high strain rate loading, *Appl. Phys. Lett.*, **104** (2014) p. 212901.

12. J. Baird and S. Shkuratov, Ferroelectric energy generator, system, and method. U.S. Patent 7,560,855. Issued July 14, 2009 to Loki Incorporated.

13. S.I. Shkuratov, E.F. Talantsev, L. Menon, H. Temkin, J. Baird and L.L. Altgilbers, Compact high-voltage generator of primary power based on shock wave depolarization of lead zirconate titanate piezoelectric ceramics, *Rev. Sci. Instrum.* **75**(8) (2004) pp. 2766–2769 .

14. S.I. Shkuratov, E.F. Talantsev, J. Baird, H. Temkin, L.L. Altgilbers and A.H. Stults, Longitudinal shock wave depolarization of Pb(Zr$_{0.52}$Ti$_{0.48}$)O$_3$ polycrystalline ferroelectrics and their utilization in explosive pulsed power, in *Proceedings of 14th APS Topical Conference on Shock Compression of Condensed Matter*, Baltimore, MD, July 2005, Vol. 845, Part 2 (AIP Conference Proceedings, 2006, eds. M.D. Furnish, M. Elert, T.P. Russel, and C.T. White), pp. 1169–1172.

15. S.I. Shkuratov, E.F. Talantsev, J. Baird, H. Temkin, Y. Tkach, L.L. Altgilbers, and A.H. Stults, The depolarization of a Pb(Zr$_{0.52}$Ti$_{0.48}$)O$_3$ polycrystalline piezoelectric energy-carrying element of compact pulsed power generator by a longitudinal shock wave, in *Digest of Technical Papers — 16th IEEE International Pulsed Power Conference*, Albuquerque, NM, 2007, Vol. 2, Article number 4084268, pp. 529–532.

16. S.I. Shkuratov, E.F. Talantsev, J. Baird, M.F. Rose, Z. Shotts, L.L. Altgilbers and A.H. Stults, Completely explosive ultracompact high-voltage nanosecond pulse-generating gystem, *Rev. Sci. Instrum.* **77**(4) (2006) p. 043904.

17. S.I. Shkuratov, E.F. Talantsev, J. Baird, A.V. Ponomarev, L.L. Altgilbers and A.H. Stults, Operation of the longitudinal shock wave ferroelectric generator charging a capacitor bank: experiments and digital model, in *Digest of Technical Papers — 16th IEEE International Pulsed Power Conference*, Albuquerque, NM, 2007, Vol. 2, Article number 4652390, pp. 1146–1150.

18. S.I. Shkuratov, J. Baird, E.F. Talantsev, Y. Tkach, L.L. Altgilbers, A.H. Stults and S.V. Kolossenok, Pulsed charging of capacitor bank by compact explosive-driven high-voltage primary power source based on longitudinal shock wave depolarization of ferroelectric ceramics, in *Digest of Technical Papers — 16th IEEE International Pulsed Power Conference*, Albuquerque, NM, 2007, Vol. 2, Article number 4084270, pp. 537–540.

19. S.I. Shkuratov, J. Baird, E.F. Talantsev, M.F. Rose, Z. Shotts, L.L. Altgilbers, A.H. Stults and S.V. Kolossenok, Completely explosive ultracompact high-voltage pulse generating system, in *Digest of Technical Papers — 16th IEEE International Pulsed Power Conference*, Albuquerque, NM, 2007, Vol. 2, Article number 4084247, pp. 445–448.

20. S.I. Shkuratov, J. Baird, E.F. Talantsev, A.V. Ponomarev, L.L. Altgilbers and A.H. Stults, High-voltage charging of a capacitor bank, *IEEE Trans. Plasma Sci.* **36**(1) (2008) pp. 44–51.

21. S.I. Shkuratov, J. Baird, V.A. Antipov, E.F. Talantsev, C.S. Lynch and L.L. Altgilbers, PZT 52/48 depolarization: quasi-static thermal heating versus longitudinal explosive shock, *IEEE Trans. Plasma Sci.* **38**(8) (2010) pp. 1856–1863.
22. R.E. Setchell and M.U. Anderson, Shock-compression response of an alumina-filled epoxy, *J. Appl. Phys.* **97** (2005) p. 083518.
23. R.E. Setchell, M.U. Anderson, and S.T. Montgomery, Compositional effects on the shock-compression response of alumina-filled epoxy, *J. Appl. Phys.* **101** (2007) p. 083527.

Chapter 6

Mechanisms of Transverse Shock
Depolarization of PZT 95/5 and PZT 52/48

6.1 Introduction

The shock compression of a ferroelectric element within an explosive
FEG induces phase transformations, domain wall motion and domain
disorientation. The amount of electric charge released by a ferroelectric element due to the shock-induced depolarization determines the
energy that can be delivered by an FEG in a load circuit. Due to their
small FEG size, the shock waves generated in explosive generators
can be not perfectly planar. This results in a more complex electrical
response by the shock-compressed ferroelectric elements in miniature
FEGs in comparison with that obtained in those experiments in
which planar shock waves are generated by gas gun facilities.

The results of a comparative experimental study of changes to the
remanent polarization of two poled ferroelectrics, PZT 95/5 and PZT
52/48, subjected to transverse shock compression within miniature
explosive FEGs, are presented in this chapter. To enable a direct
comparison of the results obtained with the PZT 52/48 and PZT
95/5, identical specimen geometry for both ferroelectrics was used in
the experiments. The results of investigations of the thermal-induced
depolarization of the ferroelectrics along with high-resolution X-ray
diffraction are also presented. The obtained results indicate that the
mechanisms of transverse shock depolarization are different for PZT
95/5 and PZT 52/48.

6.2 Transverse Shock Compression of PZT 52/48 and PZT 95/5

Schematics of the miniature transverse shock wave ferroelectric generator and the measuring circuit used in the shock depolarization experiments are shown in Figure 6.1. The diameter of the generators was 38 mm and the HE charge was 11 grams of RDX [1–3]. More details on an FEG of this type can be found in Chapter 5 of this book.

The HE detonation shock propagated through a urethane potting material and into the ferroelectric material (Figure 6.1). The shock propagation direction was perpendicular to the polarization vector (transverse shock). The output terminals of the FEG were short-circuited. The stress-induced current was monitored with a Pearson 411 current probe. Additional experimental details are described elsewhere [4, 5].

PZT 52/48 and PZT 95/5 ceramic specimens have been supplied by ITT Corporation and TRS Technologies Inc., respectively. Tables 2.1 and 2.2 in Chapter 2 of this book list the physical properties for the two ferroelectrics. The sizes and geometric shapes of the specimens affect the stress-induced depolarization behavior [5]. Identical specimen geometry was used (12.7-mm thick × 12.7-mm wide × 50.8-mm long) to enable a direct comparison of experimental results obtained from PZT 52/48 and PZT 95/5. Each specimen was poled along its thickness dimension.

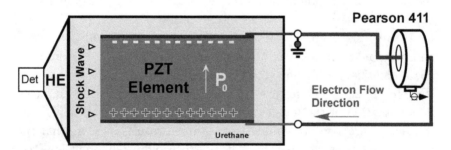

Fig. 6.1. Schematics of the ferroelectric generator and measuring circuit used in the transverse shock depolarization experiments. P_0 is the remanent polarization vector. The polarity of the surface charge is shown by positive (+) and negative (−) signs.

6.3 Uniaxial Stress Distribution in Shock-Compressed Ferroelectric Elements

The shock wave uniaxially compresses the material in the shock propagation direction with no lateral expansion (Poisson effect), as expansion is prevented by the inertia of the specimen. The material behind the adiabatic compressive stress wave has undergone a compressive stress jump in the shock propagation direction.

To determine the uniaxial stress amplitude in the ferroelectric specimens a simulation of high strain rate loading by using the CALE computer code was performed. The CALE is a two-dimensional Arbitrary Lagrange/Eulerian second order accurate hydrodynamics program [6].

We ran the simulation in cylindrical coordinates, assuming rotational symmetry around the axis and zero material rotation. In order for the numerical integration to proceed, the space was divided into "zones" by a mesh of coaxial cylindrical surfaces and a set of planes. Each zone was washer-shaped, with a nearly square section in the R-Z plane, with the coordinates R (cylindrical radius from the symmetry axis), Z (axial position), and angle (around the symmetry axis). Figure 6.2 shows CALE results for the uniaxial stress distribution in the middle of a PZT 95/5 element of transverse FEG at 8 μs after the compression front entered the element. The uniaxial stress is practically uniform in the cross-section. In the middle of the ferroelectric element the uniaxial stress amplitude was found to be 1.7 ± 0.1 GPa.

6.4 Transverse Shock Depolarization of PZT 52/48 and PZT 95/5

The shock loading produced current versus time profiles. Figure 6.3 shows typical waveforms of the stress-induced current and charge for the two ferroelectric materials.

The stress-induced electric charge, Q_{SW}, is the time integral of the stress-induced current:

$$Q_{SW}(t) = \int_0^{\tau_{SW}} I_{SW}(t) \cdot dt \qquad (6.1)$$

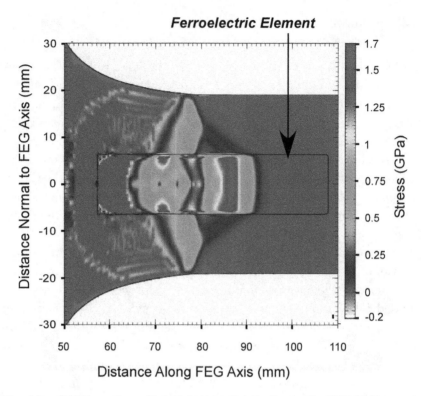

Fig. 6.2.   CALE results — Uniaxial stress distribution in the PZT 95/5 ceramic element of transverse FEG at 8 μs after the compression front entered the element.

where I_{SW} is the stress-induced current, and τ_{SW} is the shock wave transit time.

The polarity of the charge released by transversely-shocked PZT 52/48 was identical to that for PZT 95/5 (see Figure 6.3) and identical to the polarity of the surface charge that balanced the initial remanent polarization.

The stress-induced charge released by PZT 95/5 was $Q_{SW95/5} = 204.5 \pm 1.8 \mu C$ and the corresponding charge density was $\omega_{SW95/5} = 32 \pm 1 \mu C/cm^2$. The stress-induced charge released by PZT 52/48, $Q_{SW52/48}$, and the corresponding charge density, $\omega_{SW52/48}$, were more than two times lower than those of PZT 95/5 (see Table 6.1).

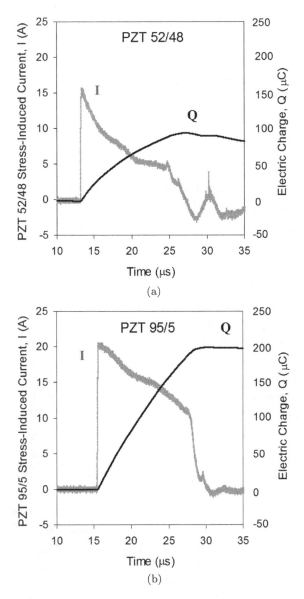

Fig. 6.3. Typical waveforms of stress-induced current and evolution of electric charge released by transversely shock-compressed (a) PZT 52/48 and (b) PZT 95/5.

Table 6.1. Summary of results obtained with PZT 52/48 and PZT 95/5 ferroelectric ceramic specimens in transverse shock depolarization experiments and thermal depolarization experiments.

Property	PZT 52/48	PZT 95/5
Stress-induced charge density, $\omega_{SW}\,(\mu C/cm^2)$	14 ± 2	32 ± 1
Thermal-induced depolarization charge density, $\omega_{therm}\,(\mu C/cm^2)$	31 ± 1	32 ± 1

6.5 Thermal-Induced Depolarization of PZT 52/48 and PZT 95/5

The remanent polarization of each material was measured by performing thermal depolarization experiments. PZT 52/48 and PZT 95/5 specimens were heated above their Curie points using a Thermodyne 47900 furnace.

A schematic diagram of the experimental setup used for thermal depolarization experiments is shown in Figure 6.4. A ferroelectric specimen was placed in a thermal bath with ultra-fine sand. The temperature was controlled using a K-type thermocouple connected to a SPER Scientific thermometer (Model 800005). Current from the ferroelectric specimen was monitored using a Keithly 2400 pico-ampere meter. Each specimen was subjected to a heating from room temperature to above its Curie point ($T_C = 320°C$ for PZT 52/48 and $T_C = 230°C$ for PZT 95/5) at 0.8 degree/min, was kept at the highest temperature for three hours, and then was cooled to room temperature at 0.8 degree/min.

One of the difficulties with measuring the thermal depolarization current was to provide reliable electrical contacts to the ferroelectric samples while we tested them, because the contacts had to maintain their electrical connections to the samples at temperatures of up to 370°C. Due to the nature of the tests, it was not acceptable to heat the specimens beforehand, therefore the contacts could not be soldered. In addition, spring-type contacts could not be used because of the piezoelectric response of the sample to any mechanical stress. So, the electrical connections to the ferroelectric specimens were made by gluing copper foil strips with a silver-based conductive epoxy

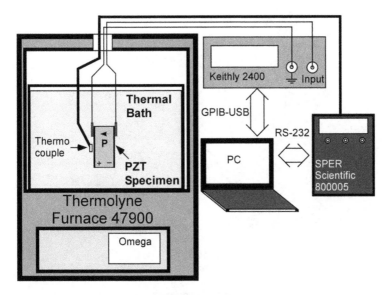

Fig. 6.4.   Schematic of the experimental setup used for the measurement of the thermal depolarization of ferroelectric specimens.

adhesive (Chemtronic CW2400) that cures at room temperature. Additional experimental details can be found elsewhere [5].

Typical plots of thermal depolarization current, $I_{therm}(T(t))$ and charge released, $Q(T(t))$ for PZT 52/48 and 95/5 are shown in Figure 6.5. There are well-resolved peaks in the current plots that occurred at $T_{52/48} = 147.2 \pm 2.9°C$ and $T_{95/5} = 225.4 \pm 2.3°C$.

The low-temperature spike in the PZT 95/5 thermal-induced current plot (Figure 6.5(b)) occurred at the transition from the low-temperature to the high-temperature FE rhombohedral phase [7, 8].

The depolarization charge density, ω_{therm}, is the integral of the experimentally measured thermal-induced current, $I_{therm}(T(t))$, divided by the electrode area, A:

$$\omega_{therm}(T(t)) = \left(\int_0^{t_F} I_{therm}(T(t)) \cdot dt \right) \Big/ A \qquad (6.2)$$

where $t_F = 25000$ s is the heating time.

The thermal-induced charge released by the PZT 52/48 and PZT 95/5 and the corresponding charge densities were very close,

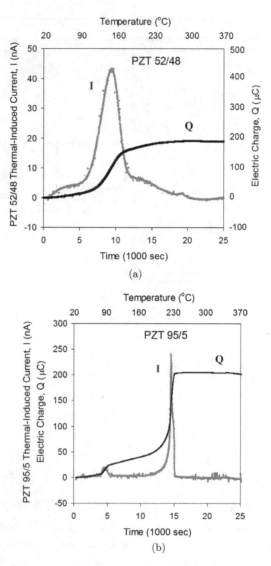

Fig. 6.5. Typical plots of thermal depolarization current and charge for (a) PZT 52/48 and (b) PZT 95/5.

$\omega_{therm52/48} = 31 \pm 1\mu\,C/cm^2$ and $\omega_{therm95/5} = 32 \pm 1\mu\,C/cm^2$. These results indicate similar initial remanent polarization of PZT 52/48 and PZT 95/5. Table 6.1 summarizes the shock and thermal depolarization experimental results.

6.6 Mechanisms of Transverse Shock Depolarization of PZT 52/48 and PZT 95/5

A comparison of the charge density released under the shock loading and thermal heating of PZT 95/5 presented in Table 6.1 shows that the PZT 95/5 was completely depolarized by the transverse shock loading. This is in good agreement with the results of the transverse shock depolarization of PZT 95/5 obtained using gas gun facilities [9–11].

A similar comparison for PZT 52/48 indicates only a 45% transverse shock depolarization (Table 6.1). Apparently, a significant difference in the amount of stress-induced charge for the two ferroelectrics is caused by the different depolarization mechanisms.

Polarization can be changed by lattice distortion that gives rise to the piezoelectric effect, polarization reorientation can be induced by differences between the stress components through domain wall motion, and polarization can be eliminated by a phase transition from a polar phase to a non-polar phase.

The composition of PZT 95/5 lies very close to a boundary between ferroelectric and antiferroelectric phases in the phase diagram. It has been demonstrated [12] that in hydrostatically loaded PZT 95/5 the FE (rhombohedral) to AFE (orthorhombic) phase transition occurs abruptly at a pressure of 0.32 GPa. This pressure-induced FE-to-AFE transition results in a decrease of the volume of the material [12] and a corresponding complete depolarization of the material.

It was also shown [12] that under the uniaxial quasi-static compression of PZT 95/5 with the stress applied in the polarization direction, domain reorientation occurred at low stress levels at nearly constant volume and the phase transformation with a corresponding volume reduction occurred at higher stress levels beginning at 0.2 GPa and remained incomplete at 0.6 GPa when the specimens broke. Domain reorientation may contribute to the shock depolarization of PZT 95/5.

There do not appear to be other studies of the depolarization of PZT 52/48 when compressed by a transverse shock. Reynolds and Seay [13] reported on detailed experimental studies of PZT

52/48 shock compressed in the direction parallel to the polarization (longitudinal shock). It was shown that the charge released by longitudinally shock-compressed PZT 52/48 was practically linear, increasing with pressure up to 1.1 GPa [13]. The authors concluded in [13] that the mechanism of charge release by longitudinally shocked PZT 52/48 is not a phase transition or domain switching but a reduction of the dipole moment by axial compression (i.e., the piezoelectric effect). It should be noted that in [13], because of longitudinal shock compression, the polarity of the charge released by specimens due to the piezoelectric effect was identical to the polarity of the charge that would be released due to domain-switching depolarization.

The direction of current flow observed here for transverse shock loading of PZT 52/48 specimens was inconsistent with the linear piezoelectric effect. A typical assumption for shock loading is that the shock uniaxially compresses the material in the shock propagation direction with no lateral expansion. Under this state of stress, the linear piezoelectric effect should increase the material's polarization, but this would cause a current in the direction opposite to what was observed (Figure 6.3).

To confirm the linear piezoelectric behavior, we subjected PZT 52/48 specimens to low-pressure impulse loading in the Z-direction (perpendicular to the polarization vector, see XYZ-axes in Figure 6.6). In these experiments the PZT 52/48 released an electric charge opposite in polarity to that of the charge released by PZT 52/48 under transverse shock loading. These results confirm that the direction of charge flow observed in the transverse shock experiments indicates depolarization of PZT 52/48.

Depolarization of PZT 52/48 under transverse shock compression cannot be explained by a transition from a polar to a non-polar phase. PZT 52/48 elements were in the tetragonal ferroelectric phase. PZT 52/48 does not undergo a phase transition under shock loading, it remains in the tetragonal phase under pressure at experimental conditions within the explosive FEG.

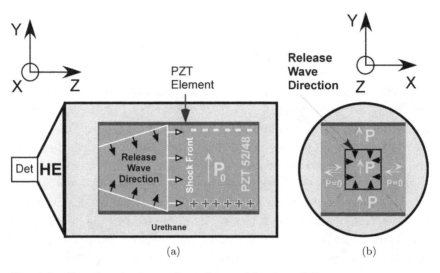

Fig. 6.6. Snapshot in time, where the combination of the stresses creates an inhomogeneous stress field in the ferroelectric element. (a) — Side view. (b) — Top view. P_0 and P are polarization vectors. The polarity of the surface charge is shown by $(+)$ and $(-)$ signs.

The mechanism of shock depolarization is not immediately apparent. Compression in a direction perpendicular to the polarization direction would not be expected to induce domain reorientation.

The observed results are explained by considering the multiaxial time-dependent state of stress in the specimen. Consider the cross-sections of the shock-loaded specimen shown in Figure 6.6. The shock wave is propagating in the Z-direction and the polarization is in the Y-direction. Using an elastic approximation, the transverse $(X$ and $Y)$ stress components are given by:

$$\sigma_{xx} = \sigma_{yy} = \frac{\sigma_{zz}\nu}{1 - \nu} \tag{6.3}$$

where σ_{zz}, σ_{yy}, and σ_{xx} are components of stress in the Z-, Y-, and X-directions respectively, and ν is the Poisson's ratio. The material behind the adiabatic compressive stress wave has undergone a compressive stress jump to σ_{zz}. Lateral expansion (the Poisson effect) is prevented by the inertia of the specimen, resulting in the compressive stresses $\sigma_{xx} = \sigma_{yy}$ given by Eq. (6.3).

As the shock wave moves into the specimen, the stress field changes along the specimen sides (Figure 6.6). Because the surrounding material (urethane) has a lower acoustic impedance than that of the PZT, a release wave forms at the interface of the compressive shock and the interface between the PZT and urethane. This release wave propagates into the specimen behind the shock wave, reducing the lateral stress components σ_{xx} and σ_{yy}.

Figure 6.6 shows a snapshot in time, where the combination of the stresses creates an inhomogeneous stress field. Polarization reorientation occurs when the component of stress in the polarization direction is more compressive than a component of stress in a direction transverse to the polarization direction.

In the wedge-shaped regions that surround the square center region (Figure 6.6(b)), the stress component normal to the side surface has decreased due to the interface boundary conditions. In the right and left wedges this has little effect on the polarization because σ_{xx} is more compressive than σ_{yy} and provides a driving force for keeping the polarization in the Y-direction.

In the top and bottom wedges, however, σ_{yy} is more compressive than σ_{xx}. This provides a driving force for rotating the polarization from the Y-direction to the X-direction through the ferroelastic effect (domain reorientation). As the shock propagates through the entire crystal in the Z-direction, the effect of the unloading wave coming in from the X-faces will depolarize half of the material. This is consistent with the observation of close to 50% depolarization of the PZT 52/48 specimens.

6.7   High-Resolution X-Ray Diffraction of PZT 95/5 and PZT 52/48

High-resolution X-ray diffraction (XRD) was performed on each material using a Bragg-Brentano diffractometer (PANalytical X'Pert Pro). Specimens were crushed, ground and sieved using a 100 mesh to produce a powder. The XRD data were collected in a broad 2θ range

(0 through 140°) using Cu $K\alpha$ radiation ($\lambda = 0.154060$ nm). The continuous scanning mode was used and the data were recorded with a 0.033° 2θ step.

All the diffraction patterns were subjected to the fitting algorithm provided by the PANalytical X'Pert Highscore Plus software to determine the position, intensity, broadening and shape of each peak. X'Pert HighScore Plus uses the Pseudo-Voigt profile function, which is the weighted mean between a Lorentz and a Gauss function. The d-spacing was performed using the X'Pert Highscore FWHM single peak fitting routine. The results were used to confirm the ABO_3 perovskite structure of the PZT 52/48 and PZT 95/5 specimens, to determine the lattice parameters, and to calculate the volume of the unit cell.

Figure 6.7 shows the most informative part (2θ from 10 to 100°) of the X-ray diffraction patterns for the PZT 52/48 and PZT 95/5 specimens in the as-received and after-shocked condition.

Analysis of the XRD patterns indicates that all specimens have a single-phase dominant perovskite structure. No significant presence of a second phase was traced. The peak splitting of the (100) and (200) reflections in the diffraction pattern of PZT 52/48 (see inset in Figure 6.7(a)) indicates a predominantly tetragonal distortion. This peak splitting of the (100) and (200) peaks does not appear in the PZT 95/5 patterns; instead, splitting in the (111) peak indicates a predominantly rhombohedral distortion (see insert in Figure 6.7(b)).

The lattice parameters for the PZT 52/48 and PZT 95/5 specimens were determined from the d-spacing values of the high-angle peaks in the diffraction pattern to reduce the systematic error. The (400) ($2\theta = 99.21°$) peaks were used for PZT 52/48 and the (222) ($2\theta = 63.57°$) and (400) ($2\theta = 96.00°$) peaks were used for PZT 95/5. Although the XRD system has an error of 0.005° in 2θ, the systematic error in the calculation of d-spacing values was estimated to be 0.04° [14]. The calculated lattice parameters with the margins of error are listed in Table 6.2.

The PZT 95/5 rhombohedral lattice constants were found to be $a = 4.146 \pm 0.014$Å and $\alpha = 89.75°$ in the polarized state, and $a = 4.148 \pm 0.014$Å and $\alpha = 89.75°$ after high strain rate loading.

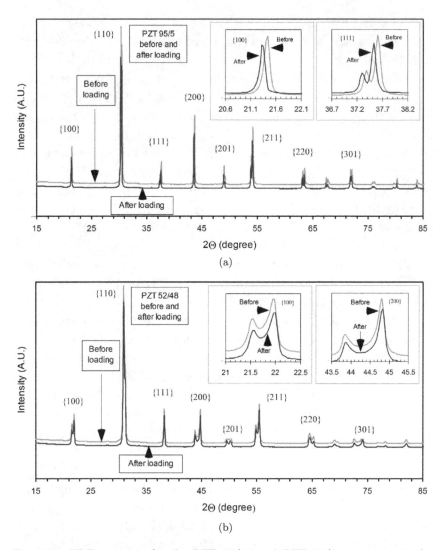

Fig. 6.7. XRD patterns for the PZT 52/48 and PZT 95/5 specimens in the as-received and after-shocked condition. (a) PZT 95/5 and (b) PZT 52/48.

Although there appears to be a peak shift between poled and high strain rate depoled PZT 95/5 specimens, the amount of peak shift is only slightly larger than the margin of error in the d-spacing. These results are in good agreement with Jaffe, Cooke, and Jaffe [7] for rhombohedral PZT 95/5.

Table 6.2.   Lattice constants and unit cell volumes for PZT 52/48 and PZT 95/5 specimens determined from the XRD measurements.

Composition	Structure	Lattice constant (Å)	Unit cell volume (Å^3)
Poled PZT 52/48	Tetragonal	$a = 4.043 \pm 0.014$ $c = 4.125 \pm 0.013$ $c/a = 1.0202 \pm 0.0062$	67.436 ± 0.065
PZT 52/48 after high strain rate loading	Tetragonal	$a = 4.042 \pm 0.014$ $c = 4.126 \pm 0.013$ $c/a = 1.0206 \pm 0.0062$	67.415 ± 0.065
Poled PZT 95/5	Rhombohedral	$a = 4.146 \pm 0.014,$ $\alpha = 89.75°$	71.269 ± 0.066
PZT 95/5 after high strain rate loading	Rhombohedral	$a = 4.148 \pm 0.014,$ $\alpha = 89.75°$	71.370 ± 0.066

The X-ray diffraction results indicate that PZT 95/5 returns to the rhombohedral state after stress-induced depolarization (Table 6.2). This observation is consistent with the results of hydrostatic studies of PZT 95/5 [12]. Fritz [12] showed that when the pressure for hydrostatically loaded PZT 95/5 is decreased from 0.32 to 0.14 GPa, it undergoes a transformation from the AFE orthorhombic state back to the FE rhombohedral state. He also indicated that at atmospheric pressure the specimen was slightly larger after hydrostatic pressure cycling than before. Based on the results of Fritz, we conclude that PZT 95/5 was in a pressure-stabilized antiferroelectric state when it was subjected to stress in our experiments, and that the FE to AFE transition is the dominant depolarization mechanism for PZT 95/5.

The poled PZT 52/48 specimens have lattice constants, a and c, and a ratio c/a that are close to those reported by Jaffe [7] for tetragonal PZT 52/48. The lattice constants and unit cell volume for PZT 52/48 specimens that have been subjected to high strain rate loading are similar to those of the poled specimens (Table 6.2). The XRD results indicate that there are no changes in the PZT 52/48 lattice parameters before and after high strain rate loading. Apparently, the PZT 52/48 does not undergo a phase transition and remains in the tetragonal phase under adiabatic shock compression.

These XRD results confirm the conclusion (see Section 6.6 above) that the stress-induced charge produced by PZT 52/48 under transverse shock compression is not caused by a transition from a polar to non-polar phase, but is the result of domain reorientation caused by release waves propagating behind the shock wave. Under 1.7 GPa transverse stress, PZT 52/48 is not completely depolarized. It retains half of its initial remanent polarization and the corresponding piezoelectric properties.

6.8 Summary

- Changes to the remanent polarization of two ferroelectrics, PZT 95/5 and PZT 52/48, subjected to transverse shock compression within miniature explosive generators, have been investigated experimentally.
- The measurements of thermal-induced depolarization and high-resolution X-ray diffraction of PZT 95/5 and PZT 52/48 specimens have been performed.
- The experimental results indicate that the mechanisms of transverse shock depolarization are different for PZT 95/5 and PZT 52/48.
- Transverse shock compression within miniature FEGs completely depolarizes the PZT 95/5. PZT 95/5 undergoes a pressure-induced transition from a polar (rhombohedral) phase to a non-polar antiferroelectric (orthorhombic) phase. These results are in good agreement with the results of shock depolarization studies of PZT 95/5 obtained using perfectly planar shock waves generated by gas gun facilities.
- PZT 95/5 elements transversely shock-compressed within miniature FEGs release electric charge with density equal to their remanent polarization. This stress-induced charge can be used to produce current and electric potential in the FEG load circuit.
- The polarity of electric charge released by PZT 52/48 under transverse shock loading is opposite to the polarity of the charge generated due to the piezoelectric effect, but it is identical to the polarity of the surface charge that balances the initial remanent

polarization. Based on these results the conclusion can be reached that PZT 52/48 is depolarized under transverse shock compression within the FEG.

- The depolarization of PZT 52/48 under transverse shock loading is not caused by a transition from a polar to a non-polar phase. The PZT 52/48 elements were in the tetragonal ferroelectric phase. PZT 52/48 does not undergo a phase transition under shock loading, it remains in the tetragonal phase under pressure at experimental conditions within explosive FEGs.

- Based on the obtained results, the conclusion was reached that the transverse shock depolarization of PZT 52/48 is the result of non-180° domain reorientation caused by release waves propagating behind the shock wave front.

- The stress-induced charge density released by transversely shock-compressed PZT 52/48 is about half of its remanent polarization. This charge can be utilized for the generation of high voltage and current pulses in the FEG load circuit.

Bibliography

1. L.L. Altgilbers, J. Baird, B. Freeman, C.S. Lynch and S.I. Shkuratov, *Explosive Pulsed Power* (Imperial College Press, London, 2010).
2. S.I. Shkuratov, J. Baird and E.F. Talantsev, Miniature 120-kV autonomous generator based on transverse shock-wave depolarization of Pb(Zr0.52Ti0.48)O3 ferroelectric, *Rev. Sci. Instrum.* **82** (2011) p. 086107.
3. S.I. Shkuratov, J. Baird, V.G. Antipov, E.F. Talantsev, W.S. Hackenberger, A.H. Stults and L.L. Altgilbers, High voltage generation with transversely shock-compressed ferroelectrics: breakdown field on thickness dependence, *IEEE Trans. Plasma Sci.* **44**(10) (2016) pp. 1919–1927.
4. S.I. Shkuratov, J. Baird, V.G. Antipov and E.F. Talantsev, Depolarization mechanisms of PbZr0.52Ti0.48O3 and PbZr0.95Ti0.05O3 poled ferroelectrics under high strain rate loading, *Appl. Phys. Lett.* **104** (2014) p. 212901.
5. S.I. Shkuratov, J. Baird, V.G. Antipov, E.F. Talantsev, C.S. Lynch and L.L. Altgilbers, PZT 52/48 depolarization: Quasi-static thermal heating versus longitudinal explosive shock, *IEEE Trans. Plasma Sci.* **38**(8) (2010) pp. 1856–1863.
6. R.E. Tipton, A 2D Lagrange MHD code, in *Proc. 4th Int. Conf. Megagauss Magnetic Field Generation and Related Topics* (Editors: C.M. Fowler, R.S. Caird and D.J. Erickson) (Plenum Press, New York, U.S.A. 1987) pp. 299–302.

7. Z. Ujima and J. Handerek, Phase transition and spontaneous polarization in PbZrO3, *Physics Status Solidi* **28**(2) (1975) pp. 489–496.

8. B. Jaffe, W.R. Cook and H. Jaffe, *Piezoelectric Ceramics* (Academic, London, 1971).

9. P. C. Lysne and C.M. Percival, Electric energy generation by shock compression of ferroelectric ceramics: normal-mode response of PZT 95/5, *J. Appl. Phys.* **46**(4) (1975) pp. 1519–1525.

10. P. C. Lysne, Dielectric properties of shock-compressed ferroelectric ceramic, *J. Appl. Phys.* **48**(3) (1977) pp. 1020–1023.

11. R.E. Setchell, Shock wave compression of the ferroelectric ceramic Pb0.99(Zr0.95Ti0.05) 0.98Nb0.02O3: depoling currents, *J. Appl. Phys.* **97** (2005) p. 013507.

12. I.J. Fritz, Uniaxial stress effects in a 95/5 lead zirconate titanate ceramic, *J. Appl. Phys.* **49**(9) (1978) pp. 4922–4928.

13. C. E. Reynolds and G.E. Seay, Two-wave shock structures in the ferroelectric ceramics: barium titanate and lead zirconate titanate, *J. Appl. Phys.* **33** (1962) pp. 2234–2239.

14. W. Parrish, The precision determination of lattice parameters, *Acta Cryst.* **13** (1960) pp. 838–850.

Chapter 7

High-Current Generation by
Shock-Compressed Ferroelectric Ceramics

7.1 Introduction

The electric charge released by a ferroelectric element in the FEG due
to shock depolarization is one of the main parameters that determine
the amplitude of the current generated in the FEG load circuit. It
is shown in the previous chapter that transversely shock-compressed
PZT 95/5 and PZT 52/48 elements with volumes of $8\,cm^3$ are capable
of producing currents with amplitudes of a few tens of amperes.
The results of the study of high-current generation by PZT 95/5
and PZT 52/48 ceramic elements transversely and longitudinally
shock-compressed within miniature FEGs are presented in this
chapter. It was experimentally demonstrated that shock-compressed
ferroelectric elements with volumes of 0.4 to $8\,cm^3$ are capable
of producing current pulses with amplitudes of several hundreds
of amperes. The obtained results indicate that the direction of
propagation of shock waves relative to the direction of polarization
and design of ferroelectric modules has a significant effect on the
amount and dynamics of stress-induced charge, output current
amplitude and waveform. The mechanisms of longitudinal shock
depolarization of the two ferroelectrics are discussed.

7.2 Multi-Element PZT 95/5 Modules

It is shown in the previous chapter that PZT 95/5 elements transversely shock-compressed within miniature FEGs become completely depolarized and release electric charge with density equal to their remanent polarization. A possible way to increase stress-induced charge and, correspondingly, current and energy delivered from an FEG into the load, is to increase the remanent polarization of the ferroelectric element. However, there are fundamental limits to this parameter due to the nature of ferroelectric materials. For instance, the highest remanent polarization for PZT 95/5 was found to be $P_0 = 38\,\mu C/cm^2$ [1].

Another possibility to increase the stress-induced charge is to design the ferroelectric element of the FEG as a module containing several ceramic elements connected in parallel. This approach could result in the multiplication of the stress-induced charge and the amplitude of the current could be produced by miniature generators.

However, this approach raises a question about mechanical and electrical interference between the elements within the module during the propagation of shock waves through the elements and the shock depolarization of ferroelectric material. To answer this question, a systematic experimental study of shock depolarization and generation of current by multi-element PZT 95/5 modules was performed.

Multi-element PZT 95/5 modules were designed and fabricated by TRS Technologies [1]. A schematic diagram of a four-element ferroelectric module is shown in Figure 7.1.

The multi-element PZT 95/5 modules were constructed with rectangular PZT 95/5 ceramic elements (25-mm long x 19-mm wide x 2-mm thick) connected in parallel. Silver electrodes were deposited on the desired surfaces of each ceramic element after sintering. The plates were assembled to form a multi-element module and co-fired to make it a whole unit. All the elements of the module were connected in parallel and polarized along their thickness dimensions to the remanent polarization $P_0 = 32\,\mu C/cm^2$.

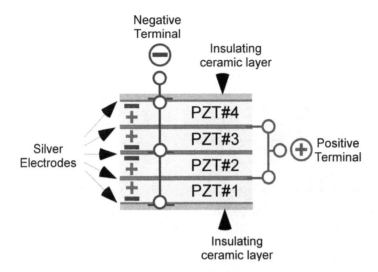

Fig. 7.1. Schematic diagram of a multi-element ferroelectric module containing four PZT 95/5 ceramic elements connected in parallel. The polarity of the surface charge is shown by (+) and (−) signs.

The module shown in Figure 7.1 contains four PZT 95/5 elements. Two positive electrodes of the bottom pair of elements (PZT#1 and PZT#2) were attached to each other and electrically connected. The top pair of elements (PZT#3 and PZT#4) was assembled in the same way as the bottom one. The negative electrode of PZT#2 was glued with conductive adhesive and electrically connected to the negative electrode of PZT#3. All the negative electrodes and positive electrodes of the ferroelectric elements were connected in parallel by wires.

Figure 7.2 shows schematics of the experimental setup and the measuring circuit used in the experiments. The output terminals of the FEG were connected to a 1 Ω resistive load, i.e. it was practically a short-circuit operation of the FEG. The shock compression induced a flow of electric current in the circuit. The load current was monitored with a Pearson Electronics 411 current probe. Additional experimental details are described elsewhere [2–4].

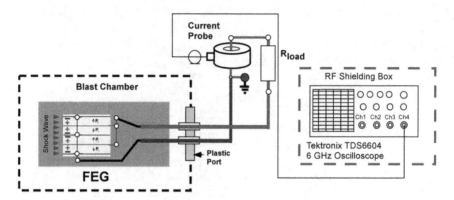

Fig. 7.2.   Schematics of the experimental setup and measuring circuit used in the experiments with multi-element PZT 95/5 modules.

7.3   Depolarization of PZT 95/5 within Transverse Shock Wave FEGs Based on Different Designs

The number of elements in the multi-element PZT 95/5 modules varied from one to eight. Each ferroelectric element of the module had an electrode area of $25 \times 19 \, \text{mm}^2$ and a thickness of 2 mm. The total thickness of the multi-element modules ranged from 2 mm (one-element module) to 16 mm (eight-element module).

The uniaxial stress distribution in multi-element modules can vary with the number of elements (module thickness) because of the difference in the shock impedance of encapsulating material (urethane) and ferroelectric ceramics that could cause a distortion of the shock wave front within the FEG. This may have an effect on the dynamics of stress-induced charge and, correspondingly, the amplitude and waveform of the current generated by modules with different numbers of elements.

Before conducting the experiments with multi-element modules, the investigation of the depolarization of PZT 95/5 single elements with identical electrode areas $(25 \times 19 \, \text{mm}^2)$ and different thicknesses (2 and 16 mm) shock-compressed within miniature FEGs based on two different designs was performed.

One of the two FEG designs was based on transverse planar shock waves (Figure 7.3) and the other one was based on transverse

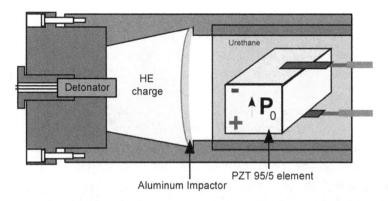

Fig. 7.3. Schematic of transverse planar shock wave FEG with PZT 95/5 ferroelectric element. P_0 is the remanent polarization vector. The polarity of the surface charge is shown by $(+)$ and $(-)$ signs.

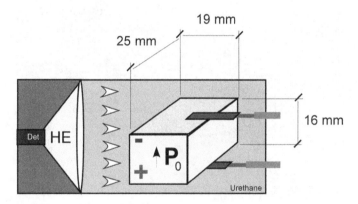

Fig. 7.4. Schematic of transverse spherically expanding shock wave FEG with PZT 95/5 ferroelectric element.

spherically expanding shock waves (Figure 7.4). In the design shown in Figure 7.3, planar shock waves in the urethane body of the FEG were initiated by an explosively accelerated metallic impactor (flyer plate). Planar shock waves initiated in solids with impacting projectiles have certain advantages over non-planar shock waves, i.e. stable shock wave front geometry, constant pressure along the front, and constant velocity of the shock wave front. Each of these advantages makes it possible to simplify theoretical models used for the analysis of the operation of shock wave ferroelectric generators. However, the presence of a flyer plate increases the amount of HE and the size and weight of the generators, and could become a limiting factor in developing miniature prime power sources. The other FEG design (Figure 7.4) was based on the initiation of shock waves in the ferroelectric element directly from the detonation of HE.

Figure 7.5 shows the typical waveforms of the current and the dynamics of the stress-induced charge released by single 16-mm-thick PZT 95/5 elements transversely shock-compressed within FEGs based on the two designs.

The difference between the waveforms of current produced by the two FEGs (Figure 7.5) is negligible. Both waveforms are close to rectangular, with rise time $0.7\,\mu$s and pulse duration $\tau_p = 5.3\,\mu$s. The amplitudes of current generated by the FEG based on spherically expanding shock waves and planar shock waves were 34.8 and 34.4 A, respectively.

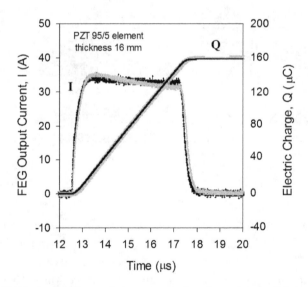

Fig. 7.5. Typical waveforms of the stress-induced current and dynamics of the electric charge released by single 16-mm-thick PZT 95/5 elements (electrode area $25 \times 19\,\text{mm}^2$) transversely shock-compressed within a planar shock wave FEG (black plots) and within FEGs based on spherically expanding shock waves (gray plots).

These results allow one to reach the conclusion that FEGs of both types provide complete depolarization of PZT 95/5 elements and produce output currents with practically identical waveforms.

Figure 7.6 shows a typical waveform of the stress-induced current and the dynamics of the charge released by a single 2-mm-thick PZT 95/5 element transversely shock-compressed within the FEG based on spherically expanding shock waves. The current waveform was very close to that for a 16-mm-thick PZT 95/5 element (see Figure 7.5). The current rise time was $0.7\,\mu s$, the pulse duration was $\tau_p = 5.4\,\mu s$ and the current amplitude was 34.7 A. The stress-induced charge was $Q = 155.7\,\mu C$ (charge density $32\,\mu C/cm^2$). The element was completely depolarized under shock compression. The experimental results indicate that the element thickness has no effect on the transverse shock depolarization and the waveform of the current pulses produced by PZT 95/5 elements shock-compressed within miniature FEGs.

Based on the obtained results, the decision was made to use transverse spherically expanding shock wave FEGs (Figure 7.4) for

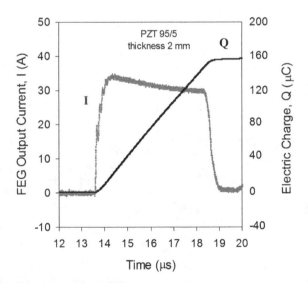

Fig. 7.6.   Typical waveform of output current and evolution of electric charge released by a single 2-mm-thick PZT 95/5 element (electrode area $25 \times 19\,mm^2$) shock-compressed within FEG based on spherically expanding shock waves.

experiments with multi-element ferroelectric modules. The results of these investigations are presented in the following sections of this chapter. The advantages of the FEG based on the initiation of the shock wave directly from the detonation of HE are its simplicity, its very few components, the small amount of HE used, and the small size and weight of the generator.

The FEGs based on spherically expanding shock waves were also used for investigations of ultrahigh-voltage generation by transversely shock-compressed ferroelectrics, and for transverse shock depolarization studies of PIN-PMN-PT ferroelectric single crystals and PZT 95/5 films. The results of these investigations are presented in the following chapters of this book.

7.4 Multi-Element PZT 95/5 Modules: Transverse Shock Depolarization and High-Current Generation

A schematic diagram of a transverse shock wave FEG with a four-element PZT 95/5 module is shown in Figure 7.7. The direction of propagation of the shock wave was perpendicular to the remanent polarization vectors in the elements. The shock wave was initiated at the long side of the ferroelectric module.

The waveforms of stress-induced current and the evolutions of electric charge released by one-, two-, four- and eight-element PZT 95/5 modules are shown in Figure 7.8. The waveforms of current

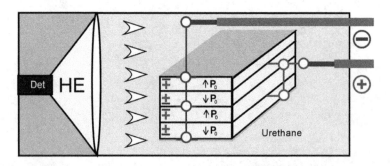

Fig. 7.7. Schematic of a transverse shock wave FEG with a four-element PZT 95/5 module.

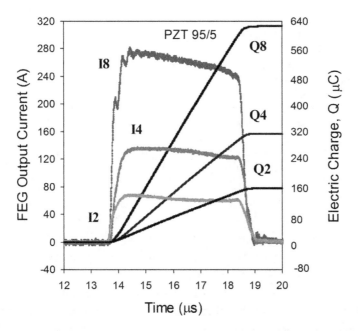

Fig. 7.8.   The typical waveforms of stress-induced current and the evolution of the electric charge released by two-element (I2 and Q2), four-element (I4 and Q4) and eight-element (I8 and Q8) PZT 95/5 modules.

generated by multi-element modules were close to rectangular with a rise time of 0.7 to 1.0 μs.

The amount of stress-induced charge and the amplitude of the current generated by multi-element modules were directly proportional to the number of elements. The current produced by the eight-element module (282 A) was higher by a factor of eight than that for the one-element module (35 A). The stress-induced charge density released by the PZT 95/5 modules, $\omega_{SW} = 32\,\mu C/cm^2$, was equal to the remanent polarization of the elements (see Table 7.1).

A complete depolarization of all the elements of the multi-element PZT 95/5 modules and the similarity of waveforms of stress-induced current generated by modules with those generated by single elements (see Figures 7.5 and 7.6) allow one to reach the conclusion that there is no indication of mechanical or electrical interference between the elements of the multi-element modules during transverse

Table 7.1.   Experimental results obtained with multi-element PZT 95/5 modules (electrode area of each element $25 \times 19\,\text{mm}^2$ and thickness 2 mm).

Number of elements	Module thickness (mm)	Stress-induced current (A)	Current pulse duration, τ_p (μs)	Stress-induced charge (μC)	Charge density (μC/cm^2)
One	2	34.7 ± 1.2	5.3 ± 0.1	150.6 ± 3.3	31.7 ± 0.07
Two	4	70.2 ± 2.1	5.3 ± 0.1	303.1 ± 2.4	31.9 ± 0.1
Four	8	141.1 ± 2.7	5.3 ± 0.1	608.0 ± 3.2	32.0 ± 0.1
Eight	16	281.5 ± 3.5	5.3 ± 0.1	1208.4 ± 5.1	31.8 ± 0.1

shock depolarization within miniature FEGs. Table 7.1 summarizes the experimental results obtained with PZT 95/5 modules.

Obviously, a complete depolarization of transversely shock-compressed PZT 95/5 multi-element modules is the result of the pressure-induced phase transition from a ferroelectric rhombohedral phase to a non-polar antiferroelectric orthorhombic phase similar to that which occurred in the shock-compressed single PZT 95/5 elements (see details in Chapter 6).

The duration of the current pulses produced by PZT 95/5 multi-element modules of all types (Table 7.1) was very reproducible, $\tau_p = 5.3 \pm 0.1\,\mu$s. Apparently, the pulse duration was determined by the shock wave transit time across the modules. Assuming that the velocity of the shock wave is constant for a PZT 95/5 module, the velocity can be derived from:

$$U_s = \frac{L_{sw}}{\tau_p}, \qquad (7.1)$$

where U_s is the velocity of the shock wave in the module, L_{sw} is the geometrical dimension of the PZT 95/5 module in the direction of the shock wave propagation (shock wave path), and τ_p is the duration of the current pulse produced by the FEG.

The shock wave velocity determined from the experimental data obtained for multi-element modules (Table 7.1) was $U_s = 3.8 \pm 0.1\,\text{mm}/\mu$s. Thus, the duration of the current pulse produced by a multi-element module can be controlled through the module geometry:

$$\tau_p = \frac{L_{sw}}{U_s} \tag{7.2}$$

An increase in the width of the module (L_{sw}) results in increasing the current pulse duration and vice versa.

Based on the results described above the conclusion was reached that the multi-element module approach can be successfully used for the generation of high-current pulses by miniature FEGs.

Figure 7.9 presents the current amplitude versus the number of elements in a multi-element PZT 95/5 module. In accordance with the extrapolated plot, a kiloampere current could be produced by a 27-element module. The thickness of the high-density ferroelectric ceramic elements can be as small as 0.5 mm. One can expect that a miniature transversely shock-compressed 45-element PZT 95/5

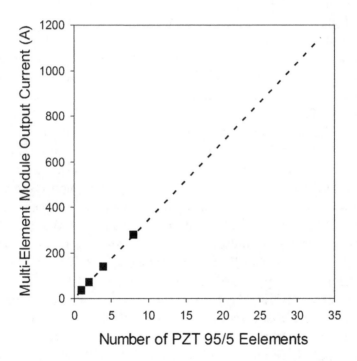

Fig. 7.9. Amplitudes of current produced by transversely shock-compressed multi-element PZT 95/5 modules as a function of the number of elements in the modules.

module with total thickness of 23 mm would be capable of producing a current pulse with an amplitude of 1.5 kA.

The experimental results (Table 7.1 and Figure 7.9) demonstrate high efficiency in the generation of current by multi-element PZT 95/5 ceramic modules. It is shown above that the ferroelectric element thickness has no effect on transverse shock depolarization and on the stress-induced current waveform. Thus, the multi-layer approach can be used for the generation of high current by ferroelectric film elements. The experimental results obtained with multi-layer PZT 95/5 films are presented in Chapter 10 of this book.

7.5 Longitudinal Shock Depolarization and High-Current Generation by PZT 95/5

Halpin [5] reported on detailed studies of depolarization and the generation of stress-induced current by PZT 95/5 ceramic specimens longitudinally shock-compressed by planar shock waves initiated by impacting projectiles accelerated in the gas gun system. The shock depolarization of PZT 95/5 ceramic disks with diameter 12.7 mm and thickness 0.76 mm was investigated in the wide range of longitudinal stress from 0.4 to 3.3 GPa.

Over the full range of stress investigated it was found that there was a wide variation in the waveforms of the stress-induced current (current peak magnitude and position, pulse duration) and the amount of stress-induced charge released by ferroelectric specimens, which suggest the very complicated behavior of the longitudinally shocked PZT 95/5. The observation was made that the duration of stress-induced current pulses was greater than the stress wave transit time. It was shown that longitudinally shocked PZT 95/5 specimens were completely depolarized in the stress range of 1.0 to 2.5 GPa and released charge with density equal to their remanent polarization, $P_0 = 33 \, \mu C/cm^2$. At stress levels below 1.0 GPa and above 2.5 GPa a fraction of the initial remanent polarization was retained.

The highest amplitude of the current (470 A) generated by longitudinally shock-compressed PZT 95/5 specimens was recorded at a stress level of 1.9 GPa [5]. The experimental results indicated that both an increase and decrease of stress above and below 1.9 GPa resulted in a decrease in the amplitude of the current pulse with a corresponding increase in its duration.

Based on the obtained experimental results Halpin reached the conclusion that the complex behavior of longitudinally shocked PZT 95/5 specimens reflects the presence of stress wave front tilt which influences the overall character of the current pulse.

Setchell [6] reported on experimental studies of the depolarization of PZT 95/5 ceramic disks with diameter 25.4 mm and thickness 4.0 mm under longitudinal stress ranging from 0.9 to 2.4 GPa. The experiments were conducted on a 63.5-mm diameter compressed-gas gun capable of achieving impact velocities from 0.03 to 1.3 km/s. Similar to the results obtained by Halpin [5], wide variations in the stress-induced current amplitude, pulse duration, and stress-induced electric charge were observed over the full range of longitudinal stress [6].

The PZT 95/5 specimens were completely depolarized, released stress-induced charge density equal to their remanent polarization ($P_0 = 30 \,\mu C/cm^2$) and generated the highest current, with an amplitude of 170 A at the stress level of 0.9 GPa. Progressively smaller electric charges and amplitudes of the current pulse were recorded as stress was increased from 0.9 to 2.4 GPa.

It was concluded in [5, 6] that the complete depolarization of longitudinally shock-compressed PZT 95/5 occurs due to the pressure-induced FE-to-AFE phase transition. The hydrostatic compression component of the stress tensor induces the phase transformation from the ferroelectric rhombohedral to a non-polar anti-ferroelectric orthorhombic phase that causes a complete depolarization of PZT 95/5 under longitudinal shock loading.

In the experiments conducted by Halpin, the highest amplitude of stress-induced current was generated by PZT 95/5 under 1.9 GPa longitudinal stress, while in the experiments conducted by Setchell the maximum current was generated by PZT 95/5 under 0.9 GPa longitudinal stress.

The only difference in these two studies was the thickness and diameter of the ferroelectric specimen. Based on these results the conclusion can be reached that the magnitude of stress resulting in the highest amplitude of current generated by longitudinally shock-compressed PZT 95/5 depends on the geometrical dimensions of the ferroelectric specimen.

7.6 Longitudinal Shock Depolarization and High-Current Generation by PZT 52/48

In this section, the results of the experimental investigations of shock depolarization and high-current generation by PZT 52/48 ceramic elements longitudinally shock-compressed within miniature FEGs are presented. A schematic of the longitudinal shock wave FEG used in these experiments is shown in Figure 7.10.

The shock wave propagation direction was anti-parallel to the remanent polarization. The uniaxial stress magnitude in the longitudinally shocked PZPT 52/48 ferroelectrics was 1.7 ± 0.1 GPa. The experimental setup and measuring circuit used in the experiments were identical to those in the experiments with PZT 95/5 elements (see Figure 7.2).

PZT 52/48 ceramic disks with thickness ranging from 0.65 to 6.5 mm were used in the investigations. The PZT 52/48 element sizes are listed in Table 7.2. Typical waveforms of stress-induced current and the dynamics of electric charge density released by

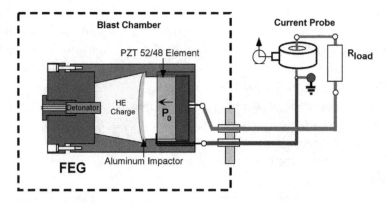

Fig. 7.10. Schematics of the longitudinal shock wave FEG and the measuring circuit used in the experiments.

Table 7.2. Stress-induced current amplitudes and electric charge densities released by PZT 52/48 elements under longitudinal shock compression and thermal-induced electric charge densities.

Disk diameter (mm)	26.2	27.0	25.0	25.0	25.0
Disk thickness (mm)	0.65	2.1	2.5	5.1	6.5
Stress-induced current amplitude (A)	352.3 ± 19.2	205.5 ± 11.3	183.7 ± 13.1	138.0 ± 12.0	108.3 ± 13.7
Stress-induced charge, $Q_{SW}(\mu C)$	160.1 ± 1.1	160.2 ± 1.2	133.0 ± 1.2	126.1 ± 1.1	119.2 ± 1.1
Stress-induced charge density, $\omega_{SW}(\mu C/cm^2)$	29.7 ± 2.2	27.5 ± 2.1	27.1 ± 2.1	26.4 ± 2.0	25.6 ± 2.1
Thermal depolarization charge density, $\omega_{therm}(\mu C/cm^2)$	27.9 ± 1.9	27.8 ± 1.7	27.4 ± 1.7	27.6 ± 1.7	27.6 ± 1.8

longitudinally shocked 0.65-mm-thick PZT 52/48 elements are shown in Figure 7.11.

The current amplitude was $I(1.6\,\mu s) = 362$ A and it had a full width at half maximum (FWHM) of 0.5 μs. The density of stress-induced charge released by PZT 52/48 in this experiment was $\omega_{SW} = 31\,\mu C/cm^2$. It is comparable with stress-induced charge densities released by PZT 95/5 under transverse and longitudinal shock-compression (see previous sections).

Figures 7.12 and 7.13 present the typical waveforms of stress-induced current and the dynamics of the electric charge densities released by longitudinally shock-compressed PZT 52/48 elements with thicknesses 2.1 and 5.1 mm. The amplitudes of stress-induced current generated by 2.1-mm-thick and 5.1-mm-thick PZT 52/48 elements were $I(1.8\,\mu s) = 223$ A and $I(1.9\,\mu s) = 149$ A, respectively. The experimental results for ferroelectric elements of all sizes are summarized in Table 7.2.

The experimental results indicate that longitudinally shock-compressed PZT 52/48 elements with volumes of 0.35 to 2.5 cm^3 are capable of producing current pulses with amplitudes of several hundreds of amperes.

An increase in the PZT 52/48 element thickness results in a progressively smaller amplitude of the stress-induced current and a

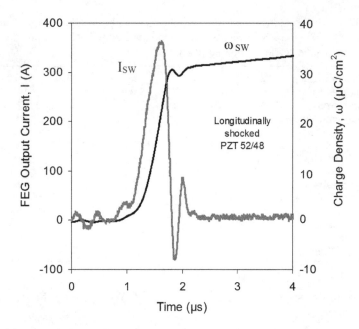

Fig. 7.11. Typical waveform of the stress-induced current and the dynamics of the electric charge density released by a longitudinally shock-compressed 0.65-mm-thick PZT 52/48 element.

longer current pulse duration. Apparently, the latter is the result of a longer shock wave transit time through the thicker ferroelectric elements. The thickness of the PZT 52/48 element can be used to control the amplitude of the generated current and current pulse duration.

Figure 7.14 shows the amplitudes of stress-induced current generated by longitudinally shocked PZT 52/48 elements as a function of the element thickness. The dependence of the current on the element thickness is non-linear. It assumes the complex behavior of longitudinally shock-compressed PZT 52/48. In accordance with the fitted and extrapolated curves (Figure 7.14), a PZT 52/48 element with thickness 0.2 mm and total volume less than a cubic centimeter would be capable of producing current pulses with amplitude exceeding one kiloampere.

In conclusion, longitudinally shock-compressed PZT 52/48 and PZT 95/5 ferroelectrics provide high dynamics of stress-induced

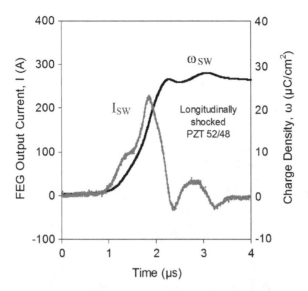

Fig. 7.12. Typical waveform of the stress-induced current and the dynamics of the electric charge density released by a longitudinally shock-compressed 2.1-mm-thick PZT 52/48 element.

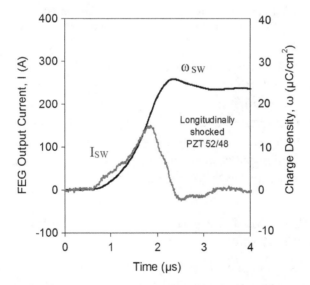

Fig. 7.13. Typical waveform of the stress-induced current and the dynamics of the electric charge density released by a longitudinally shock-compressed 5.1-mm-thick PZT 52/48 element.

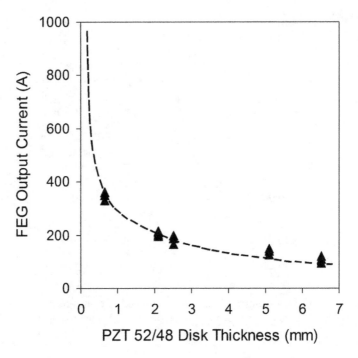

Fig. 7.14. Amplitudes of stress-induced current generated by longitudinally shock-compressed PZT 52/48 elements as a function of the element thickness.

charge and can be considered as efficient high-current sources. Miniature longitudinal shock wave FEGs containing ferroelectric elements with volume not exceeding a cubic centimeter are capable of producing current pulses with the amplitude of several hundreds of amperes. However, the short duration of the current pulse generated by longitudinally shocked ferroelectrics could create a problem for applications dealing with high-inductance loads.

7.7 Mechanism of Longitudinal Shock Depolarization of PZT 52/48

The remanent polarization of PZT 52/48 specimens used in the shock depolarization study was measured by performing thermal depolarization experiments. PZT 52/48 elements were heated above their Curie points using a Thermodyne 47900 furnace. The thermal depolarization experimental setup and procedure are described in Chapter 6.

Typical plots of the thermally induced current, $I_{therm}(T)$, for PZT 52/48 ceramic specimens with thickness 0.65 and 5.1 mm are shown in Figures 7.15 and 7.16. The $I_{therm}(T)$ is not a monotonic function of temperature, and it has a few well-resolved peaks. The amplitude of the peaks and their width varies with PZT 52/48 specimen thickness.

Each PZT 52/48 specimen investigated, no matter the size, had the most pronounced peak of $I_{therm}(T)$ at $T = 145 \pm 5°$C. This temperature is in good agreement with the temperature at which the elements were polarized by the manufacturer. Commercial PZT 52/48 specimens are polarized by ITT Corp at temperatures ranging from 125 to 150°C [7]. Heating the PZT 52/48 ferroelectric elements

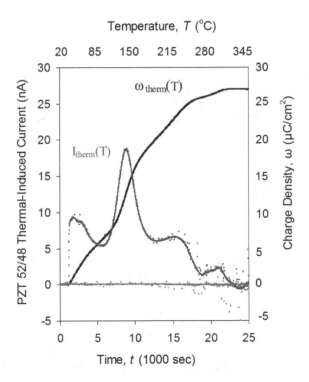

Fig. 7.15. Typical thermal depolarization current and electric charge density curves for a 0.65-mm-thick PZT 52/48 specimen.

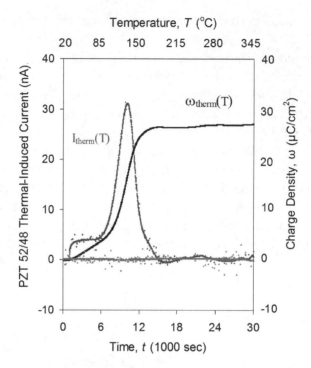

Fig. 7.16. Typical thermal depolarization current and electric charge density curves for a 5.1-mm-thick PZT 52/48 specimen.

higher than 155°C during the polarization procedure can lead to an increase in leakage currents to a level that could result in thermal runaway and electrical breakdown [8].

The experimental results indicate that the second heating cycle of the PZT 52/48 specimens caused no current flow in the circuit (Figures 7.15 and 7.16). In this case, the current recorded from each investigated element was practically zero. This is direct experimental evidence that during the first heating cycle, PZT 52/48 undergoes a structural phase transition into the cubic phase on the PZT composition-temperature phase diagram [9]. The PZT 52/48 elements were completely depolarized during the first heating cycle, and all the surface electric charge that balanced the polarization was released to the external circuit. The thermal depolarization results obtained with PZT 52/48 specimens are summarized in Table 7.2.

The experimental results for thermal-induced electric charge densities for PZT 52/48 specimens and stress-induced charge densities released by PZT 52/48 specimens with different thicknesses under longitudinal shock loading are shown in Figure 7.17. The stress-induced charge densities released by PZT 52/48 are practically equal to the thermal-induced charge densities (i.e. remanent polarization of the specimens). Based on these results the conclusion can be reached that PZT 52/48 elements were completely depolarized under longitudinal shock compression.

These results are different from the PZT 52/48 transverse shock depolarization results (see Chapter 6 of this book). The stress-induced charge density released by transversely shock-compressed PZT 52/48 was just a fraction (about 50%) of the remanent polarization (see Table 6.1 in Chapter 6). Apparently, the mechanisms of longitudinal and transverse shock depolarization of PZT 52/48 are different.

The results obtained in the experiments described above are in good agreement with the results obtained by Reynolds and

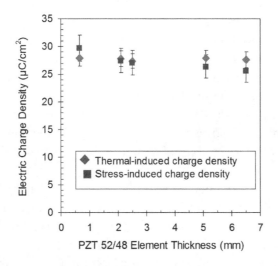

Fig. 7.17.   Experimental results for thermal-induced electric charge densities for PZT 52/48 specimens (diamonds) and stress-induced electric charge densities (squares) released by PZT 52/48 specimens under longitudinal shock compression.

Seay [10], which were reported in an experimental study of the shock depolarization of PZT 52/48 ceramic specimens longitudinally shock-compressed by planar shock waves initiated by a specially designed planar shock wave generator. It was shown [10] that the charge released by longitudinally shocked PZT 52/48 was practically linear, increasing with the pressure up to 1.1 GPa. Above 1.1 GPa, the stress-induced charge density released by PZT 52/48 specimens was equal to their remanent polarization. The authors concluded in [10] that the mechanism of depolarization of longitudinally shocked PZT 52/48 is not a phase transition or due to domain switching, but a reduction of the dipole moment by longitudinal compression.

Under ambient conditions, PZT 52/48 elements are in the tetragonal ferroelectric phase. The PZT 52/48 does not undergo a phase transition under longitudinal shock compression, it remains in the tetragonal phase under pressure at experimental conditions in the longitudinal shock wave ferroelectric generators (see previous section) and in the experiments conducted by Reynolds and Seay [10]. The complete depolarization of longitudinally shocked PZT 52/48 cannot be explained by a pressure-induced phase transition from a polar (tetragonal) to a non-polar (cubic) phase.

Under transverse shock loading, the dipoles are compressed perpendicular to their polar axes, while under longitudinal shock loading, dipoles are compressed along their polar axes. Therefore, opposite to the transverse mode of shock compression, the polarity of electric charge released by longitudinally shocked PZT 52/48 due to the shock depolarization is identical to the polarity of the charge generated due to the direct piezoelectric effect. This makes it difficult to identify the mechanism of longitudinal shock depolarization of PZT 52/48.

A significantly higher stress-induced charge density released by PZT 52/48 under longitudinal shock in comparison with that under transverse shock can be explained by an involvement of not one, but two different mechanisms of depolarization under longitudinal shock loading, a pressure-induced reorientation of non-180° domains, and a reduction of the dipole moment caused by longitudinal compression.

7.8 Summary

- It was experimentally demonstrated that miniature generators utilizing PZT 95/5 and PZT 52/48 ferroelectric ceramic elements having volumes of 0.4 to 8 cm^3 are capable of producing current pulses with the amplitude of hundreds of amperes.

- The approach based on transverse shock depolarization of multi-element PZT 95/5 ceramic modules with elements connected in parallel can be successfully used for the generation of high current by explosive ferroelectric generators. There is no indication of electrical or mechanical interference between ferroelectric elements within the modules during shock-induced depolarization. All elements of the modules become completely depolarized under transverse shock loading and release electric charge with density equal to their remanent polarization.

- The amplitudes of the current pulses produced by multi-element modules are directly proportional to the number of elements. The current pulse duration and current amplitude can be precisely controlled through the geometrical dimensions of the modules and the number of elements.

- The experimental results indicate that the thickness of PZT 95/5 elements shock-compressed within miniature FEGs has no effect on the transverse shock depolarization, amplitude and waveform of the stress-induced current. Thus, the multi-element module approach can be used for the generation of high current by transversely shocked multi-layer ferroelectric films with layers connected in parallel (see Chapter 10 of this book).

- PZT 95/5 ceramic elements become completely depolarized under longitudinal shock compression and produce current pulses with amplitudes of hundreds of amperes. Similar to transverse shock compression, the hydrostatic compression component of the stress tensor induces the phase transformation in PZT 95/5 from the ferroelectric rhombohedral to a non-polar anti-ferroelectric orthorhombic phase that causes a complete depolarization under longitudinal shock loading. However, the behavior of longitudinally shock-compressed PZT 95/5 specimens is more complex than the behavior of the same specimens under transverse shock.

- PZT 52/48 elements with thickness ranging from 0.6 to 6 mm become completely depolarized under longitudinal shock compression and produce current pulses with amplitudes of hundreds of amperes.

- A complete depolarization of longitudinally shock-compressed PZT 52/48 is not in agreement with a partial (50%) transverse shock depolarization (see Chapter 6).

- A significantly higher stress-induced charge density released by PZT 52/48 under longitudinal shock in comparison with that under transverse shock can be explained by an involvement of not one, but two different mechanisms of depolarization under longitudinal shock loading, a pressure induced reorientation of non-180° domains and a reduction of the dipole moment caused by longitudinal compression.

- The longitudinally shock-compressed PZT 95/5 and PZT 52/48 ferroelectrics provide high dynamics of stress-induced charge and they can be considered as efficient high-current sources. Miniature longitudinal FEGs containing ferroelectric elements with volumes not exceeding a cubic centimeter would be capable of producing current pulses with a kiloampere amplitude.

- Based on obtained experimental results, the conclusion can be reached that there is no significant difference between the operation of planar transverse shock wave FEGs and generators based on spherically expanding shock waves. The FEGs of both types provide complete transverse shock depolarization of ferroelectric elements and generate stress-induced current with almost identical waveforms. The FEGs based on spherically expanding shock waves were used for investigations of depolarization and high-current generation by the multi-element ferroelectric ceramic modules described in this chapter. The advantages of the FEG based on the initiation of the shock wave directly from the detonation of HE are its simplicity, its very few components, the small amount of HE used, and the small size and weight of the generator.

- The FEGs based on spherically expanding shock waves were also used for investigations of ultrahigh-voltage generation by transversely shock-compressed ferroelectrics, for transverse shock depolarization studies of PIN-PMN-PT ferroelectric single crystals and PZT 95/5 films. The results of these investigations are presented in following chapters of this book.

Bibliography

1. http://trstechnologies.com/
2. S.I. Shkuratov, J. Baird, V.G. Antipov, E.F. Talantsev, C.S. Lynch and L.L. Altgilbers, PZT 52/48 depolarization: quasi-static thermal heating versus longitudinal explosive shock, *IEEE Trans. Plasma Sci.* **38**(8) (2010) pp. 1856–1863.
3. S.I. Shkuratov, J. Baird, V.G. Antipov, E.F. Talantsev, H.R. Jo, J.C. Valadez and C.S. Lynch, Depolarization mechanisms of $PbZr_{0.52}Ti_{0.48}O_3$ and $PbZr_{0.95}Ti_{0.05}O_3$ poled ferroelectrics under high strain rate loading, *Appl. Phys. Lett.* **104** (2014) p. 212901.
4. S.I. Shkuratov, J. Baird, V.G. Antipov, E.F. Talantsev, J.B. Chase, W.S. Hackenberger, J. Luo, H.R. Jo and C.S. Lynch, Mechanism of complete stress-induced depolarization of relaxor ferroelectric single crystals without transition through non-polar phase, *Appl. Phys. Lett.* **112** (2018) p. 122903.
5. W.J. Halpin, Current from a shock-loaded short-circuited ferroelectric ceramic disk, *J. Appl. Phys.* **37**(1) (1966) pp. 153–163.
6. R.E. Setchell, Shock wave compression of the ferroelectric ceramic Pb0.99(Zr0.95Ti0.05)0.98Nb0.02O3: depoling currents, *J. Appl. Phys.* **97** (2005) p. 013507.
7. http://itt.com/
8. A.J. Moulson and J.M. Herbert, *Electroceramics: Materials, Properties, Applications.* 2nd Edition (John Wiley & Sons Ltd., West Sussex, 2003).
9. B. Jaffe, W.R. Cook and H. Jaffe, *Piezoelectric Ceramics* (Academic Press, London, 1971).
10. C. E. Reynolds and G.E. Seay, Two-wave shock structures in the ferroelectric ceramics: barium titanate and lead zirconate titanate, *J. Appl. Phys.* **33** (1962) pp. 2234–2239.

Chapter 8

Shock Depolarization of Ferroelectrics in High-Voltage Mode

8.1 Introduction

Before a shock wave is initiated, a net electric field within the ferro-electric element is equal to zero because the remanent polarization is balanced by the surface charge density. The shock wave travels through the element and depolarizes it, and surface charges are released on the electrodes of the element. When the output terminals of the FEG are connected to a low-resistance load, the ferroelectric element is operating in the high-current depolarization mode. In this mode of operation the shock depolarization results in the generation of a high current in the load circuit (see Chapters 6, 7 and 10). When the output terminals of the FEG are open or connected to a high-resistance load, the ferroelectric element is operating in the high-voltage depolarization mode. In this mode of operation the stress-induced charge is utilized for charging the ferroelectric element itself, resulting in a high electric field and a high electric potential across the element.

In this chapter, the results of investigations of the operation of ferroelectric generators in the high-voltage mode are presented. The experimental results indicate that transversely and longitudinally shock-compressed PZT 95/5 and PZT 52/48 elements are capable of producing high-voltage pulses with amplitudes directly proportional to the element thickness. The important observation is that internal breakdown within shocked ferroelectrics has a significant effect on their operation in the high-voltage mode. Energy losses and energy densities generated by the two ferroelectrics in the high-voltage mode are discussed.

8.2 High-Voltage Generation by Shock-Compressed PZT 52/48

8.2.1 *High-voltage generation by longitudinally shock-compressed PZT 52/48*

The schematics of the experimental setup and the measuring circuit used in the study of longitudinal shock depolarization of ferroelectrics in the high-voltage mode are shown in Figure 8.1. The ferroelectric generator was placed in the blast chamber. The FEG output terminals were connected to a Tektronix P6015A high-voltage probe that was placed outside the blast chamber. It was practically an open circuit operation of the FEG. The stress-induced charge was not transferred from the FEG into the external circuit but was utilized for charging the ferroelectric element itself. In this mode of operation shock-compressed ferroelectric elements produced voltage pulses with the highest amplitude possible. Additional experimental details are described elsewhere [1–4].

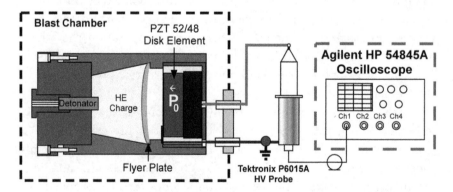

Fig. 8.1. Schematics of the longitudinal planar shock wave ferroelectric generator and the measuring circuit used in the high-voltage experiments. P_0 is the remanent polarization vector.

Ferroelectric generators based on longitudinal planar shock waves were used in the high-voltage experiments with PZT 52/48 (Figure 8.1). The direction of shock wave propagation was anti-parallel to the direction of polarization. More details about this FEG design can be found in Chapter 5 of this book.

The PZT 52/48 ceramic disk elements used in this study had an order of magnitude range of thicknesses from 0.65 to 6.5 mm. All the elements were polarized across their thickness dimensions to their remanent polarization by the manufacturer. The sizes of PZT 52/48 elements are listed in Table 8.1.

Typical waveforms of voltage generated by longitudinally shock-compressed PZT 52/48 elements are shown in Figure 8.2. The high-voltage waveforms were very reproducible. The voltage wave-forms produced by elements with different thicknesses look similar (Figure 8.2): the voltage increased to its maximum during a microsecond interval of time and then rapidly decreased to zero.

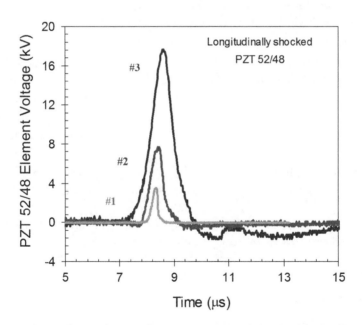

Fig. 8.2. Typical waveforms of voltage generated by longitudinally shock-compressed PZT 52/48 elements with thicknesses 0.65 mm (#1), 2.1 mm (#2) and 5.1 mm (#3).

The experimental results indicate that the element thickness has a significant effect on the amplitude of generated voltage and its rise time. An increase in the PZT 52/48 element thickness resulted in an increase in both the voltage pulse amplitude and its rise time.

The amplitude of voltage generated by the longitudinally shock-compressed 0.65-mm-thick PZT 52/48 element was 3.4 kV with the rise time 0.3 μs. The amplitudes of voltage generated by longitudinally shocked elements with thicknesses 2.1 and 5.1 mm were 8.6 and 17.6 kV, respectively (see Table 8.1).

The rapid rise of voltage from zero to its peak value is the result of shock depolarization of the ferroelectric element. In this mode of operation, the stress-induced charge was not transferred into an external circuit but it charges the element itself (which is initially a capacitor).

After reaching the maximum, the high voltage rapidly decreased to zero with no flat part in the voltage waveform (Figure 8.2). Apparently, this rapid decrease of the voltage is the result of a significant increase in the electrical conductivity of the shock-compressed ceramic material at the moment when the voltage reached its peak value.

The experimental results indicate that the peak electric field generated within shocked PZT 52/48 elements was not constant for all element sizes. The highest electric field, 5.5 kV/mm, was generated by elements with thickness of 0.65 mm. The lowest electric field, 3.1 kV/mm, was generated by elements with thickness of 6.5 mm (see Table 8.1).

The longitudinal shock depolarization of PZT 52/48 elements which have geometrical dimensions identical to those studied in the

Table 8.1. Experimental results obtained with longitudinally shock-compressed PZT 52/48 elements operating in the high-voltage mode.

Disk diameter (mm)	26	27	25	25	25
Disk thickness (mm)	0.65	2.1	2.5	5.1	6.5
Voltage, V_{FEG} (kV)	3.6 ± 0.3	7.8 ± 0.6	8.9 ± 0.9	17.1 ± 1.2	21.1 ± 1.8
Electric field (kV/mm)	5.5 ± 0.5	3.8 ± 0.2	3.7 ± 0.3	3.3 ± 0.3	3.1 ± 0.3
Rise time (μs)	0.23 ± 0.04	0.63 ± 0.1	0.64 ± 0.1	1.23 ± 0.2	1.64 ± 0.3

high-voltage mode has been investigated in the short-circuit mode (see Chapter 7 of this book). A comparison of the experimental results obtained in the high voltage and short-circuit depolarization modes raises a few questions.

The rise time of high-voltage pulses was significantly shorter than the shock wave transit time. Obviously, the generation of the high voltage was interrupted long before the shock wave passed through the PZT 52/48 elements.

The appearance of a high electric field and a high electric potential across a shock-compressed ferroelectric element makes the process of shock depolarization fundamentally different from that in the short-circuit mode. A high electric field is applied to both the shock-compressed portion and uncompressed portion of the element. A rapid decrease in the voltage could be due to a few factors: (1) the mechanical destruction of the ferroelectric element could be caused by the detonation of HE; (2) the electrical conductivity of the ferroelectric material could be increased due to the shock compression; (3) the high electric field could cause re-polarization of the shock depolarized portion of the element, resulting in significant energy losses; (4) an electric breakdown could occur within the ferroelectric element. These factors are discussed below in Section 8.4.

Figure 8.3 shows amplitudes of high voltage generated by longitudinally shock-compressed PZT 52/48 and voltage rise times as a function of the element thickness [1, 2]. The important result is that the voltage amplitude was directly proportional to the element thickness in the full range investigated. The experimental results obtained with longitudinally shock-compressed PZT 52/48 operating in the high-voltage mode are summarized in Table 8.1.

8.2.2   *High-voltage generation by transversely shock-compressed PZT 52/48*

A schematic diagram of the transverse planar shock wave FEG used for investigations of the generation of high voltage by transversely shock-compressed PZT 52/48 is shown in Figure 8.4. The explosive part of this FEG was identical to that for the longitudinal FEG used in the experiments described above (see Figure 8.1). The only difference between the two designs was the positioning of

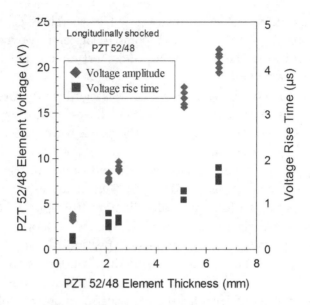

Fig. 8.3. Experimentally obtained amplitudes of high-voltage pulses (diamonds) generated by longitudinally shock-compressed PZT 52/48 elements and voltage rise times (squares) as a function of the element thickness.

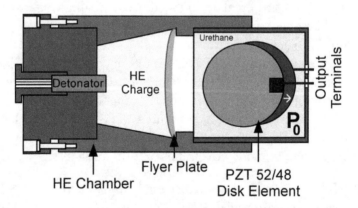

Fig. 8.4. Schematic diagram of a transverse planar shock wave ferroelectric generator used in the high-voltage experiments with PZT 52/48. P_0 is the remanent polarization vector.

the ferroelectric elements within the FEG. In the transverse FEG, the direction of shock wave propagation was perpendicular to the remanent polarization vector (Figure 8.4). More details about this FEG design can be found in Chapter 5. The experimental setup and measuring circuit used in these experiments were identical to those shown in Figure 8.1.

Identical geometries of PZT 52/48 elements were used in the investigations of high-voltage generation by longitudinal and transverse shock wave FEGs to enable a direct comparison of experimental results obtained for the two modes of shock compression.

Figure 8.5 shows typical waveforms of high voltage generated by transversely shock-compressed PZT 52/48 [3]. Comparing the results obtained with transversely and longitudinally shocked PZT 52/48 (Figures 8.5 and 8.2, respectively) the conclusion can be reached that ferroelectric elements having identical geometrical dimensions generate voltage with practically equal amplitudes in the two modes of shock compression. Similar to longitudinally shocked PZT 52/48, an increase in the thickness of a transversely shocked PZT 52/48

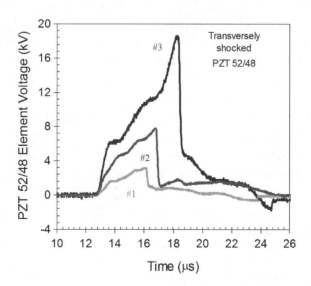

Fig. 8.5. Typical waveforms of voltage generated by transversely shock-compressed PZT 52/48 elements with thickness 0.65 mm (#1), 2.1 mm (#2) and 5.1 mm (#3).

element from 0.65 to 5.1 mm resulted in an increase in the amplitude of generated voltage from 3.3 kV to 17 kV (Figure 8.5).

At the same time, the rise times of voltage pulses generated by PZT 52/48 elements under transverse shock loading (Figure 8.5) were significantly longer than those generated under longitudinal shock. This difference could be caused by the dynamics of the stress-induced charge released by PZT 52/48 elements under transverse shock compression being lower than those released under longitudinal shock (see Chapter 7 of this book).

The rise times of high voltage generated by transversely shock-compressed PZT 52/48 were always longer than the shock wave transit time. The shock wave velocity in PZT 52/48 was determined to be $U_s = 3.8 \pm 0.2$ mm/μs [4]. The shock wave transit time through the element with diameter 26 mm was 6.8 μs. However, the rise time of the high-voltage pulse generated by a transversely shocked PZT 52/48 element (see Table 8.2 and Figure 8.5) was significantly shorter. It is obvious that the process of generation of a high voltage by PZT 52/48 elements under transverse shock loading was interrupted before the shock wave front passed through the element. A part of the stress-induced charge was not utilized for charging the ferroelectric element. These results are similar to those obtained in the experiments with high-voltage generation by PZT 52/48 under longitudinal shock loading described in the previous section.

Figure 8.6 shows the amplitudes of high voltage generated by PZT 52/48 elements under transverse shock compression and voltage rise times as a function of the element thickness. The important result is that the amplitude of voltage generated by transversely shocked PZT 52/48 is directly proportional to the element thickness. An

Table 8.2. Experimental results obtained with transversely shock-compressed PZT 52/48 elements operating in the high-voltage mode.

Disk diameter (mm)	26.2	27	25	25	25
Disk thickness (mm)	0.65	2.1	2.5	5.1	6.5
Voltage, V_{FEG} (kV)	3.6 ± 0.4	7.9 ± 1.1	9.1 ± 0.4	17.5 ± 1.2	22.3 ± 1.3
Electric field (kV/mm)	5.5 ± 0.6	3.9 ± 0.4	3.8 ± 0.2	3.3 ± 0.3	3.1 ± 0.3
Rise time (μs)	2.7 ± 0.5	4.1 ± 0.3	4.2 ± 0.5	5.6 ± 0.6	6.0 ± 0.3

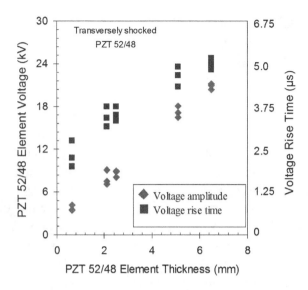

Fig. 8.6. Experimentally obtained amplitudes of high-voltage pulses (diamonds) generated by transversely shock-compressed PZT 52/48 elements and voltage rise times (squares) as a function of the element thickness.

increase in the PZT 52/48 element thickness from 0.65 to 6.5 mm resulted in an increase in the voltage pulse amplitude from 3.6 to 22 kV. These results are close to those obtained for longitudinally shock-compressed PZT 52/48 (Figure 8.3). The experimental results for transversely shocked PZT 52/48 elements operating in the high-voltage mode are summarized in Table 8.2.

8.3 High-Voltage Generation by Shock-Compressed PZT 95/5

There do not appear to be studies of high-voltage generation by longitudinally shock-compressed PZT 95/5. There are very few studies of the operation of transversely shock-compressed PZT 95/5 with high ohmic loads and capacitive loads.

Lysne and Percival [5] reported on investigations of the operation of transversely shock-compressed PZT 95/5 with high ohmic loads. Experiments were conducted on a compressed-gas gun system. It was shown that transversely shock-compressed PZT 95/5 rectangular specimens with thickness of 9.5 mm were capable of producing a 48 kV pulse across the 15 kΩ load. A decrease of the load resistance to 2.95 kΩ resulted in the reduction of the amplitude of generated voltage to 9.3 kV. Also, the operation of transversely shocked PZT 95/5 with high ohmic loads has been investigated experimentally in [6, 7].

To obtain systematic experimental data for the operation of transversely shock-compressed PZT 95/5 in the high-voltage mode, a series of high-voltage shock depolarization experiments was conducted with rectangular PZT 95/5 elements having different thicknesses and identical electrode areas.

A schematic of transverse FEG used in these experiments is shown in Figure 8.7. This FEG was based on a spherically expanding shock wave. Investigated rectangular PZT 95/5 specimens had electrode area 12.7×50.8 mm^2 and thicknesses ranging from 2 to 6 mm (see Table 8.3). All the elements were polarized across their thickness dimensions to the remanent polarization of $32\,\mu C/cm^2$. The shock wave travelled parallel to the long side of the ferroelectric specimens (Figure 8.7). The experimental setup and measuring circuit used in these experiments were similar to those used for PZT 52/48 (see Figure 8.1).

Figure 8.8 shows a typical waveform of voltage generated by transversely shock-compressed 6-mm-thick PZT 95/5 element. The voltage amplitude was 41 kV and the rise time was $1.4\,\mu s$. After reaching the maximum the voltage rapidly reduced to almost zero level during several hundreds of nanoseconds.

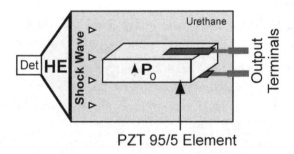

Fig. 8.7. Schematic of transverse FEG used in the experimental investigation of high-voltage generation by PZT 95/5 ferroelectrics. P_0 is the remanent polarization vector.

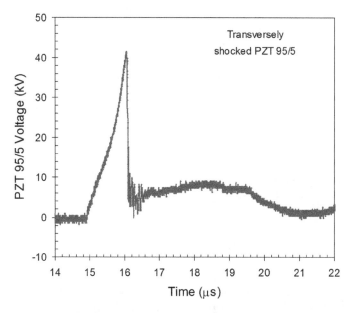

Fig. 8.8. Typical waveform of voltage generated by transversely shock-compressed PZT 95/5 element (6.0-mm thick × 12.7-mm wide × 50.8-mm long).

PZT 95/5 rectangular elements of identical size were investigated in the short-circuit depolarization mode (see Chapter 6). In the short-circuit depolarization mode the stress-induced electric charge was continually released on the electrodes of the element for a time interval of 14.6 μs. This is a much longer time than the rise time of the high-voltage pulse shown in Figure 8.8 (1.8 μs). It is obvious that the process of high-voltage generation by the PZT 95/5 element was abruptly interrupted. Similar results were observed with shock-compressed PZT 52/48 (see sections above).

Figure 8.9 shows experimentally-obtained amplitudes of high voltage generated by transversely shock-compressed PZT 95/5 elements as a function of the element thickness. The important result is that the amplitude of high voltage generated by a transversely shock-compressed PZT 95/5 element is directly proportional to the element thickness. Similar results were obtained with longitudinally and transversely shocked PZT 52/48. An increase of the PZT 95/5 element thickness from 2 to 6 mm resulted in an increase in the

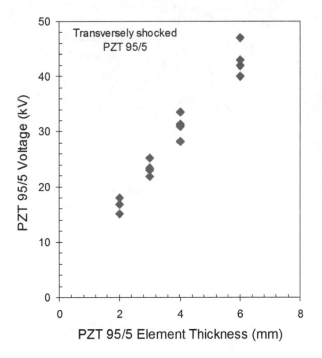

Fig. 8.9. Experimentally-obtained amplitudes of high voltage generated by transversely shock-compressed PZT 95/5 elements as a function of the element thickness.

voltage pulse amplitude from 17 to 44 kV. All experimental results for high-voltage generation by transversely shocked PZT 95/5 are summarized in Table 8.3.

Figure 8.9 shows experimentally-obtained amplitudes of high voltage generated by transversely shock-compressed PZT 95/5 elements as a function of the element thickness. The important result is that the amplitude of high voltage generated by a transversely shock-compressed PZT 95/5 element is directly proportional to the element thickness. Similar results were obtained with longitudinally and transversely shocked PZT 52/48. An increase of the PZT 95/5 element thickness from 2 to 6 mm resulted in an increase in the voltage pulse amplitude from 17 to 44 kV. All experimental results

Table 8.3. Experimental results obtained with transversely shock-compressed PZT 95/5 elements operating in the high-voltage mode.

PZT 95/5 thickness (mm)	2	3	4	6
PZT 95/5 electrode area (mm^2)	12.7×50.8	12.7×50.8	12.4×50.0	12.7×50.8
Voltage, V_{FEG} (kV)	16.8 ± 1.9	23.5 ± 3.1	30.7 ± 3.2	43.5 ± 4.4
Electric field (kV/mm)	8.3 ± 0.9	7.8 ± 0.6	7.7 ± 0.6	7.3 ± 0.6

for high-voltage generation by transversely shocked PZT 95/5 are summarized in Table 8.3.

Comparison of the results obtained for transversely shocked PZT 95/5 and PZT 52/48 (Tables 8.3 and 8.2) shows that the amplitudes of high voltage generated by PZT 95/5 elements are higher by a factor of two than those for PZT 52/48 elements having the same thicknesses. Apparently, this is caused by different physical and electrical properties of the two ferroelectrics.

Experimental results (Table 8.3) indicate that the peak electric field generated by transversely shocked PZT 95/5 elements is inversely proportional to the element thickness.

8.4 Electric Breakdown in Shock-Compressed Ferroelectrics

As is described above, the waveforms of the high voltage generated by shock-compressed PZT 52/48 and PZT 95/5 ferroelectrics were similar: the voltage increased to its maximum and then rapidly decreased to almost zero level. An increase in the voltage pulse to its maximum was the result of the depolarization of the ferroelectric element due to the shock compression. In the high-voltage mode, the stress-induced charge is not transferred from the ferroelectric to the external circuit because of the high resistance and low capacitance of the voltage probe used in these experiments, but it is utilized for charging the ferroelectric element itself.

The obtained experimental results indicate that the high voltage reached its maximum during a time interval that is significantly shorter than the time of shock wave transit in the PZT 95/5 and

PZT 52/48 elements. An abrupt and rapid decrease of the voltage from its maximum could be due to a few factors:

(1) Effect of the load circuit.
(2) Mechanical destruction of the ferroelectric element caused by high-pressure gas expansion from the HE detonation, followed by short-circuiting of the electrodes of the element.
(3) A significant increase of electrical conductivity of the shock-compressed ferroelectric material and the corresponding electric current in the element.
(4) An internal electric breakdown within the ferroelectric element.

Consider the effect of the load circuit. Figure 8.10 shows equivalent circuits of longitudinal and transverse ferroelectric generators operating in the high-voltage mode. The capacitance of the unstressed zone and the capacitance of the stressed zone are shown as C_1 and C_2, respectively. The capacitance of a ferroelectric element is denoted as C_{FEG}. The resistance and capacitance of the high-voltage probe are shown as R_{probe} and C_{probe}, respectively. The capacitance of the probe ($3\,\text{pF}$) is negligible in comparison with the capacitance of the ferroelectric element which varies from $300\,\text{pF}$ to $1\,\text{nF}$. The time constant of the FEG circuit in these experiments can be estimated as follows:

$$\tau_{circuit} = R_{probe} \cdot C_{FEG} \tag{8.1}$$

Substitution of the resistance of the voltage probe ($10^8\,\Omega$) and the lowest capacitance of a ferroelectric element used in this study

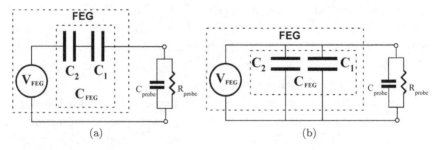

(a)   (b)

Fig. 8.10. Equivalent circuits of (a) longitudinal FEG and (b) transverse FEG operating in the high-voltage mode.

($3 \cdot 10^{-10}$ F) gives us the time constant $\tau_{circuit} = 3 \cdot 10^{-2}$ s. The $\tau_{circuit}$ is several orders of magnitude longer than the time for the voltage pulse duration (see Figures 8.2, 8.5 and 8.8). It is evident that the rapid decrease of the voltage across shocked ferroelectric elements is not related to discharging the energy generated in the ferroelectric element into the load circuit.

The time required for mechanical destruction of the ceramics behind the shock front, t_{dest}, can be estimated as [8]:

$$t_{dest} = \frac{L_{sw}}{u_p} \tag{8.2}$$

where L_{sw} is the shock wave travel distance in the ferroelectric element, and u_p is the particle velocity. The latter can be determined as [8]:

$$u_p = \frac{p_{sw}}{\rho_0 U_s} \tag{8.3}$$

where p_{sw} is the shock pressure, ρ_0 is the density of ceramics before the shock action, and U_s is the shock wave velocity. In our experiments, $U_s = 3.8 \pm 0.2$ mm/μs [4], and, correspondingly, $u_p = 0.05$ mm/μs. The lower bound of the destruction time is $t_{dest} = 13 \, \mu$s. It is about an order of magnitude longer than the time taken to reduce the voltage in the experiments (Figures 8.2, 8.5 and 8.8), so mechanical destruction does not appear to be a factor in the rapid voltage decrease.

As regards the shock-induced electrical conductivity factor, above, it follows from recent experimental studies of the electrical conductivity of shock-compressed piezoelectric crystals (quartz) [9] and polarized lead zirconate titanate ferroelectric ceramics [10] at $p_{sw} = 12$ GPa (nearly an order of magnitude higher than the shock pressure in this study) that there is no conductivity increase in shocked PZT ceramic materials or any electric charge leakage at electric field strength up to $E = 4.4$ kV/mm [10]. So, shock-induced electric conductivity is not a factor in the rapid voltage decrease, either.

Based on the consideration of all the factors above, unless some unknown factor is at play, one can conclude that the experimentally detected rapid decrease of the voltage generated by shocked ferroelectrics is the result of internal electric breakdown. Therefore, the experimentally obtained maximum electric field generated within a shock-compressed ferroelectic element can be considered the *breakdown field, E_{break}*.

8.5 Power Law for Breakdown Field of Shock-Compressed Ferroelectrics

Studies of electric breakdown in shock-compressed solid dielectrics (including ferroelectrics) were started in the 1960s (a historical review and a complete list of references can be found in Ref. [11]). The kinetic model of electric breakdown of shocked ferroelectrics developed by Lysne [5, 12, 13] allows one to predict the time, t_{break}, from initial shock compression to breakdown. This model and its later modifications have been used to analyze t_{break} as a function of shock pressure, porosity, remanent polarization, and other factors for lead zirconate titanate ceramic compositions.

We decided to use a different approach from the kinetic model developed by Lysne [5, 12, 13] for analysis of the experimental data for shock-compressed ferroelectrics. Since the 1950s, electric breakdown in solid dielectrics at ambient conditions has been extensively studied [14–19]. One of the results of these studies is the thickness-dependent dielectric breakdown law, experimentally proved for a variety of dielectric materials investigated at ambient conditions:

$$E_{break}(d) = \gamma \cdot d^{-\xi}, \tag{8.4}$$

where E_{break} is the breakdown field, γ is the material-dependent constant, d is the thickness of the dielectric, and ξ is a coefficient that is justified by the mechanism of electric breakdown, namely the injection of electrons and electron-phonon scattering. Recently, discussions have begun on the role of space charge in the breakdown event.

To analyze the results of shock experiments, we utilized a procedure that was developed earlier for the analysis of the breakdown of dielectrics at ambient conditions, Eq. (8.4). An important question presents itself — can experimental data for the breakdown electric field, $E_{break}(d)$, obtained under explosive shock conditions, be described in accordance with the law (Eq. (8.4)) that was experimentally proven for the breakdown of dielectric materials under ambient conditions [14–19]? It should be mentioned that this approach for shocked ferroelectrics data analysis was used for the first time by Shkuratov, Talantsev and Baird in Ref. [2].

Figure 8.11 presents a plot of the breakdown electric field, $\log(E_{break})$, of a shocked PZT 52/48 element as a function of the element thickness, $\log(d)$. These data can be represented by a straight line, and the plot is similar to plots experimentally obtained for the breakdown electric field of dielectrics at ambient conditions [14–19]. Experimental data for the breakdown electric field, E_{break}, for shocked PZT 52/48 elements with different thicknesses are summarized in Tables 8.1 and 8.2.

The slope of the curve in Figure 8.11 is 0.248 ± 0.020, and this slope corresponds to the coefficient ξ in Eq. (8.4). In accordance

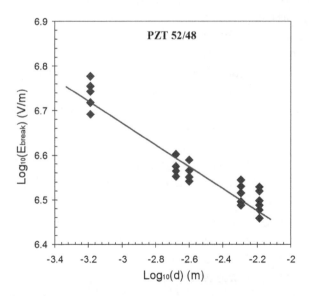

Fig. 8.11. Breakdown electric field of shock-compressed PZT 52/48 as a function of the element thickness.

with theoretical analysis [16, 19], a coefficient of $\xi = 0.25$ implies
the presence of a tunnel mechanism of the injection of electrons
from the negative electrode into the dielectrics, and strong electron-
phonon scattering in an adiabatically compressed material. The
current density, j, of the prime electrons injected into the dielectric
material due to the tunnel effect is given by the following expression
[16, 19]:

$$j(\phi_{\text{eff}} E_{el}) = \frac{q_e}{16\,\pi^2 h} \cdot \frac{(q_e E_{el})^2}{\phi_{\text{eff}}} \cdot \exp\left[-\frac{4(2m)^{3/2}\phi_{\text{eff}}^{3/2}}{3hq_e E_{el}}\right], \qquad (8.5)$$

where E_{el} is the electric field strength at the negative electrode, ϕ_{eff} is
the effective height of the potential barrier at the negative electrode-
dielectric interface, h is Planck's constant, and q_e and m are the
charge and the effective mass of the electron, respectively.

It follows from Eq. (8.5) that E_{el} and ϕ_{eff} are the main parameters
that determine the current density of the injected electrons and,
correspondingly, the probability of electric breakdown. Apparently,
a reduction of the electric field, E_{el}, through control/minimization
of micro-protrusions at the negative electrode surface facing the
dielectric material and an increase of the effective height of the
potential barrier, ϕ_{eff}, through the use of various materials in
the negative electrode, could suppress the tunnel current density
and increase the breakdown voltage. The physical properties of
PZT-electrode interfaces and PZT specimens with electrodes made
of different materials were recently studied [20].

Figure 8.12 presents a plot of the breakdown electric field,
$\log(E_{break})$, of a shocked PZT 95/5 element as a function of the
element thickness, $\log(d)$. Similar to PZT 52/48 (Figure 8.11) this
data can be represented by a straight line, and the plot is similar
to plots experimentally obtained for the breakdown electric field of
dielectrics at ambient conditions [14–19]. The slope of the PZT 95/5
curve in Figure 8.12 is 0.21 ± 0.03.

The obtained relationship between the breakdown field, E_{break},
and the thickness of the ferroelectric element, d, (Figures 8.11

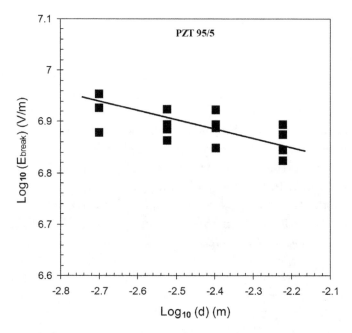

Fig. 8.12.   Breakdown electric field of shock-compressed PZT 95/5 as a function of the element thickness.

and 8.12) allows one to predict the breakdown field in shocked ferroelectrics and, correspondingly, the amplitude of the high voltage produced by the ferroelectric generator.

8.6   Energy Density Generated by Shock-Compressed Ferroelectrics in High-Voltage Mode

In the high-voltage mode, the ferroelectric element combines three stages of conventional pulsed power system in one, i.e. a prime power source, a high-voltage generator, and a capacitive energy storage (see equivalent circuits in Figure 8.10). The energy density of a capacitive energy storage device can be determined as follows:

$$W_{FEG} = \frac{\varepsilon_0 \varepsilon_{\mathrm{r}} A V_{break}^2}{2\, d \cdot vol} = \frac{1}{2} \varepsilon_0 \varepsilon_{\mathrm{r}} E_{break}^2 \qquad (8.6)$$

where ε_r is the relative dielectric permittivity of the ferroelectric, ε_0 is the dielectric permittivity of free space, d is the distance between the electrodes of the ferroelectric element (element thickness), A is the area of the electrodes, V_{break} is the breakdown voltage of the element, *vol* is the volume of the element, and E_{break} is the breakdown electric field of the element.

In accordance with Eq. (8.6), the energy density generated by shock-compressed ferroelectrics in the high-voltage mode is directly proportional to the permittivity of ferroelectric material and to the breakdown field to the power of two.

Estimate the energy density generated by transversely shock-compressed PZT 52/48 in the high-voltage mode. The PZT 52/48 breakdown field is $E_{break52/48}$ = 4.1 kV/mm (see Table 8.2). The relative permittivity for polarized and for depolarized PZT 52/48 is 1300 and 1140, respectively (see Table 2.2 in Chapter 2 of this book). A substitution of the PZT 52/48 parameters into Eq. (8.6) gives us the energy density $W_{52/48}$ = 101 mJ/cm^3.

Estimate the energy density generated by transversely shock-compressed PZT 95/5 in the high-voltage mode. The PZT 95/5 breakdown field is $E_{break95/5}$ = 9.2 kV/mm (see Table 8.3). The relative permittivity for polarized and depolarized PZT 95/5 is 295 and 225, respectively (see Table 2.1 in Chapter 2 of this book). Substitution of PZT 95/5 parameters into Eq. (8.6) gives us the energy density $W_{95/5}$ = 110 mJ/cm^3.

The important result is that the energy density generated by transversely shocked PZT 52/48 in the high-voltage mode is close to that for PZT 95/5. The electric field generated by shocked PZT 95/5 elements is twice of that for PZT 52/48. However, the energy density is directly proportional to the permittivity of the ferroelectric material (Eq. (8.6)), which is higher for PZT 52/48.

The breakdown field is one of the main parameters that have an effect on energy density generated by shocked ferroelectrics in the high-voltage mode. An increase in the breakdown field of shock-compressed ferroelectrics is important for the development of high energy density FEG-based high-voltage systems.

Another important parameter that has an effect on the energy density generated by shocked ferroelectrics in the high-voltage mode is the permittivity of ferroelectric materials (Eq. (8.6)).

The energy densities generated by shocked PZT 95/5 and PZT 52/48 discussed above were determined with permittivities obtained from low electric field measurements. The dielectric properties of ferroelectric materials can change under shock compression and high electric field. A brief review of the results of the measurements of the permittivity of ferroelectrics during shock wave transit and under high electric field is presented in the next section. The results obtained by using different experimental methods vary significantly [22–28]. However, these results indicate that the permittivities of unstressed and stressed ferroelectrics under high electric field are higher than those under low field. Based on these results one can conclude that the energy density generated by an FEG in the high-voltage mode determined with low field permittivity should be considered the lower bound of this parameter.

8.7   Measurements of Dielectric Properties of Ferroelectrics During Shock Wave Transit and under High Electric Field

The measurement of the dielectric properties of ferroelectrics during shock wave transit and under high electric field is a challenging task. Lysne and Percival [22] and Lysne [23] reported on investigations of the dielectric properties of PZT 95/5 during transverse shock wave transit and under high electric field. Their approach was based on the analysis of waveforms of stress-induced current and stress-induced voltage generated by transversely shock-compressed PZT 95/5 specimens operating with resistive loads. It was found that depolarized material in the stressed zone was a linear dielectric with an absolute permittivity (dielectric constant) of 9 nF/m. In addition, it was noted that for stress states above 1.6 GPa, the permittivity of material in the stressed zone was independent of both the stress and the electric field. The permittivity of polarized

PZT 95/5 determined from the hysteresis loops recorded at ambient conditions was represented as a linear function of the electric field and it was found to be higher than the permittivity of shock depolarized material.

Setchell *et al.* [24] developed new experimental methods to better isolate dielectric properties in both unstressed and stressed zones of PZT 95/5 specimens. These methods are based on shock-driven circuits containing multiple, small PZT 95/5 elements that are displaced both parallel and perpendicular to the shock motion. For the experimental data analysis the model was used that assumed that the displacement current is governed by the Debye approximation for a relaxing dielectric. The absolute permittivity of polarized PZT 95/5 in the unstressed zone was obtained at three different levels of electric field: $\varepsilon(1.58\,\mathrm{kV/mm}) = 15.8\,\mathrm{nF/m}$, $\varepsilon(3.04\,\mathrm{kV/mm}) = 14.8\,\mathrm{nF/m}$ and $\varepsilon(3.38\,\mathrm{kV/mm}) = 13.1\,\mathrm{nF/m}$. The experimental results indicated that the permittivity of depolarized material in the stressed zone was significantly lower than that for polarized material in the unstressed zone.

Wu *et al.* [25] developed a new experimental technique for investigations of the dynamic dielectric properties of PZT 95/5 during shock compression and under high electric field. The dynamic permittivity was determined from oscillating periods in currents generated in the external oscillating circuits. This technique made it possible to measure the relative dielectric permittivity of PZT 95/5 in the stressed and unstressed zone under high electric fields. It was found that the relative permittivity of polarized PZT 95/5 in the unstressed zone is influenced heavily by the electric field. The permittivity increased from $\varepsilon_{\mathrm{r}FE} = 198$ at zero field to $\varepsilon_{\mathrm{r}FE} = 481$ at an electric field of $1.3\,\mathrm{kV/mm}$. Experimental results indicated that the permittivity of depolarized material in the stressed zone, on the contrary, had no correlation with the electric field. The relative permittivity of depolarized PZT 95/5 was found to be $\varepsilon_{\mathrm{r}} = 227$ under an electric field ranging from zero to $3\,\mathrm{kV/mm}$.

Fritz and Keck [26] reported on investigations of the dielectric properties of PZT 95/5 under hydrostatic stress at low electric field. The relative dielectric permittivity was studied as a function of pressure. It was known from separate dielectric charge loss measurements on polarized specimens that at 0.3 GPa hydrostatic stress a transition from a ferroelectric to a non-polar anti-ferroelectric state occurred. The experimental results indicated that there was a drop in permittivity from its initial value, $\varepsilon_{\mathrm{r}FE} = 290$ to $\varepsilon_{\mathrm{r}AFE} = 205$ near 0.3 GPa stress. The conclusion was reached that the low electric field relative permittivity of PZT 95/5 in the anti-ferroelectric state is $\varepsilon_{\mathrm{r}AFE} = 205$.

Hwan and Lynch [27] and Valadez *et al.* [28] reported on studies of dielectric properties of polarized PZT 95/5 under hydrostatic stress and high electric fields. In accordance with their investigations the relative permittivity of polarized PZT 95/5 at low field was found to be $\varepsilon_{\mathrm{r}FE} = 305$. The relative permittivity of polarized PZT 95/5 increased to $\varepsilon_{\mathrm{r}FE} = 1119$ at an electric field of 3 kV/mm and zero stress.

The simultaneous application of hydrostatic stress and high electric field to polarized PZT 95/5 makes the ferroelectric-to-antiferroelectric transformation complicated. The FE-AFE transformation is a coupled field-driven phase transformation. Field-driven phase transformations in ferroelectric materials involve the driving forces of stress and electric field. If the phase transformation involves a strain change, it can be driven by stress; and if the phase transformation involves an electric displacement change, it can be driven by electric field. The relative dielectric permittivity of PZT 95/5 subjected to hydrostatic compression and high electric field was found to be $\varepsilon_{\mathrm{r}AFE} = 1384$ [27, 28].

In conclusion, the experimental results indicate that the dielectric permittivity of PZT 95/5 during shock wave transit and under high electric field vary significantly when determined using different methods. However, the results of these measurements indicate that even the lowest values of permittivity under high electric field are greater than the permittivity under low field. As was mentioned in the previous section, the energy density generated by an FEG in the high-voltage mode determined with low field permittivity should be considered as the lower bound of this parameter.

To determine actual energy density generated by shocked ferro-electrics operating in the high-voltage mode it is necessary to obtain complete information about the dielectric properties of unstressed and stressed ferroelectrics in a wide range of electric fields up to the breakdown point.

There do not appear to be studies of dielectric properties of PZT 52/48 or other ferroelectrics during shock wave transit and under a high electric field. There is a high probability that the high electric field has a significant effect on the dielectric properties of polarized and shock depolarized PZT 52/48 and other ferroelectric materials as well.

8.8 Energy Losses in Shock-Compressed Ferroelectrics in High-Voltage Mode

It is shown in Sections 8.2 and 8.3 above that the generation of high voltage by shock-compressed ferroelectric elements is interrupted by electric breakdown long before the shock wave passed through the elements and shock depolarization was complete.

Consider energy losses in PZT 52/48 and PZT 95/5 elements operating in the high-voltage mode. Assume that all electric charge released by a ferroelectric element due to the shock depolarization is utilized for charging the element itself. In this case, the electric field generated by the element can be calculated using the following expression:

$$E_{calc} = \frac{Q_{sw}}{C_{FEG}d}, \tag{8.7}$$

where E_{calc} is the calculated electric field within the ferroelectric element, Q_{SW} is the electric charge released by the element due to shock depolarization, C_{FEG} is the capacitance of the element, and d is the element thickness. Eq. (8.7) can be re-written in the following form:

$$E_{calc} = \frac{\omega_{sw}}{\varepsilon_0 \varepsilon_r}, \tag{8.8}$$

where ω_{sw} is the stress-induced charge density, ε_r is the relative permittivity of the ferroelectric material, and ε_0 is the dielectric permittivity of free space.

Calculate the electric field generated within transversely shock-compressed PZT 52/48. The stress-induced charge density released by PZT 52/48 under transverse shock loading was measured in the short-circuit depolarization mode (Chapter 6). Transversely shocked PZT 52/48 elements were partially (about 50%) depolarized and released electric charge with density $\omega_{SW52/48} = 14\,\mu C/cm^2$ (see Table 6.1 in Chapter 6). The substitution of PZT 52/48 parameters into Eq. (8.8) gives us the electric field strength generated within a PZT 52/48 element $E_{calc52/48} = 11.7\,kV/mm$. The electric field values calculated by this simple model (Eqs. (8.7) and (8.8)) are in good agreement with the results of the calculations obtained with the more complicated model developed in [21].

The experimental results obtained for transversely shock-compressed PZT 52/48 operating in the high-voltage mode are summarized in Table 8.2. The electric field generated by shocked PZT 52/48 elements, $E_{break52/48} = 4.1\,kV/mm$, is lower by a factor of 3 than the calculated field.

The difference between the calculated and experimental electric field strength generated within transversely shock-compressed PZT 52/48 is an indication of significant electric charge and energy losses. Only a fraction of the electric charge released by a shocked ferroelectric element is utilized for charging the element itself because the high-voltage generation is interrupted by electric breakdown within the shocked element, resulting in the short-circuiting of the element electrodes.

Calculate the electric field generated within transversely shock-compressed PZT 95/5. The depolarization of transversely shocked PZT 95/5 was investigated in the short-circuit mode (Chapter 6). The PZT 95/5 elements were completely depolarized under shock compression and released electric charge with density $\omega_{SW95/5} = 32\,\mu C/cm^2$ (Table 6.1 in Chapter 6). The substitution of PZT 95/5 parameters into Eq. (8.8) gives us $E_{calc95/5} = 122\,kV/mm$.

The experimental results obtained for transversely shocked PZT 95/5 operating in the high-voltage mode are summarized in Table 8.3. The electric field strength generated within transversely shocked PZT 95/5 elements did not exceed $E_{break95/5} = 9.2\,kV/mm$. It is more than an order of magnitude lower than the calculated electric field. This difference is caused by significant electric charge and energy losses due to the electric breakdown within PZT 95/5 elements. In the high-voltage mode, only a fraction of stress-induced charge is utilized for charging the PZT 95/5 element itself.

The permittivity of ferroelectrics obtained from low field measurements was used in the calculation of electric field (Eq. (8.8)) discussed above. It is shown in the previous section that the permittivity of stressed and unstressed ferroelectrics under high electric field could be higher by a factor of 2 to 3 than if they were at low electric field. Taking this into consideration, the conclusion can be reached that the results of electric field calculations made with low field dielectric permittivity, and the corresponding estimation of electric charge and energy losses in the shocked ferroelectrics operating in the high-voltage mode are the upper bonds of these parameters.

In conclusion, an increase in the breakdown field of ferroelectric ceramic materials through the improvement of the technology of their fabrication, electrode materials, electrode preparation techniques and other means is very important for increasing the energy density of high-voltage ferroelectric generators.

8.9 Summary

- It was experimentally demonstrated that shock-compressed PZT 95/5 and PZT 52/48 ferroelectrics provide reliable and reproducible generation of high-voltage pulses with amplitudes of tens of kilovolts. This ability of ferroelectrics to produce high voltage under shock compression can be used in explosive ferroelectric generators.
- The experimental results indicate that the amplitude of voltage generated by the two ferroelectrics under shock loading is directly proportional to the ferroelectric element thickness.
- Shock-compressed PZT 95/5 elements generate high-voltage pulses with amplitudes a factor of two higher than those generated by PZT 52/48 elements having the same thicknesses. This can be

caused by the different physical and electrical properties of the two ferroelectrics.

- The presence of a high electric field and a high electric potential across shock-compressed ferroelectrics makes the shock depolarization in the high-voltage mode different from that in the short-circuit mode. The electric breakdown within shock-compressed PZT 95/5 and PZT 52/48 elements is a fundamental phenomenon that has a significant effect on energy losses and energy density generated in the high-voltage mode.

- It was found that the electric breakdown in transversely shock-compressed PZT 95/5 and PZT 52/48 elements having thicknesses of 1 to 6 mm obeys the thickness-dependent breakdown law that represents the breakdown of dielectrics at ambient conditions. This law was experimentally proved for a variety of solid dielectrics at ambient conditions. It can be extended to transversely shock-compressed ferroelectric materials.

- It was found that the electric breakdown in longitudinally shock-compressed PZT 52/48 elements having thicknesses of 0.6 to 6 mm obeys the thickness-dependent breakdown law that represents the breakdown of dielectrics at ambient conditions.

- The energy density generated by shock-compressed PZT 52/48 in the high-voltage mode is close to that for PZT 95/5. Both ferroelectrics can be utilized for the development of high-voltage FEG-based systems.

- The dielectric properties of ferroelectrics change significantly during shock wave transit and under high electric field. The permittivities of the shock-compressed zone and uncompressed zone of ferroelectric specimens under high electric field were found to be higher than those obtained from standard low electric field measurements. Based on these results one can conclude that the energy and energy density generated by shock-compressed ferro-electrics in the high-voltage mode determined with permittivities obtained from low field measurements should be considered as the lower bound of these parameters. To determine the actual energy density generated by shocked ferroelectrics in the high-voltage mode it is necessary to obtain complete information about the dielectric properties of unstressed and stressed ferroelectric materials in a wide range of electric field strength up to the breakdown point.

- The experimental results indicate that the generation of high voltage by shock-compressed ferroelectric elements is interrupted by electric breakdown long before the shock wave passes through the elements and shock depolarization is complete. Only a fraction of electric charge released by shock-compressed ferroelectric elements is utilized for charging the elements themselves because of electric breakdown within the shocked elements resulting in the short-circuiting of the element electrodes.

- An increase in the breakdown field of ferroelectric ceramic materials through the improvement of the technology of their fabrication, electrode materials, electrode preparation techniques and other means is very important for increasing the energy density of high-voltage ferroelectric generators.

Bibliography

1. S.I. Shkuratov, E.F. Talantsev, L. Menon, H. Temkin, J. Baird and L.L. Altgilbers, Compact high-voltage generator of primary power based on shock wave depolarization of lead zirconate titanate piezoelectric ceramics, *Rev. Sci. Instrum.* **75**(8) (2004) pp. 2766–2769.

2. S.I. Shkuratov, E.F. Talantsev and J. Baird, Electric breakdown of longitudinally shocked $Pb(Zr_{0.52}Ti_{0.48})O_3$ ceramics, *J. Appl. Phys.* **110** (2011) p. 024113.

3. S.I. Shkuratov, J. Baird, V.G. Antipov and E.F. Talantsev, Autonomous pulsed power generator based on transverse shock wave depolarization of ferroelectric ceramics, *Rev. Sci. Instrum.* **81** (2010) p. 126102.

4. S.I. Shkuratov, J. Baird, V.G. Antipov, E.F. Talantsev, C.S. Lynch and L.L. Altgilbers, PZT 52/48 depolarization: quasi-static thermal heating versus longitudinal explosive shock, *IEEE Trans. Plasma Sci.* **38**(8) (2010) pp. 1856–1863.

5. P.C. Lysne and C.M. Percival, Electric energy generation by shock compression of ferroelectric ceramics: normal-mode response of PZT 95/5, *J. Appl. Phys.* **46**(4) (1975) pp. 1519–1525.

6. R. E. Setchell, Shock wave compression of the ferroelectric ceramic Pb0.99.Zr0.95Ti0.05.0.98Nb0.02O3: Depoling currents, *J. Appl. Phys.* **97** (2005) p. 013507.

7. S.I. Shkuratov, J. Baird, V.G. Antipov and E.F. Talantsev, Utilizing Pb(Zr0.95Ti0.05)O3 ferroelectric ceramics to scale down autonomous

explosive-driven shock-wave ferroelectric generators, *Rev. Sci. Instrum.* **83** (2012) p. 076104.

8. L. Davison, *Fundamentals of Shock Wave Propagation in Solids* (Springer, Heidelberg, 2008).

9. V.A. Borisenok, V.A. Kruchinin, V. A. Bragunets, S.V. Borisenok, V.G. Simakov and M.V. Zhernokletov, Measuring shock-induced electrical conductivity in piezoelectrics and ferroelectrics: single-crystal quartz, *Combustion, Explosion, and Shock Waves* **43** (2007) pp. 96–103.

10. V.A. Bragunets, V.G. Simakov, V.A. Borisenok, S.V. Borisenok and V.A. Kruchinin, Shock-induced electrical conductivity in some ferroelectrics, *Combustion, Explosion, and Shock Waves* **46** (2010) pp. 231–236.

11. L.L. Altgilbers, J. Baird, B. Freeman, C.S. Lynch and S.I. Shkuratov, *Explosive Pulsed Power* (Imperial College Press, London, 2010).

12. P.C. Lysne, Dielectric breakdown of shock-loaded PZT 65/35, *J. Appl. Phys.* **44**(2) (1973) pp. 577–582.

13. P.C. Lysne, Prediction of dielectric breakdown in shock-loaded ferroelectric ceramics, *J. Appl. Phys.* **46**(1) (1975) pp. 230–232.

14. G. A. Vorob'ev, Dependence of the dielectric strength of some alkali halide monocrystals on the duration of the applied voltage, *Sov. Phys. — JETP* **3**(2) (1956) pp. 225–229.

15. R. Gerson and T.C. Marshall, Dielectric breakdown of porous ceramics, *J. Appl. Phys.* **30**(11) (1959) pp. 1650–1655.

16. F. Forlani and N. Minnaja, Thickness influence in breakdown phenomena of thin dielectric films, *Phys. Status Solidi B* **4**(2) (1969) pp. 311–324.

17. J.J. O'Dwyer, Theory of double charge ejection from a dielectric, *J. Appl. Phys.* **39**(9) (1968) pp. 4356–4362.

18. W.S. Nicol, Thickness variation of breakdown field strength in plasma oxidized aluminum films, *Proc. IEEE* **56**(1), (1968) pp. 109–110.

19. F. Forlani and N. Minnaja, Electrical breakdown in thin dielectric films, *J. Vac. Sci. Technol.* **6**(4) (1969) pp. 518–526.

20. F. Chen, R. Schafranek, A. Wachau, S. Zhukov, J. Glaum, T. Granzow, H. von Seggern and A. Klein, Barrier heights, polarization switching, and electrical fatigue in Pb(Zr,Ti)O3 ceramics with different electrodes, *J. Appl. Phys.* **108** (2010) p. 104106.

21. J.A. Mazzie, Simplified model of ferroelectric energy generation by shock compression, *J. Appl. Phys.* **48**(3) (1977) pp. 1368–1369.

22. P. C. Lysne and C.M. Percival, Electric energy generation by shock compression of ferroelectric ceramics: normal-mode response of PZT 95/5, *J. Appl. Phys.* **46**(4) (1975) pp. 1519–525.

23. P. C. Lysne, Dielectric properties of shock-wave compressed PZT 95/5, *J. Appl. Phys.* **48**(3) (1977) pp. 1020–1023.

24. R.R.E. Setchell, S.T. Montgomery, D.E. Cox and M.U. Anderson, Delectric properties of PZT 95/5 during shock compression under high electric filed, *AIP Conf. Proc. CP845, Shock Compression of Condensed Matter — 2005*, eds. M.D. Furnish, M. Elert, T.P. Russell and C.T. White (American Institute of Physics, 2006), pp. 278–281.

25. Y. Wu, G. Liu, Z. Gao, H. He and J. Deng, Dynamic dielectric properties of the ferroelctric ceramic Pb(Zr0.95Ti0.05)O3 in shock compression under high electrical field, *J. Appl. Phys.* **123** (2018) p. 244102.

26. I.J. Fritz and J. D. Keck, Pressure-temperature phase diagrams for several modified lead zirconate ceramics, *Journal of Physics and Chemistry of Solids* **39** (1978) pp. 1163–1167.

27. H.R. Jo and C.S. Lynch, Effect of composition on the pressure-driven ferro-electric to antiferroelectric phase transformation behavior of (Pb0.97La0.02) (Zr1−x−ySnxTiy)O3 ceramics, *J. Appl. Phys.* **116** (2014) p. 074107.

28. J.C. Valadez, R. Sahul, E. Alberta, W. Hackenberger and C.S. Lynch, The effect of a hydrostatic pressure induced phase transformation on the unipolar electrical response of Nb modified 95/5 lead zirconate titanate, *J. Appl. Phys.* **111** (2012) p. 024109.

Chapter 9

Ultrahigh-Voltage Generation
by Shock-Compressed Ferroelectrics

9.1 Introduction

The generation of ultrahigh (100 kV and up) voltage by compact autonomous pulsed power systems is important for modern engineering applications [1]. In this chapter, the results of systematic studies of ultrahigh-voltage generation by ferroelectrics shock-compressed within miniature FEGs are presented. The experimental results indicate that transversely shocked PZT 95/5 and PZT 52/48 are capable of producing hundreds of kilovolts of electric potential. The experimentally obtained thickness-dependent dielectric breakdown law for shock-compressed ferroelectrics makes it possible to predict the FEG output voltage up to 500 kV and forms the basis for design of ultrahigh-voltage FEG-based systems. High-speed photography of the explosive operation of high voltage FEG along with a recording of electrical signals produced by the generator helped to identify the limits of the operation of ferroelectric generators in the ultrahigh-voltage mode.

9.2 PZT 95/5 Ultrahigh-Voltage Generation:
Transverse versus Longitudinal Shock Loading

It is shown in the previous chapter that transversely and longitudinally shock-compressed PZT 95/5 and PZT 52/48 elements with thicknesses lying in the range of a few millimetres are capable of generating high-voltage pulses with amplitude directly proportional to the element thickness. These results allow one to assume that ferroelectric elements having the thickness of a few tens of millimetres could produce hundreds of kilovolts of electric potential. This was the starting point of investigations of ultrahigh-voltage generation by shock-compressed ferroelectrics.

A schematic of the experimental setup and measuring circuit used for investigations of the operation of ultrahigh-voltage ferroelectric generators is shown in Figure 9.1. The FEG output terminals were connected to a North Star PVM-5 high-voltage probe (resistance 400 MΩ, capacitance 12 pF, transition time 8 ns) or to an ultrahigh-voltage capacitive probe (capacitance 3 pF, transition time 4 ns). The ferroelectric generator was placed in the blast chamber and the FEG output terminals were connected to the high-voltage probe placed outside the chamber.

It is experimentally demonstrated in Chapter 8 that longitudinally and transversely shock-compressed PZT 52/48 elements generate high-voltage pulses with practically equal amplitude. Based on these results the decision was made to investigate high-voltage generation by transversely and longitudinally shock-compressed PZT 95/5 elements.

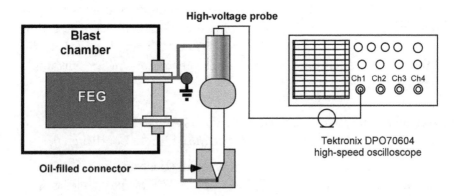

Fig. 9.1.   Schematics of the experimental setup and measuring circuit used for investigations of the operation of ultrahigh-voltage ferroelectric generators.

Ferroelectric Element

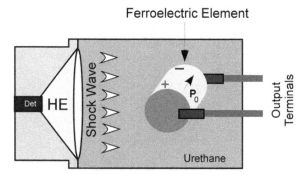

Fig. 9.2.   Schematic of transverse FEG used in investigations of ultrahigh-voltage generation by shock-compressed ferroelectrics. P_0 is the remanent polarization vector. The polarity of the surface charge in the ferroelectric element is shown by (+) and (−) signs.

Schematics of transverse and longitudinal ferroelectric generators used in the ultrahigh-voltage experiments are shown in Figures 9.2 and 9.3, respectively. The diameter of generators of both types was 38 mm, and the distance between the top of the plastic body and the ferroelectric element varied from 20 to 30 mm. The 11 grams of HE charge along with the RP-501 detonator were used in these experiments. More details can be found in Chapter 5 of this book.

In the ultrahigh-voltage mode, longitudinal shock wave ferroelectric generators could have an advantage over transverse FEGs. In a transverse FEG, the polarization vector of the ferroelectric element is perpendicular to the FEG axis (Figure 9.2). Therefore, in order to produce ultrahigh voltage the element thickness should be increased with a corresponding increase in the generator diameter and amount of HE.

Opposite to the transverse FEG, in the longitudinal FEG the polarization vector of the ferroelectric element and inter-electrode distance are positioned along the axis of the device (Figure 9.3). Correspondingly, in the longitudinal FEG, an increase in the ferroelectric element thickness does not result in an increase in the diameter of the FEG and the amount of HE.

The PZT 95/5 cylindrical elements with identical geometrical dimensions were used in transverse and longitudinal high-voltage

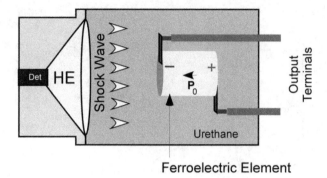

Ferroelectric Element

Fig. 9.3. Schematic of longitudinal FEG used in investigations of ultrahigh-voltage generation by shock-compressed ferroelectrics.

experiments to enable a direct comparison of the results obtained in the two modes of shock compression. The diameter of the PZT 95/5 cylindrical elements was 19 mm and the thickness was 23.1 mm.

A typical waveform of the voltage generated by a transversely shock-compressed PZT 95/5 element is shown in Figure 9.4. The voltage amplitude was $V_{FEG} = 120$ kV and the voltage rise time was 2.2 μs.

Figure 9.5 shows a typical waveform of voltage generated by a longitudinally shock-compressed PZT 95/5 element identical to that shown in Figure 9.4. The voltage generated by the longitudinally shocked PZT 95/5 cylinder, $V_{FEG} = 27$ kV, was lower by a factor of 4.5 than that generated by a transverse FEG with an identical ferroelectric element (Figure 9.4). The voltage rise time was 5.9 μs.

We performed three experiments with each type of ferroelectric generator. The mean voltages generated by transversely and longitudinally shock-compressed PZT 95/5 were 123.3 ± 3.2 kV and 29.3 ± 6.1 kV, respectively.

The experimental results obtained with transversely and longitudinally shocked PZT 95/5 elements in the high-voltage mode (Figures 9.4 and 9.5) are not in agreement with the results obtained for PZT 52/48 elements having thicknesses ranging from 0.6 to 6 mm. It was experimentally demonstrated that longitudinally and transversely shock-compressed PZT 52/48 elements having identical

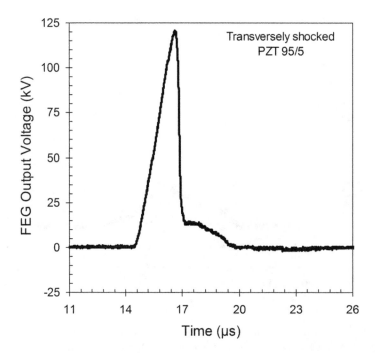

Fig. 9.4. Typical waveform of voltage generated by transversely shock-compressed PZT 95/5 cylindrical elements (diameter 19 mm and thickness 23.1 mm).

geometrical dimensions produced high voltage with practically equal amplitudes (see Chapter 8 of this book).

Longitudinal shock depolarization of ferroelectrics in the ultrahigh-voltage mode is discussed in Section 9.6 below. It is shown that the thickness of the ferroelectric element is one of the parameters that have an effect on the generation of a high voltage by longitudinally shock-compressed ferroelectrics.

The experimental results (Figures 9.4 and 9.5) indicate that in the ultrahigh-voltage mode transversely shock-compressed PZT 95/5 elements are more efficient than longitudinally shocked ones. The efforts were focused on investigations of ultrahigh-voltage generation by transversely shock-compressed PZT 95/5. The results of these investigations are described in the following sections of this chapter.

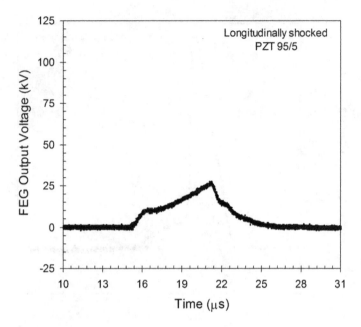

Fig. 9.5. Typical waveform of voltage generated by longitudinally shock-compressed PZT 95/5 cylindrical elements (diameter 19 mm and thickness 23.1 mm).

9.3 Extension of Thickness-Dependent Dielectric Breakdown Law on Transversely Shock-Compressed PZT 95/5 Operating in Ultrahigh-Voltage Mode

It is shown in the previous chapter that the generation of high voltages ranging from 5 to 40 kV by transversely shock-compressed PZT 95/5 elements is interrupted by the electric breakdown within the elements long before the shock depolarization is complete. One could expect that electric breakdown also plays an important role in the ultrahigh-voltage mode of the operation of transversely shocked PZT 95/5 ferroelectrics. Therefore, understanding the mechanism of the breakdown of shocked ferroelectrics is important for the development of ultrahigh-voltage ferroelectric generators.

Studies of electric breakdown in shock-compressed solid dielectrics (ferroelectrics in particular) were started at the Sandia National

Laboratories in the 1950s (a historical review and a complete list of references can be found in [1]). A breakdown in shocked ferroelectrics was always considered a different physical phenomenon from that occurring in dielectrics at ambient conditions (see [2] and references therein).

Lysne [3, 4] reported on experimental studies of breakdown for poled PZT 65/35 ferroelectrics under ambient conditions and under shock compression (the breakdown voltage ranged from 2 to 13 kV). The results of these studies [3, 4] demonstrated the difference between the breakdown of PZT 65/35 under ambient conditions and under shock compression. It was shown in [3, 4] that the remanent polarization (varied from 3 to $30 \, \mu C/cm^2$) and shock pressure (varied from 0.3 to 2.3 GPa) have a significant effect on the electric breakdown of shock-compressed PZT 65/35 ferroelectrics. These studies [3, 4] revealed the complex relationship between the breakdown field, the polarization of the specimens, and the shock pressure. For breakdowns in shocked ferroelectrics, Lysne developed the model [3–5] that makes it possible to predict the time from initial shock compression to breakdown as a function of temperature, initial polarization rate, shock pressure, and porosity of ferroelectric ceramics. This model has been used to analyze breakdown in lead zirconate titanate ceramic compositions. However, the breakdown mechanism in shock-compressed ferroelectrics is still not a phenomenon that has been completely understood.

In the previous chapter, it is shown that within the narrow range of ferroelectric element thicknesses from 1 to 6 mm, the breakdown field of transversely shock-compressed PZT 95/5 is described in accordance with the law that was experimentally proven for the breakdown of dielectric materials under ambient conditions:

$$E_{break}(d) = \gamma \cdot d^{-\xi} \qquad (9.1)$$

where E_{break} is the breakdown field, γ is the material-dependent constant (in this case it is the breakdown field for a 1.0 mm-thick PZT 95/5 element), d is the thickness of the dielectric, and ξ is a coefficient that is justified by the mechanism of electric breakdown.

These results raise an important question: can the breakdown field of explosively shocked PZT 95/5 ferroelectric elements with a thickness of a few tens of millimetres obey the same thickness-dependent dielectric breakdown law as dielectrics at ambient conditions (Eq. (9.1))?

To answer this question, systematic experimental studies of the generation of high and ultrahigh voltage by PZT 95/5 cylindrical and rectangular elements having a wide range of thicknesses were performed. The round shape of the electrodes of cylindrical ferroelectric elements provides a uniform electric field on the perimeter of the element. With rectangular elements, there is a possibility of electric field enhancement on the corners of the electrodes that can vary from element to element and increase scatter within the experimental data. The schematic of transverse FEG used in these experiments is shown in Figure 9.2. A schematic of the experimental setup is shown in Figure 9.1.

A typical waveform of the voltage generated by a transversely shocked 23-mm-thick PZT 95/5 cylindrical element is shown in Figure 9.4. The voltage increased during 2.2 μs to its maximum value $V_{FEG} = 120$ kV then started decreasing due to an internal breakdown within the element.

Figure 9.6 summarized the experimental results for the mean breakdown voltage of transversely shocked PZT 95/5 elements with thicknesses lying in the wide range from 1 to 23 mm [6–8]. The experimental results indicate that there is no significant difference between the breakdown voltages of transversely shocked rectangular and cylindrical elements having the same thickness. The scatter within the experimental data for both cylindrical and rectangular elements did not exceed 10%. The pulsed voltage generated across PZT 95/5 elements is directly proportional to the element thickness in a wide range of voltage amplitudes from 9 to 120 kV (Figure 9.6).

We performed a parametric analysis of the experimental data representing more than one order of magnitude range of PZT 95/5 element thickness and, correspondingly, a wide range of breakdown voltage using various functions. It follows from our analysis that the

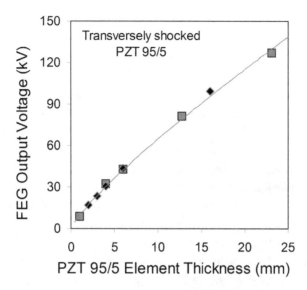

Fig. 9.6. Experimentally obtained amplitudes of high voltage generated by rectangular (squares) and cylindrical (diamonds) transversely shock-compressed PZT 95/5 elements as a function of the element thickness, and fitted curve (dot line).

best fit for the data is the function that is identical to the thickness-dependent breakdown law:

$$V_{FEG}(d) = \chi \cdot d^{1-\xi} \qquad (9.2)$$

where χ is the first fitting parameter (in this case it is voltage generated by a 1.0 mm-thick PZT 95/5 element), and ξ is the second fitting parameter. Experimental $V_{FEG}(d)$ data for PZT 95/5 along with a fitted curve are presented in Figure 9.6. For these PZT 95/5 experimental data, the best fit was achieved with $\xi = 0.243$, a value close to the parameter obtained by Forlani and Minnaja in their model of the breakdown [9] (where ξ justifies the mechanisms of breakdown).

Based on these results (Figure 9.6) the conclusion can be reached that electric breakdown in transversely shock-compressed PZT 95/5 obeys the thickness-dependent breakdown law, Eq. (9.1), in a wide range of element thicknesses and, correspondingly, in a wide range of breakdown voltages.

This conclusion does not contradict the theoretical considerations of dielectric breakdown. In fact, the works of Gerson and Marshall [10] and Forlani and Minnaja [9] did not postulate that the dielectrics under consideration were at ambient conditions. In [10], the authors based their theoretical presumption on dielectrics having porous structure. Porous structure takes place in shock-compressed ferroelectric ceramics. In [9] the only postulate the authors made was about a forbidden energy zone in the electronic structure.

At high shock pressures (above 20 GPa), some dielectrics can be transformed into their conductive states (see [11–13] and references therein). However, it was experimentally demonstrated in [11–13] that poled PZT ferroelectrics under an electric field of 4.4 kV/mm and shock-compressed at 12 GPa (eight times higher than the shock pressure in our experiments) do not posses significant conductivity. Thus, the theory of Forlani and Minnaja [9] can be applied to shock-compressed ferroelectrics in a wide range of breakdown voltages.

The experimental and analytic results shown in Figure 9.6 give the grounds to assume that it is possible to extrapolate PZT 95/5 breakdown voltages above 120 kV. Figure 9.7 shows the extrapolated curve for transversely shocked PZT 95/5. In accordance with the extrapolated curve (dot line in Figure 9.7), the FEGs utilizing 35-mm-thick PZT 95/5 elements would be capable of producing the output voltages close to 200 kV. This is an important voltage range for the application of an FEG as an ultrahigh-voltage prime power source [1].

To verify the extrapolated curve, a series of experiments was conducted with PZT 95/5 cylindrical elements having thicknesses 25.4 and 27 mm and diameter 22 mm. The FEGs built with these PZT 95/5 elements, however, generated output voltages that were significantly lower than predicted by the extrapolated curve in Figure 9.7.

Figure 9.8 shows a typical waveform of the voltage generated in these experiments by an FEG containing a 25.4-mm-thick PZT 95/5 cylinder. The amplitude of the voltage was 128 kV. After reaching the maximum, the voltage abruptly decreased (in ∼10 ns) to zero.

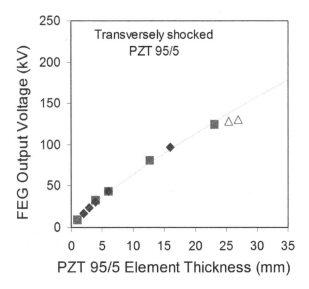

Fig. 9.7. PZT 95/5 experimental breakdown voltages (squares and diamonds) with fitted and extrapolated curve (dot line). Experimental results for 25.4-mm and 27-mm-thick PZT 95/5 cylinders are shown by triangles.

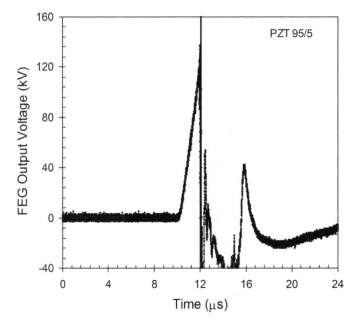

Fig. 9.8. Typical waveform of the breakdown voltage across a 25.4-mm-thick PZT 95/5 cylindrical element.

Powerful oscillations accompanied the breakdown (Figure 9.8). In one of the four experiments of this series, the keyboard and the mouse of our Tektronix TDS6604 oscilloscope were shut off. Most probably, this occurred due to radio-frequency radiation caused by electric breakdown in the FEG. We did not observe these effects in the electric breakdown of thinner PZT 95/5 elements (see Figure 9.4).

The mean voltages generated by 25.4-mm- and 27-mm-thick PZT 95/5 elements are shown in Figure 9.7 as triangles. Obviously, these values are not in agreement with the extrapolated curve. These experimental results indicate that the electric breakdown mechanism for 25.4-mm and 27-mm-thick PZT 95/5 elements is different from that for the thinner elements. A possible explanation of this difference could be a flashover along the surface of shock-compressed PZT 95/5 elements at the ferroelectric-urethane body interface.

9.4   High-Speed Photography of the High-Voltage FEG Operation

The experimental results described in the section above indicate that the electric breakdown initiation mechanism for 27-mm-thick PZT 95/5 elements is different from that for the thinner elements. Was this difference due to some shift in ferroelectric behavior? It is known that the properties of dielectric surfaces change under ultrahigh-voltage conditions [14]. Or was this difference due to some aspect of the FEG design?

The operation of the FEGs is completely different from that of the conventional pulsed power systems. An FEG generates pulsed power with the explosive shock compression of the ferroelectric element. Note, however, that the shock wave not only affects the ferroelectric element, but every part of the device. A release (tensile) wave follows the shock (compressive) front; this combination eventually destroys the device. As a result, the electrical and mechanical properties of the components change during the FEG operation. These changes might cause an electric breakdown under ultrahigh-voltage conditions.

To understand the physics and mechanics of ultrahigh-voltage breakdown in the FEG we developed a technique for high-speed photography of the operation of the ferroelectric generator while simultaneously recording the ultrahigh-voltage waveform generated across the ferroelectric element. A schematic of the experimental setup for the high-speed photography of the operation of an explosive ferroelectric generator is shown in Figure 9.9.

Figure 9.10 presents photographs of the inside and outside of the blast chamber during the test setup. A Cordin 10A framing camera at half a million frames per second was used to photograph the experiments. To observe processes within the plastic body we made it with a transparent urethane compound Crystal Clear 204.

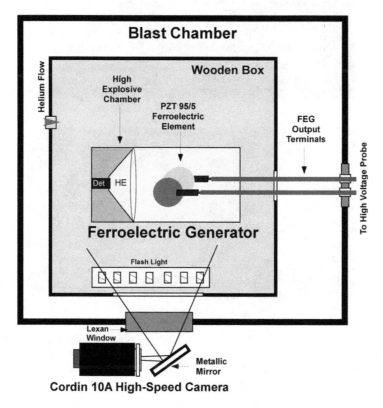

Fig. 9.9. Schematic of the experimental setup for high-speed photography of the operation of the explosive ferroelectric generator.

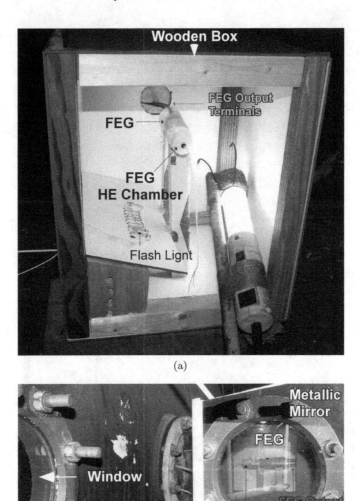

Fig. 9.10.   Experimental setup for the high-speed photography of the operation of the explosively-driven ferroelectric generator. (a) End view of the wooden box containing the FEG, with the box in position within the blast chamber. (b) Image of the FEG in the metallic mirror.

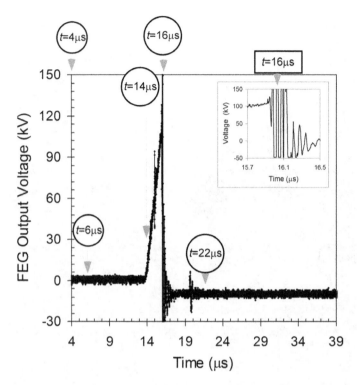

Fig. 9.11. Waveform of the voltage generated by a PZT 95/5 cylindrical element that was simultaneously recorded during the high-speed photography of the FEG explosive operation (see Figure 9.12).

The waveform of the voltage generated across the PZT 95/5 element that was simultaneously recorded during the high-speed photography of an FEG operation is shown in Figure 9.11.

Figure 9.12 presents a series of high-speed photographs ($1/2 \times 10^6$ frames per second) of the explosive operation of the FEG with a PZT 95/5 element of diameter 22 mm and thickness 27 mm [8].

The high-speed images in Figure 9.12 correspond to the marked times in the resultant voltage waveform measured at the output terminals of the FEG (Figure 9.11). Details of the experimental setup can be found herein [8].

At $t = 0\,\mu$s, the EBW detonator was fired and the HE charge detonated; subsequently, over the next 6 μs the detonation wave

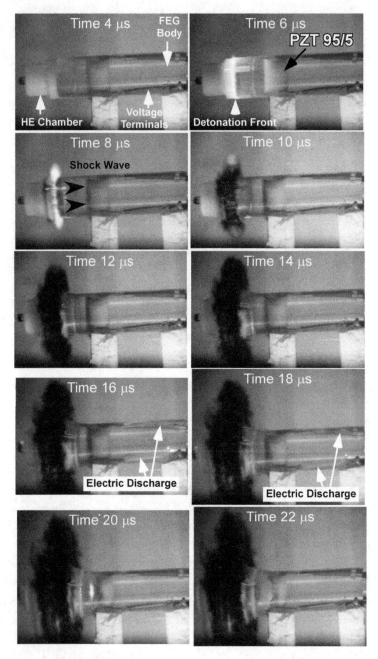

Fig. 9.12.   High-speed photographs of a ferroelectric generator in an explosive operation.

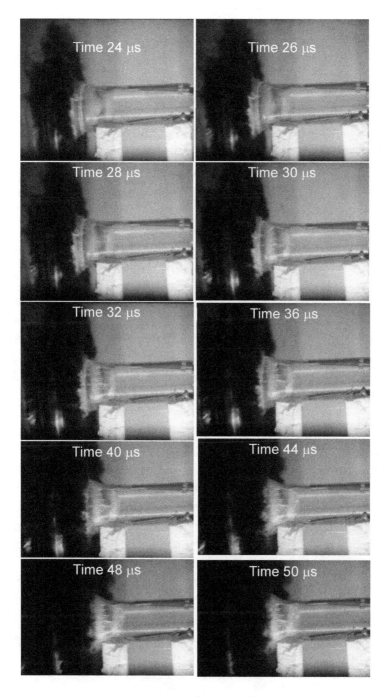

Fig. 9.12. (*Continued*)

propagated in the HE charge rises to the top of the plastic body. The first high-speed photograph was taken at $t = 4\,\mu s$ (Figure 9.12).

In Figure 9.12, at $t = 6\,\mu s$, there was a clearly visible bright light corresponding to the detonation-generated plasma at the HE/urethane body interface when the detonation front reached the top of the urethane body of the FEG. This was the beginning of the shock front propagation within the FEG body, at a velocity in the urethane compound of $2.8\,mm/\mu s$.

At $t = 10\,\mu s$, there was a visible radial expansion of the very top of the FEG body at the joint between the body and the explosive chamber. This mechanical deformation was the result of high-pressure gases coming from the HE chamber.

At $t = 14\,\mu s$, the shock front reached the PZT 95/5 element, and the voltage started rising due to the shock depolarization of the ferroelectric element (see voltage waveform in Figure 9.11).

At $t = 16\,\mu s$, the voltage reached its maximum of 129 kV and intensive oscillations began (see Figure 9.11). Bright light, likely caused by gas ionization in the high-voltage wire insulation, appeared within the red high-voltage wires; $4\,\mu s$ later, the oscillations stopped, the voltage rose again, and then dropped to zero (see Figure 9.11), and the light in the high-voltage wires disappeared ($t = 20\,\mu s$ in Figure 9.12).

At $t = 22\,\mu s$, the PZT/urethane interface (facing the HE chamber) was mechanically fragmented. There were numerous cracks clearly visible in the urethane surrounding the PZT. The shock wave velocity in PZT 95/5 is $3.8\,mm/\mu s$, so by this time, the shock front had passed through the ferroelectric element and entered the urethane compound behind it. If the PZT element had produced a longer high-voltage pulse, this mechanical destruction could have caused electric breakdown.

This mechanical fragmentation of the potting material limits the duration of the the high-voltage pulse generated by a ferroelectric element. In this experiment, the shock wave front entered the ferroelectric element at $t = 14\,\mu s$ (see Figure 9.11). The mechanical fragmentation of the potting material was observed at $t = 22\,\mu s$ (see Figure 9.12). If the duration of the high-voltage pulse was longer than $8\,\mu s$, the FEG would experience the electric breakdown in the mechanically fragmented potting material.

At $t = 40\,\mu s$, the shock front had completely passed through the 100-mm-long FEG body. Mechanical fragmentation of the body had proceeded beyond the PZT element, destroying the working part of the FEG. Experimental results indicate that the destruction of the FEG due to the expanding detonation products had reached its limit at this time. There was no more visible fragmentation during the following $10\,\mu s$ ($t = 50\,\mu s$ in Figure 9.12).

One can see on the high-speed photographs that there are no visible signs of electric breakdown along the ferroelectric surface-urethane interface. Instead, the results suggest that the voltage across the PZT 95/5 electrodes was oscillating due to an electric discharge at the output terminals within the high-voltage wire. It is probable that the ultrahigh-voltage conditions initiated a gas discharge between the layers of insulation of the high-voltage wire. The development of the gas discharge was restricted within the FEG body by the 60 mm of urethane insulation. This initial gas discharge, however, could cause a bulk electric discharge in the gas outside the device.

The experimental results indicate that for an explosive charge of this size, high-pressure gas destroys the region of the urethane body located 25–50 mm from the HE/urethane body interface during the first $30\,\mu s$ of the operation of the FEG. This could cause electric breakdown in the FEGs, utilizing large ferroelectric elements and producing relatively long ($10\,\mu s$ and up) high-voltage pulses. One may avoid this destruction by positioning the ferroelectric element farther from the HE/urethane interface.

9.5 Ultrahigh-Voltage Ferroelectric Generator Design and Performance

Based on the conclusion drawn from the analysis of the high-speed photography of the FEG operation, the ferroelectric generators were completely redesigned to suppress electric discharge at the wire terminals. Figure 9.13 shows a schematic of the redesigned FEG [15].

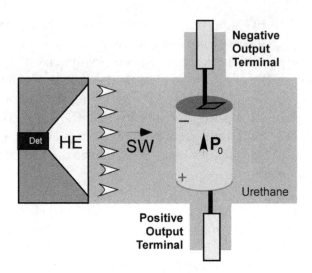

Fig. 9.13. Schematic of ultrahigh-voltage ferroelectric generator with improved electrical insulation. P_0 is the remanent polarization vector. The polarity of the surface charge in the ferroelectric specimen is shown by $(+)$ and $(-)$ signs.

The electrical insulation was improved by increasing the distance between the output terminals, and by increasing the thickness of the high-voltage wire insulation. The diameter of the FEG body was not changed. Since the same-sized HE charge was used in the redesigned FEG as in the original design, the spacing between the HE chamber/urethane body interface and the ferroelectric element was not increased.

A photograph of the redesigned ultrahigh-voltage FEG is shown in Figure 9.14. This FEG was made with Crystal Clear 204 urethane encapsulating material.

Figure 9.15 shows a typical waveform of the voltage produced by the FEG based on the design with improved electrical insulation shown in Figures 9.13 and 9.14, with a 25.4-mm-thick PZT 95/5 cylindrical element. The waveform is similar to that for the thinner PZT 95/5 elements (see Figure 9.3) and is different from that (Figure 9.8) obtained with the FEG based on the previous design shown in Figure 9.2; note that there were no powerful oscillations of the voltage during breakdown.

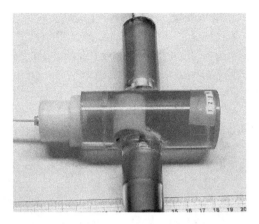

Fig. 9.14. Ultrahigh-voltage ferroelectric generator with improved electrical insulation is loaded with HE charge and prepared for explosive and electrical operation (courtesy of Loki Incorporated).

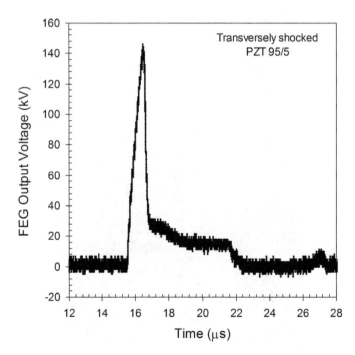

Fig. 9.15. Typical waveform of the voltage produced by ultrahigh-voltage FEG (see Figures 9.13 and 9.14) containing a 25.4-mm-thick PZT 95/5 cylindrical element.

The average breakdown voltage in several experiments using the redesigned FEG was 147.3 ± 5.6 kV. The $V_{FEG}(d)$ value for elements of this type is in good agreement with the extrapolated curve in Figure 9.7, so these results confirm the conclusions we have drawn from our analysis of the high-speed photography of the FEG operation.

A series of experiments was conducted with different encapsulating materials. A photograph of the ultrahigh-voltage FEG made with Stycast 2651-40 epoxy encapsulating material is shown in Figure 9.16.

Stycast 2651-40 has almost double the electric breakdown field in comparison with that of Crystal Clear 204. In addition to this, Stycast 2651-40 has significantly higher density and hardness than Crystal Clear 204. Stycast 2651-40 provides a better match of the mechanical properties of ceramic ferroelectric elements to those of the encapsulating material and, correspondingly, a lower distortion of the shock wave profiles within the FEGs. However, no difference was observed between the ultrahigh-voltage waveforms and the voltage amplitudes generated by FEGs made with Stycast 2651-40 and FEGs made with Crystals Clear 204.

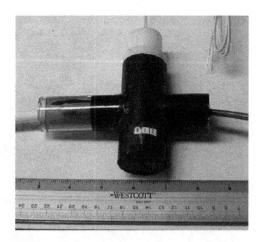

Fig. 9.16.   Ultrahigh-voltage ferroelectric generator made with Stycast 2651-40 epoxy encapsulating material is loaded with high explosives and prepared for explosive and electrical operation (courtesy of Loki Incorporated).

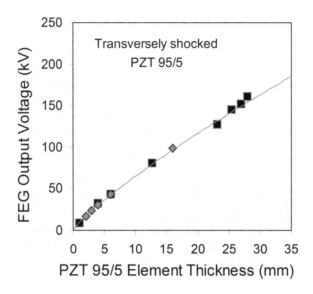

Fig. 9.17. Experimental results obtained from re-designed FEGs. High voltage generated by transversely shock-compressed PZT 95/5 cylindrical (squares) and rectangular (diamonds) elements and fitted curve.

In Figure 9.17, the results are summarized for experiments with the cylindrical PZT 95/5 elements along with experimental data for the rectangular PZT 95/5 elements [8, 15, 16].

There is a good agreement between the results obtained with the cylindrical and rectangular elements in a wide range of voltages. We did a parametric analysis of the complete set of experimental data using the $V_{FEG}(d) = \chi \cdot d^{1-\xi}$ function, and the resultant first fitting parameter (voltage generated by a 1.0 mm-thick PZT 95/5 element) $\chi_{95/5} = 10.7\,\mathrm{kV}$ and second fitting parameter $\xi_{95/5} = 0.799$ were very close to those obtained when we performed the analysis for breakdown voltages below the 100-kV level.

Based on these results, the conclusion was reached that the thickness-dependent breakdown law, Eq. (9.1), can be extended to shock-compressed PZT 95/5 ferroelectrics under ultrahigh-voltage conditions. We extrapolated the fitted curve from 170 to 500 kV level (Figure 9.18) in order to estimate the thickness of the PZT 95/5 elements that would be capable of producing the output voltage in

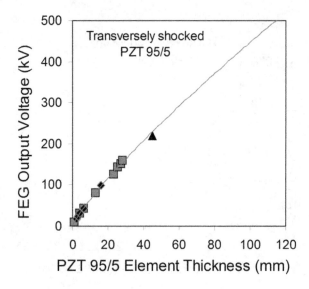

Fig. 9.18. Experimentally obtained amplitudes of voltage (squares and diamonds) generated by transversely shock-compressed PZT 95/5 with fitted and
extrapolated curve (dot line); experimental data for 46-mm-thick PZT 95/5
elements are shown by a triangle.

this voltage range that is important for powering autonomous HPM
systems.

To verify the extrapolated curve in Figure 9.18, we conducted
experiments with PZT 95/5 cylindrical elements with thickness
46 mm and diameter 22 mm. A typical waveform of the voltage
generated in these experiments is shown in Figure 9.19. The voltage
increased in 1.5 μs to its maximum value $V_{FEG} = 224$ kV.

We performed three experiments with 46-mm-thick elements,
and the resulting voltage waveforms were reproducible. The value
of $V_{FEG} = 207 \pm 19$ kV obtained in these experiments is shown
in Figure 9.18 as a triangle. This value is in agreement with the
extrapolated curve. These experimental results indicate that the
obtained law for the breakdown field on thickness dependence for
transversely shock-compressed PZT 95/5 ferroelectrics in a wide
range of ferroelectric element thicknesses and a wide range of
breakdown voltages makes it possible to predict the output voltage

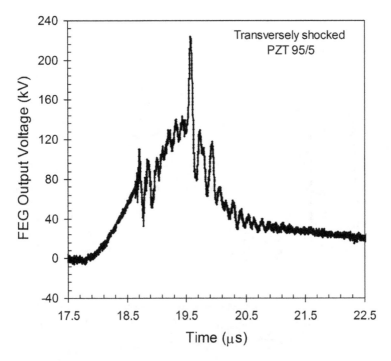

Fig. 9.19.   Waveform of voltage pulse produced by the FEG with a 46-mm-thick PZT 95/5 element.

produced by ultrahigh-voltage generators and forms the basis for the design of ultrahigh-voltage FEG systems.

9.6   PZT 52/48 Ultrahigh-Voltage Generation: Transverse versus Longitudinal Shock Loading

It is shown in the previous chapter that the energy density generated by PZT 52/48 in the high-voltage mode is close to that for PZT 95/5. Thus, both ferroelectrics can be used for the development of high-voltage FEG-based systems. The results of systematic experimental studies of ultrahigh-voltage generation by PZT 52/48 are presented in the following sections.

It is experimentally demonstrated in Section 9.3 above that there is a significant difference between amplitudes of high voltage generated by 23-mm-thick PZT 95/5 elements under longitudinal

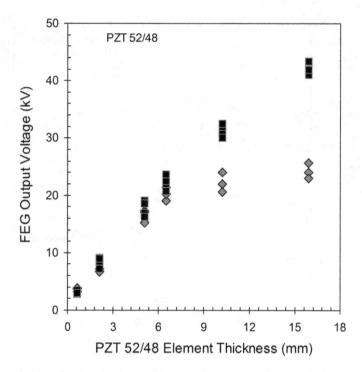

Fig. 9.20.   Amplitude of voltage pulse generated by transversely (squares) and longitudinally (diamonds) shock-compressed PZT 52/48 cylindrical elements as a function of the element thickness.

and transverse shock compression. To obtain more information about high-voltage generation by PZT 52/48 under transverse and longitudinal shock loading, the experiments were conducted with PZT 52/48 cylindrical elements in a wide range of element thicknesses.

Figure 9.20 shows amplitudes of high voltage produced by PZT 52/48 elements under transverse and longitudinal shock loading as a function of the element thickness. The experimental results indicate that high-voltage amplitudes generated by PZT 52/48 under transverse and longitudinal shock are practically equal for the elements with thicknesses ranging from 0.65 to 6.5 mm.

However, the amplitudes of high voltage generated by PZT 52/48 elements with thickness 10 and 16 mm (diameter 19 and 22 mm, respectively) under longitudinal shock compression are significantly lower than those generated under transverse shock (Figure 9.20). These results are in agreement with those obtained with longitudinally and transversely shocked 23-mm-thick PZT 95/5 elements (see Figures 9.4 and 9.5).

The high-voltage generation by 10-mm-thick and 16-mm-thick PZT 52/48 elements (Figure 9.20) is not in agreement with the breakdown field on the thickness dependence that describes the high-voltage generation for longitudinally shock-compressed PZT 52/48 elements with thicknesses ranging from 0.6 to 6 mm (see Chapter 8 for details). Based on these results, the conclusion one can come to is that the thickness of ferroelectric elements is one of the main parameters that have an effect on the mechanism of electric breakdown of longitudinally shock compressed ferroelectrics.

Consider the depolarization of ferroelectric elements under transverse and longitudinal shock loading in the high-voltage mode. When the shock wave propagates through the ferroelectric element, its volume is divided into compressed and uncompressed zones differing in their physical parameters. The compressed zone is that through which the shock wave has already passed and the uncompressed zone is that through which the shock wave has not passed. Equivalent circuits for the ferroelectric elements in the transverse and longitudinal high-voltage shock depolarization modes are shown in Figure 9.21.

The transversely shock-compressed ferroelectric element is represented by two capacitors connected in parallel (Figure 9.21(a)). The uncompressed zone is represented by capacitor C_1 and the compressed zone is represented by capacitor C_2. In the transverse shock depolarization mode, the voltage across the compressed zone is always equal to the voltage across the uncompressed zone, and equal to the FEG output voltage, V_{FEG}.

It is not the same in the longitudinally shock-compressed ferroelectric element that is represented by two capacitors connected in series (Figure 9.21(b)). The uncompressed zone is represented by capacitor C_1 and the compressed zone is represented by capacitor C_2. In the longitudinal mode, the voltage and electric field across the compressed zone and uncompressed zone depend on the dielectric properties and size of each zone. The longitudinal FEG output

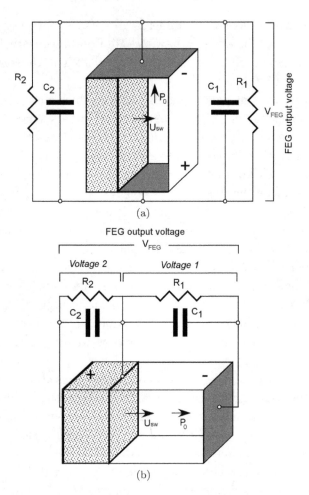

Fig. 9.21. Equivalent circuits for the ferroelectric element in (a) transverse and (b) longitudinal high-voltage shock depolarization mode. $\boldsymbol{P}_0$ is the remanent polarization vector. $\boldsymbol{U}_{SW}$ is the shock vector. The polarity of the surface charge in the ferroelectric specimen is shown by (+) and (−) signs.

voltage, V_{FEG}, is a sum of two voltages, *Voltage 1* (voltage across uncompressed zone) and *Voltage 2* (voltage across compressed zone).

The volume of each zone is under change during the shock wave transit. The dielectric properties of the compressed zone are changing due to the shock compression and the action of release waves following behind the shock wave front. Consequently, the electric field distribution within the longitudinally shocked element changes in a very complex way.

Increasing the thickness of longitudinally shocked ferroelectric elements leads to an increase in the shock wave transit time, the longer rise time of the high-voltage pulse, and a greater change in the electrical and mechanical properties of the compressed zone than in thinner elements. All these along with the complex distribution of the electric field could cause earlier electric breakdown in the longitudinally shocked thick elements.

Lysne developed a theoretical model of breakdown in longitudinally shocked PZT 65/35 [3, 5]. The breakdown voltage predicted with this model was in good agreement with the experimental results obtained for 2.5-mm-thick PZT 65/35 ferroelectric elements [3, 5]. However, to develop a comprehensive theoretical model describing the breakdown of longitudinally shock-compressed ferroelectrics in a wide range of element thicknesses and, correspondingly, a wide range of breakdown voltages, it is necessary to take into account all factors that have an effect on the electrical properties of ferroelectrics during shock wave transit and under high electric field.

In conclusion, in the ultrahigh-voltage mode, longitudinally shock-compressed PZT 95/5 and PZT 52/48 are significantly less efficient than under transverse shock loading. Based on the obtained experimental results the conclusion can be reached that the mechanism of electric breakdown within longitudinally shocked PZT 95/5 and PZT 52/48 is different from that under transverse shock compression.

9.7 Ultrahigh-Voltage Generation by Multi-Element PZT 52/48 Modules under Transverse Shock Compression

The shock-compressed ferroelectric element is a prime power source of the capacitive type. The energy generated by a ferroelectric element in the high-voltage mode is directly proportional to the amplitude of the voltage to the power of two and the capacitance of the ferroelectric element:

$$W_{FEG} = \frac{C_{FEG}V_{FEG}{}^2}{2} = \frac{\varepsilon_0\varepsilon_{\mathrm{r}}A\,V_{FEG}{}^2}{2d} \qquad (9.3)$$

where W_{FEG} is the energy produced by the ferroelectric generator in the high-voltage mode, C_{FEG} is the capacitance of the ferroelectric element in the FEG, V_{FEG} is the voltage generated by the FEG, ε_{r} is the relative dielectric permittivity of the ferroelectric material, ε_0 is the dielectric permittivity of free space, d is the distance between the electrodes of the ferroelectric element, and A is the area of the ferroelectric element electrodes.

An increase in the thickness of the ferroelectric element, d, results in an increase in the generated voltage (Figure 9.20) but it also results in a decrease in its capacitance (Eq. (9.3)). To generate ultrahigh voltage and keep the capacitance of the ferroelectric element not too low it is necessary to increase both the element thickness and its electrode area. However, an increase in the ferroelectric cylinder diameter causes an increase in the diameter of a miniature FEG and the amount of HE, which is not appropriate for a system intended to be as small in cross-section as possible. Therefore, we investigated ultrahigh-voltage generation by transversely shock-compressed PZT 52/48 elements of rectangular geometry. With rectangular elements, there is the opportunity to increase the capacitance of the FEG and, correspondingly, the FEG output energy, by lengthening the elements without increasing the FEG cross-sectional dimensions.

The fabrication of large size ceramic elements is a complex technological problem because an increase in the element thickness and its overall size can cause cracks during sintering and polarization. A possible solution is to design the ferroelectric element of a high-voltage FEG as a module containing several specimens connected in series. However, this approach raises questions about the mechanical and electrical interference of the ferroelectric elements within the module during the propagation of the shock waves through the module, the depolarization of the ferroelectric material and the generation of high voltage and high electric field within the module. To answer these questions, systematic experimental investigations of ultrahigh-voltage generation by transversely shock-compressed multi-element PZT 52/48 modules were performed.

The schematics of transverse FEGs with multi-element PZT 52/48 modules are shown in Figure 9.22 [16]. The multi-element modules were constructed with rectangular PZT 52/48 elements of

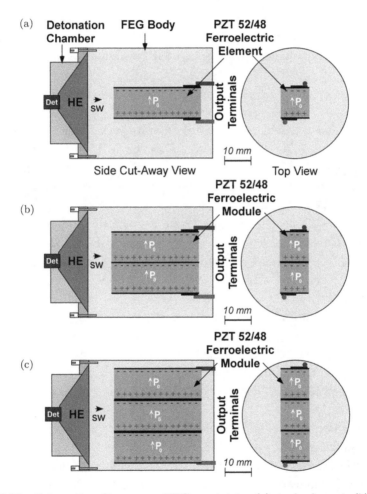

Fig. 9.22.   Schematics of transverse FEGs containing (a) single-element, (b) two-element and (c) three-element PZT 52/48 modules. SW is the shock vector. P_0 is the remanent polarization vector. The polarity of the surface charge is shown by ($+$) and ($-$) signs.

12.7 mm thick × 12.7 mm wide × 50.8 mm long. In a two-element module, the negative electrode of the first ferroelectric element is connected to the ground terminal of the FEG. The positive electrode of the first element is electrically and mechanically connected to the negative electrode of the second element. The positive electrode of the second element is connected to the high-voltage terminal of the FEG. To avoid heating the ferroelectric elements, instead of using a soldering procedure, silver epoxy (Chemtronics CW2400) is used for all connections within the FEGs. The positive electrode of the second element was connected to the high-voltage output terminal of the FEG. The serial connection in the three-element module was made in a similar fashion to that in the two-element module.

A typical waveform of the output voltage produced by the FEG with a single PZT 52/48 element is shown in Figure 9.23 (plot 1). The amplitude of the voltage pulse was 39.4 kV with rise time 1.6 μs. The mean output voltage averaged over seven experiments of this series was $V_{\text{FEG}} = 38.7 \pm 1.2$ kV.

Figure 9.23 also shows a typical waveform of the output voltage produced by the FEG with a two-element PZT 52/48 module (plot 2). The voltage amplitude was 71.0 kV with rise time 2.4 μs. The mean voltage was $V_{\text{FEG}} = 70.4 \pm 2.3$ kV. The experimental results indicate that a twofold increase in the number of elements and, correspondingly, the thickness of the PZT 52/48 module led to an increase in the FEG output voltage by a factor of 1.8. Based on these results, the conclusion can be reached that there is no significant interference among the surface charges released by PZT 52/48 elements connected in series during high-voltage shock depolarization.

The diameter of the ferroelectric generators containing one- and two-element PZT 52/48 modules was 38 mm. The amount of HE used in these devices was 11 g. For the three-element modules, FEGs with diameter 62 mm were constructed. The cone angle of the detonation chamber of this FEG was 60° and the HE mass was 59.0 ± 2.5 g.

A typical waveform of the output voltage produced by a three-element PZT 52/48 module is shown in Figure 9.23 (plot 3).

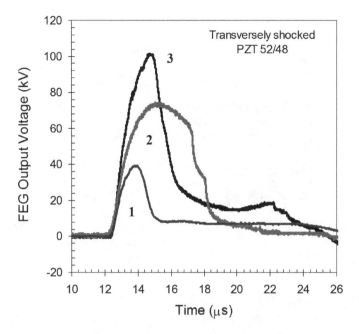

Fig. 9.23. Typical waveforms of the output voltage produced by transverse FEGs containing a single PZT 52/48 rectangular element (plot 1), and also two-element (plot 2), and three-element (plot 3) PZT 52/48 modules.

The amplitude of the voltage was $V_{FEG} = 105\,$kV with rise time $2.4\,\mu$s. The mean voltage, $V_{FEG} = 102.7 \pm 3.4\,$kV, is 2.65 times higher in voltage than that produced by a single-element PZT 52/48 with thickness 12.7 mm. No signs of electric breakdown were observed within the insulation of the FEGs.

The next ultrahigh-voltage FEG design was based on a four-element PZT 52/48 module. A schematic of this FEG is shown in Figure 9.24 [15, 16]. The elements were grouped in pairs and placed in a 62 mm diameter generator body.

For this design (Figure 9.24) we used the ultrahigh-voltage FEG design with improved electrical insulation (Figure 9.13) as the basis. In the first generators of this type, the inter-electrode distance (this is the distance between the positive and negative electrodes, "H" in Figure 9.24) was $H = 10\,$mm. A typical waveform of the output voltage produced by this FEG is shown in Figure 9.25 (plot 1). The

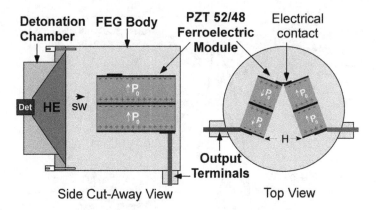

Fig. 9.24. Schematic of transverse FEGs containing a four-element PZT 52/48 module. **SW** is the shock vector. **P_0** is the remanent polarization vector. The polarity of the surface charge is shown by $(+)$ and $(-)$ signs.

amplitude of the voltage was $V_{FEG} = 100.2\,\text{kV}$ with $\tau = 1.9\,\mu\text{s}$. This voltage was lower than that produced by a three-element PZT 52/48 module (see Figure 9.23). There are clear signs of electric breakdown in the voltage waveform (Figure 9.25). Most probably it was a breakdown between the high-voltage and ground electrodes of the ferroelectric module within a plastic body of the FEG.

To suppress the internal FEG breakdown the distance between the high voltage and ground electrodes was increased to $H = 20\,\text{mm}$. A typical waveform produced by one of the FEGs of this type is shown in Figure 9.25 (plot 2). The amplitude of the output voltage was $V_{\text{FEG}} = 126\,\text{kV}$. The experimental results obtained for transversely compressed multi-element PZT 52/48 ferroelectric modules operating in the high-voltage mode are summarized in Table 9.1.

The experimental results obtained with FEGs having different distances between the top of the plastic body and the top of the ferroelectric module indicate that the 30 mm potting material layer provided reliable electrical insulation in the generators of this type.

In conclusion, it is experimentally demonstrated that transversely shock-compressed PZT 52/48 ferroelectrics can be successfully used

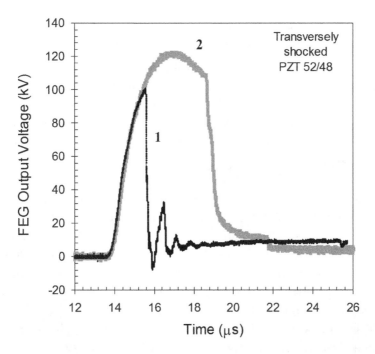

Fig. 9.25. Typical waveforms of the output voltage produced by transverse FEGs containing four-element PZT 52/48 ferroelectric modules with inter-electrode distance $H = 10\,\text{mm}$ (plot 1) and $H = 20\,\text{mm}$ (plot 2).

Table 9.1. Experimental results obtained for high-voltage generation with PZT 52/48 multi-element modules with electrode area $12.7 \times 50.8\,\text{mm}^2$.

Module thickness (mm)	12.7	25.4	38	51
Number of elements	1	2	3	4
Generated voltage (kV)	38.7 ± 1.2	70.4 ± 2.3	102.7 ± 3.4	123.3 ± 3.2

for ultrahigh-voltage generation. Multi-element PZT 52/48 modules with serial connections of elements provide the multiplication of generated voltage without parasitic interference between elements during the high-voltage shock depolarization. The generators of this type demonstrate reliable operation and reproducible parameters of ultrahigh-voltage pulses.

Explosive Ferroelectric Generators

9.8 PZT 52/48 Ultrahigh-Voltage Generation: Breakdown Field on Thickness Dependence

Figure 9.26 summarizes the experimental results for high-voltage generation by transversely shock-compressed PZT 52/48 cylindrical and rectangular elements. We performed a parametric analysis of these experimental data representing a range of two orders of magnitude of PZT 52/48 element thickness (from 0.6 to 60 mm) and a wide range of generated (breakdown) voltage (from 3 to 130 kV).

Our analysis indicates that the best fit for the data is the function that is identical to the thickness-dependent breakdown law that represents the breakdown of solid dielectrics at ambient conditions, $V_{FEG}(d) = \chi \cdot d^{1-\xi}$. Thus, the two different shock-compressed ferroelectrics, PZT 95/5 and PZT 52/48, obey this law. One can draw the conclusion that the breakdown law for dielectrics at ambient conditions (Eq. (9.2)) can be extended to shocked ferroelectrics under ultrahigh-voltage conditions.

It should be noticed that there is a significant difference in the voltages generated by transversely shock-compressed PZT 52/48

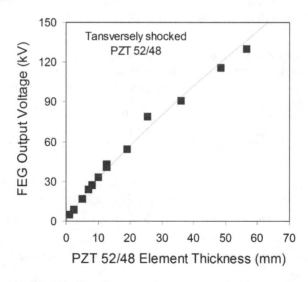

Fig. 9.26. Amplitude of voltage pulses generated by transversely shock-compressed PZT 52/48 cylindrical and rectangular elements and fitted curve.

and PZT 95/5 elements. The first fitting parameter χ (voltage generated by a 1.0 mm-thick ferroelectric element) for PZT 52/48 is $\chi_{52/48} = 4.7$ kV and the second fitting parameter is $\xi_{52/48} = 0.835$. The $\chi_{52/48}$ is about two times lower than that for PZT 95/5; $\chi_{95/5} = 10.7$ kV. A possible cause for this difference is the higher dynamics of stress-induced charge and the different dielectric properties of the compressed zone in PZT ceramics having formulations very close to the ferroelectric/antiferroelectric phase transformation boundary (PZT 95/5) compared with those having formulations near the morphotropic phase boundary (PZT 52/48).

We extrapolated the PZT 52/48 fitted curve from the 130 to the 500 kV level (Figure 9.27). To verify this extrapolated curve, we designed, constructed and experimentally investigated PZT 52/48 modules with the thickness of 150 mm. A typical waveform of the voltage generated in these experiments is shown in Figure 9.28. The

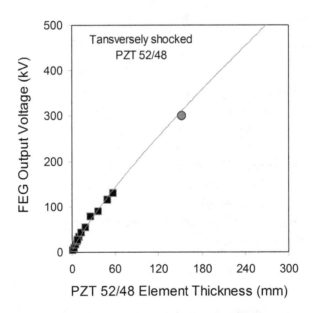

Fig. 9.27. Voltage generated by transversely shock-compressed PZT 52/48 rectangular and cylindrical elements as a function of the element thickness and corresponding fitted-extrapolated curve (dot line); experimental data for the 152-mm-thick PZT 52/48 module are shown by a circle.

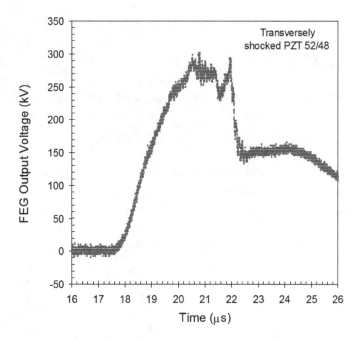

Fig. 9.28. Waveform of voltage produced by transverse FEG containing a 150-mm-thick PZT 52/48 module.

voltage increased during 2.8 μs to its maximum $V_{FEG} = 304$ kV. Data obtained in these experiments are shown as a circle in Figure 9.27. These data are in agreement with the extrapolated curve.

Based on these results (Figure 9.27) the conclusion can be reached that the experimentally-obtained law for the breakdown field on thickness dependence for transversely shock-compressed PZT 52/48 makes it possible to predict the output voltage produced by PZT 52/48 elements up to the 500 kV level and forms the basis for the design of ultrahigh-voltage FEG-based systems.

9.9 Summary

- It was experimentally demonstrated that PZT 95/5 and PZT 52/48 ferroelectrics transversely shock-compressed within minia-ture explosive FEGs provide reliable and reproducible generation of ultrahigh-voltage pulses with an amplitude of hundreds of kilovolts. Both ferroelectrics can be successfully used for the development of compact autonomous ultrahigh-voltage FEG-based systems.

- The experimental results indicate that the amplitude of ultrahigh voltage generated by the two ferroelectrics under transverse shock compression is directly proportional to the ferroelectric element thickness.

- It was experimentally proved that the thickness-dependent dielectric breakdown law, $E_{break}(d) = \gamma \cdot d^{-\xi}$, can be extended to ferroelectric ceramics under high shock pressure. This law was found to be true under ultrahigh-voltage conditions in a wide range of ferroelectric element thicknesses.

- The experimentally-obtained relationship between the breakdown field and the thickness of the ferroelectric elements in the wide range of breakdown voltages allows one to predict the breakdown field in shock-compressed ferroelectrics and the amplitude of the voltage produced by ferroelectric generators up to 500 kV. This relationship forms the basis for the development of ultrahigh-voltage FEG systems utilizing PZT 95/5, PZT 52/48 and other ferroelectric ceramic materials.

- It was experimentally demonstrated that the approach based on transverse shock depolarization of multi-element PZT 52/48 ceramic modules with elements connected in series can be successfully used for the generation of ultrahigh voltage by miniature FEGs. There was no indication of electrical or mechanical interference between ferroelectric elements within the modules during shock-induced depolarization and ultrahigh-voltage generation.

- The experimental results indicate that the amplitudes of ultrahigh-voltage pulses generated by multi-element PZT 52/48 modules under transverse shock loading are directly proportional to the number of elements and their thickness.

- High-speed photography of the operation of an explosive FEG helped to identify the causes of electrical breakdown within generators. Ultrahigh voltages within the generator place severe requirements on terminal materials, design, and layout. In addition, detonation product expansion during generator operation causes mechanical destruction of a portion of the FEG plastic body, and the destruction starts just a few microseconds after the shock front enters the FEG plastic body. This destruction may cause electrical breakdown within the PZT element, if the destruction proceeds to the location of the element.

- It was experimentally demonstrated that longitudinally shock-compressed PZT 95/5 and PZT 52/48 are significantly less efficient for ultrahigh-voltage generation than the same elements under transverse shock compression.

- The high-voltage generation by longitudinally shocked PZT 52/48 and PZT 95/5 elements having thicknesses of a few tens of millimeters is not in agreement with the breakdown law that describes high-voltage generation for longitudinally shock-compressed thin elements.

- It was experimentally demonstrated that the thickness of ferroelectric elements is one of the main parameters that have an effect on the electric breakdown mechanism of longitudinally shock-compressed ferroelectrics.

- The experimental results indicate that the mechanism of electric breakdown of longitudinally shock-compressed PZT 95/5 and PZT 52/48 is different from that under transverse shock loading.

Bibliography

1. L.L. Altgilbers, J. Baird, B. Freeman, C.S. Lynch and S.I. Shkuratov, *Explosive Pulsed Power* (Imperial College Press, London, 2010).

2. Yu.N. Vershinin, D.S. Il'ichev and P.A. Morozov, Effect of shock compression of solid insulators on the injection of valence electrons in strong magnetic fields, *Tech. Phys.* **45**(1) (2000) pp. 84–87.

3. P.C. Lysne, Dielectric breakdown of shock-loaded PZT 65/35, *J. Appl. Phys.* **44**(2) (1973) pp. 577–582.

4. P.C. Lysne, Prediction of dielectric breakdown in shock-loaded ferroelectric ceramics, *J. Appl. Phys.* **46**(1) (1975) pp. 230–232.

5. P.C. Lysne and L.C. Bartel, Electromechanical response of PZT 65/35 subjected to axial shock loading, *J. Appl. Phys.* **46**(1) (1975) pp. 222–229.

6. S.I. Shkuratov, J. Baird, E.F. Talantsev, E.F. Alberta, W.S. Hackenberger, A.H. Stults and L.L. Altgilbers, Miniature 100-kV explosive-driven prime power sources based on transverse shock-wave depolarization of PZT 95/5 ferroelectric ceramics, *IEEE Trans. Plasma Sci.* **40**(10) (2012) pp. 2512–2516.

7. S.I. Shkuratov, J. Baird, V.G. Antipov and E.F. Talantsev, Autonomous pulsed power generator based on transverse shock wave depolarization of ferroelectric ceramics, *Rev. Sci. Instrum.* **81** (2010) p. 126102.

8. S.I. Shkuratov, J. Baird and E.F. Talantsev, Extension of thickness-dependent dielectric breakdown law on adiabatically compressed ferroelectric materials, *Appl. Phys. Lett.* **102** (2013) p. 052906.

9. F. Forlani and N. Minnaja, Thickness influence in breakdown phenomena of thin dielectric films, *Phys. Status Solidi B* **4**(2) (1969) pp. 311–324.

10. R. Gerson and T.C. Marshall, Dielectric breakdown of porous ceramics, *J. Appl. Phys.* **30**(11) (1959) pp. 1650–1655.

11. V.A. Borisenok, V.A. Kruchinin, V.A. Bragunets, S.V. Borisenok, V.G. Simakov and M.V. Zhernokletov, Measuring shock-induced electrical conductivity in piezoelectrics and ferroelectrics: single-crystal quartz, *Combustion, Explosion, and Shock Waves* **43** (2007) pp. 96–103.

12. S.D. Gilev, *Combustion, Explosion, and Shock Waves* **47** (2011) p. 375.

13. V.A. Bragunets, V.G. Simakov, V.A. Borisenok, S.V. Borisenok and V.A. Kruchinin, Shock-induced electrical conductivity in some ferroelectrics, *Combustion, Explosion, and Shock Waves* **46** (2010) pp. 231–236.

14. G.A. Mesyats, *Pulsed Power* (Kluwer/Plenum, New York, 2005).

15. S.I. Shkuratov, J. Baird, V.G. Antipov, E.F. Talantsev, W.S. Hackenberger, A.H. Stults and L.L. Altgilbers, High voltage generation with transversely shock-compressed ferroelectrics: breakdown field on thickness dependence, *IEEE Trans. Plasma Sci.* **44**(10) (2016) pp. 1919–1927.

16. S.I. Shkuratov, J. Baird and E.F. Talantsev, Miniature 120-kV autonomous generator based on transverse shock-wave depolarization of Pb(Zr0.52Ti0.48)O3 ferroelectric, *Rev. Sci. Instrum.* **82** (2011) p. 086107.

Chapter 10

PZT 95/5 Films: Depolarization and High-Current Generation under Transverse and Longitudinal Shock Compression

10.1 Introduction

Ferroelectric films are used in the electronics industry for non-volatile ferroelectric random access memories, ferroelectric film capacitors, microelectromechanical systems and different types of ferroelectric transducers that operate in the low-power mode [1–4]. Modern technologies make it possible to fabricate high-quality ferroelectric films with precise control of chemical composition and ferroelectric properties. There is a significant interest in expanding the usage of ferroelectric films to ultrahigh-power systems. Opposite to the operation of ferroelectric films in the low-strain mode that is piezoelectric in nature with no large-scale disorientation of the ferroelectric domains, the operation of ferroelectric films in explosive ferroelectric generators under high mechanical stress induces phase transitions and domain reorientation with possible loss of the initial remanent polarization. In this chapter, the results of experimental studies of depolarization and high-current generation by single-layer and multi-layer PZT 95/5 films under transverse and longitudinal shock compression are presented. The obtained results indicate that ferroelectric film technology is very promising for ultrahigh-power applications.

10.2   Fabrication of PZT 95/5 Films

Single-layer and multi-layer PZT 95/5 films were fabricated by TRS Technologies [5]. PZT 95/5 powder was prepared by mixing stoichiometric amounts of PbO, ZrO_2, and TiO_2 raw material with dopants in a water-based slurry. The powder mixture was calcined at high temperature to form the PZT perovskite phase. The powder was then remilled to reduce particle size and dried. After drying, the powder was mixed with non-aqueous solvents and binders for tape casting using a ball mill. After milling, the slurry was dispensed in a thin layer on a moving steel belt using a doctor blade. This tape-casting system produced films that were $32\,\mu m$ thick. The film was continuously rolled onto a spool during casting. The films were next blanked into $12.5 \times 12.5\,cm^2$ squares in preparation for printing and stacking.

Films were printed with platinum ink to form the internal conductive layers for the multi-layer elements. The multi-layer stack was formed on the printer. After printing, the stacks were sealed in a vacuum bag and laminated using a warm isostatic press. The laminated stacks were then cut into individual elements using a guillotine cutter. The elements were slowly heat-treated to remove the organic binder materials and then sintered (co-fired) in a sealed crucible with PbO source powder.

After firing, the elements were terminated with fired silver electrodes to facilitate contact to the internal Pt electrode layers. Finally, the single-layer and multi-layer elements were polarized by applying a high voltage at elevated temperature. The film specimens were encapsulated within PZT 95/5 non-polarized ceramic bodies.

10.3   Transverse and Longitudinal Shock Wave FEGs with PZT 95/5 Film Elements

Transverse and longitudinal ferroelectric generators based on spherically expanding shock waves were used in the depolarization studies. Schematics of transverse and longitudinal FEGs with ferroelectric film specimens are shown in Figures 10.1 and 10.2, respectively.

The film specimens were encapsulated within urethane bodies of the FEGs. The potting material was in direct contact with high

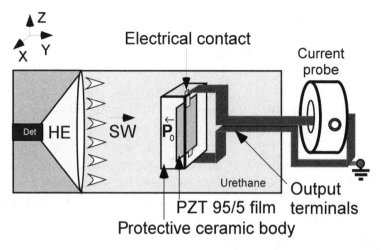

Fig. 10.1. Schematics of transverse shock wave FEG with a PZT 95/5 film specimen and the measuring circuit. P_0 is the film remanent polarization vector. SW is the shock wave vector. The direction of polarization of the film is anti-parallel to the X-axis. The direction of shock wave propagation is parallel to the Y-axis.

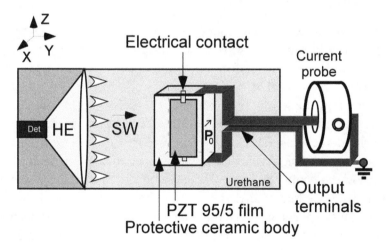

Fig. 10.2. Schematics of longitudinal shock wave ferroelectric generator with a PZT 95/5 film specimen and the measuring circuit. P_0 is the film remanent polarization vector. SW is the shock wave vector. The direction of polarization of the film is anti-parallel to the Y-axis. The direction of shock wave propagation is parallel to the Y-axis.

explosives. After initiation of the detonator, the detonation wave propagates through a conical HE charge. The HE detonation shock propagates through a urethane potting material and into the ceramic body of the film specimen.

The stress wave propagation direction was perpendicular (Figure 10.1) and anti-parallel (Figure 10.2) to the film polarization direction (transverse and longitudinal shock, respectively).

The diameter of the generators was 38 mm and their length was 55 mm. The FEG output terminals were made as a strip transmission line in order to reduce the measuring circuit inductance to a negligible level. The FEG output terminals were short-circuited. The stress-induced current was monitored with a Pearson Electronics 411 current probe and/or a Prodyn I-265 current probe, which have transition times of 10 and 0.2 ns, respectively.

10.4 Uniaxial Stress Distribution in Transversely and Longitudinally Shock-Compressed Ferroelectric Films

An adiabatic stress wave in a semi-infinite solid uniaxially compresses the material in the stress wave propagation direction with no lateral expansion (the Poisson effect does not apply). In finite geometry specimens the lateral expansion is resisted by the inertia of the specimen surrounded by polymer, and the Poisson effect applies.

In these studies the stress wave in the ferroelectric films originated from the detonating HE charge (see Figures 10.1 and 10.2) and the stress wave parameters affecting the ferroelectric films were determined by the geometry of the HE chamber, the initiation point, the geometry of the ferroelectric generator, the shock impedance matching among the parts of the system and the ferroelectric film specimen.

The uniaxial stress distribution in the ferroelectric films was determined by performing a simulation of the high strain rate loading using the CALE computer code. CALE is a two-dimensional Arbitrary Lagrange/Eulerian second-order accurate hydrodynamics program [6]. The simulation was run in cylindrical coordinates, assuming rotational symmetry around the axis and zero material rotation. In order for the numerical integration to proceed, the space was divided into "zones" by a mesh of coaxial cylindrical surfaces and a set of planes. Each zone was washer-shaped, with a nearly square section in the *R-Y* plane, with the coordinates *R* (cylindrical radius from the symmetry axis), *Y* (axial position), and angle (around the symmetry axis).

Figures 10.3 and 10.4 show the CALE results for the uniaxial stress distribution in the cross-section of transversely and longitudinally shock-compressed PZT 95/5 films, respectively. The simulation results indicate that the shock waves generated in transversely and longitudinally shocked films may not have perfectly planar geometry,

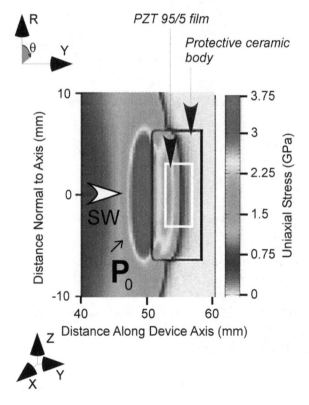

Fig. 10.3. CALE results for the uniaxial stress distribution in the cross-section of a PZT 95/5 film specimen under transverse shock compression (see FEG schematic in Figure 10.1). P_0 is the film remanent polarization vector. The direction of polarization of the film is anti-parallel to the X-axis and perpendicular to the direction of shock wave propagation. The direction of shock wave propagation is parallel to the Y-axis.

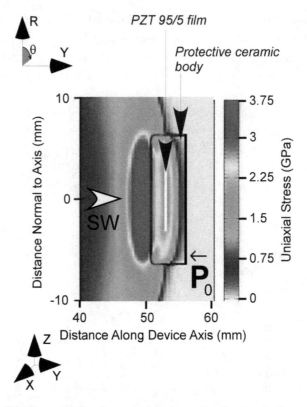

Fig. 10.4. CALE results for the uniaxial stress distribution in the cross-section of a PZT 95/5 film specimen under longitudinal shock compression (see FEG schematic in Figure 10.2). P_0 is the film remanent polarization vector. The direction of polarization of the film is anti-parallel to the Y-axis and anti-parallel to the direction of shock wave propagation. The direction of shock wave propagation is parallel to the Y-axis.

but instead have slightly hemispherical geometry. The uniaxial stress magnitude was 2.4 ± 0.14 GPa in both the transversely and longitudinally shocked films.

10.5 Transverse Shock Depolarization of Single-Layer PZT 95/5 Films

All single-layer film specimens subjected to transverse and longitudinal shock depolarization studies had identical geometrical dimensions to enable a direct comparison of the results obtained in the two modes of high strain rate loading. The film thickness was $32\,\mu$m and the electrode area was $4 \times 6.3\,\text{mm}^2$. All films were poled across their thicknesses to the remanent polarization $P_0 = 32 \pm 1\,\mu\text{C/cm}^2$. The films were encapsulated within non-polarized PZT 95/5 protective ceramic bodies of $10 \times 8 \times 4\,\text{mm}^3$.

When the high-amplitude stress wave propagates through the ferroelectric film it produces a current versus time profile. The stress-induced electric charge is the time integral of the stress-induced current (see Eq. (6.1)). The stress-induced current waveforms and the amount of stress-induced charge released by PZT 95/5 films were reproducible for both the longitudinal and transverse shock loading.

A typical waveform of the current and the dynamics of the electric charge released by PZT 95/5 film under transverse shock compression are shown in Figure 10.5 (see position of the film within FEG in Figure 10.1). The peak amplitude of the current generated under transverse shock loading was $I_{TSW} = 11.1\,\text{A}$, the pulse duration was

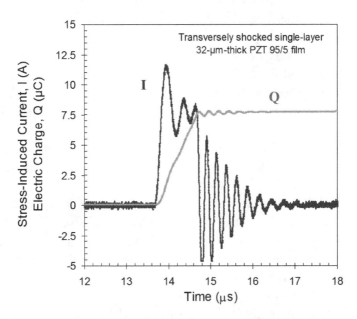

Fig. 10.5. Typical waveform of stress-induced current and the dynamics of the electric charge released by a single-layer PZT 95/5 film under transverse shock compression. The film thickness was $32\,\mu$m and the electrode area was $25\,\text{mm}^2$.

$\tau_p = 1.0\,\mu s$ and the rise time was $0.2\,\mu s$. The stress-induced charge was $Q_{TSW} = 7.8\,\mu C$.

The direction of the stress-induced current provides information about the origin of the electric charge generated by ferroelectrics under high stress. Before loading, there is no electric potential across the films because the initial remanent polarization is balanced by the surface charge density. When the polarization is decreased or completely lost under high stress, the surface charge is no longer in balance. This results in an electric potential, causing electric current to flow in the measuring circuit.

The piezoelectric effect would cause a current in the direction opposite to what was observed in the experiments with transversely shocked films (Figure 10.5). This indicates that the charge released by the films under transverse stress was caused not by the piezoelectric effect but by depolarization.

The stress-induced charge density, $31.2\,\mu C/cm^2$, released by transversely shocked PZT 95/5 films was very close to their initial remanent polarization, i.e. the PZT 95/5 films were completely depolarized under high strain rate loading.

The waveforms of the stress-induced current and the complete depolarization of PZT 95/5 films transversely shocked within the FEGs are in agreement with the results obtained with bulk PZT 95/5 ceramic specimens transversely shocked within explosive FEGs (see [7] and Chapters 6 and 7 of this book), and in agreement with the results obtained with bulk ceramic specimens compressed by planar shock waves generated by projectiles accelerated with multi-stage gas gun systems [8–10].

In the FEG, the explosive-generated shock travel was not purely uniaxial (due to the shock wave's slightly hemispherical shape as shown by the CALE simulation in Figure 10.3) within the films. The obtained results indicate that the hemispherical geometry of the stress wave in the films does not have a significant effect on the film depolarization under transverse shock. Apparently, the relatively long shock wave travel distance through the films (4 mm) negated most of the effect of the off-axis vector of the shock travel.

Based on the experimental results the conclusion can be reached that transversely shocked PZT 95/5 films similar to bulk PZT 95/5 ceramic specimens underwent a pressure-induced phase transition from the ferroelectric rhombohedral to a non-polar antiferroelectric orthorhombic phase [7–10]. This transformation was observed in hydrostatic studies [11] where the FE-to-AFE phase transformation occurs abruptly at a pressure of 0.32 GPa in PZT 95/5. Under uniaxial adiabatic compression, the hydrostatic component of the stress tensor induces the FE-to-AFE phase transformation in the PZT 95/5 film that results in its complete depolarization.

The important result is that shock-compressed PZT 95/5 films released a specific electric charge per unit of volume, $10^4 \, \mu C/cm^3$, that is an order of magnitude higher than that for a 0.5-mm-thick bulk PZT 95/5 ceramic specimen, $640 \, \mu C/cm^3$.

10.6 Longitudinal Shock Depolarization of Single-Layer PZT 95/5 Films

Figure 10.6 presents a typical waveform of stress-induced current and dynamics of electric charge released by a single-layer 32-μm-thick PZT 95/5 film under longitudinal stress. The film was loaded across its thickness anti-parallel to the direction of polarization (see position of the film within FEG in Figure 10.2). The longitudinally shocked films had geometrical dimensions identical to those for films studied under transverse shock compression (see previous section).

The direction of stress-induced current produced by longitudinally shocked films (Figure 10.6) was identical to that under transverse shock (Figure 10.5). The charge released by the film was $Q_{LSW} = 8.1 \, \mu C$ (corresponding charge density $32.4 \, \mu C/cm^2$). Similar to transversely shocked films, single-layer PZT 95/5 films were completely depolarized under longitudinal shock compression.

Table 10.1 summarizes experimental results obtained with transversely and longitudinally shocked 32-μm-thick PZT 95/5 films. Based on the experimental results the conclusion can be reached that

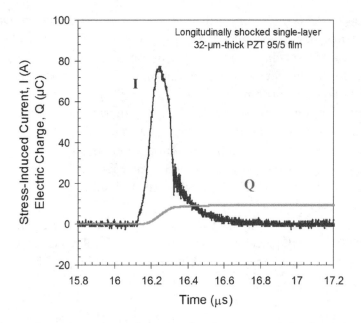

Fig. 10.6.   Typical waveform of stress-induced current and the dynamics of the electric charge released by a single-layer PZT 95/5 film under longitudinal shock compression. The film thickness was $32\,\mu$m and the electrode area was $25\,\mathrm{mm}^2$.

Table 10.1.   Experimental results obtained with transversely and longitudinally shock-compressed single-layer PZT 95/5 films (film thickness 32 μm and electrode area $25\,\mathrm{mm}^2$).

Type of shock loading	Film area (cm^2)	Current (A)	Current pulse FWHM (μs)	Electric charge (μC)	Charge density $(\mu C/\mathrm{cm}^2)$
Transverse shock	0.25	11.2 ± 1.4	1.0 ± 0.04	7.8 ± 0.2	31.1 ± 2.2
Longitudinal shock	0.25	87.1 ± 7.5	0.1 ± 0.01	8.0 ± 0.2	31.9 ± 1.9

transversely and longitudinally shocked single-layer PZT 95/5 films were completely depolarized within FEGs due to the stress-induced FE-to-AFE phase transition and released a charge density equal to their remanent polarization.

10.7   Complex Behavior of Single-Layer PZT 95/5 Films under Longitudinal Shock Loading

The density of stress-induced current (current per unit of the film electrode area) generated by shock-compressed ferroelectric films is an important parameter for ultrahigh-power applications. The stress-induced current density generated by PZT 95/5 film under longitudinal stress, $360\,\mathrm{A/cm^2}$, is almost an order of magnitude higher than that under transverse shock.

Consider factors that may have an effect on the parameters of stress-induced current pulses produced by PZT 95/5 films. Assume that the stress-induced phase transformations and depolarization of PZT 95/5 occur immediately in the shock wave front. In this case, the stress-induced charge dynamics and, correspondingly, the stress-induced current density is directly proportional to the amount of electric charge released by the shocked ferroelectric film and inversely proportional to the shock wave transit time:

$$j_{sw} = \frac{\omega_{sw}}{\tau_{sw}} \tag{10.1}$$

where j_{sw} is the stress-induced current density, ω_{sw} is the stress-induced charge density, and τ_{sw} is the shock wave transit time.

Assuming that the velocity of the shock wave is constant for PZT 95/5 film, the shock wave transit time can be derived from:

$$\tau_{sw} = \frac{L_{sw}}{U_s} \tag{10.2}$$

where L_{sw} is the geometric dimension of the film in the direction of the shock wave propagation (the shock wave transit distance in the film), and U_s is the shock wave velocity.

Experimental results indicate that PZT 95/5 films were completely depolarized under transverse and longitudinal shock compression (see Table 10.1). With Eq. (10.2), Eq. (10.1) can be rewritten in the following form:

$$j_{sw} = \frac{P_0 U_s}{L_{sw}} \tag{10.3}$$

where P_0 is the remanent polarization of a ferroelectric film. In accordance with Eq. (10.3), the current density produced by a shocked ferroelectric film is directly proportional to the remanent polarization of the film and inversely proportional to the shock wave transit distance.

The shortest shock wave transit distance appears to be in the longitudinally shock-compressed films because shock waves propagate across the thickness of the films. In accordance with Eq. (10.3), longitudinally shock-compressed films should be capable of producing the highest density of stress-induced current possible.

Substitute the parameters of a transversely shock-compressed PZT 95/5 film ($P_0 = 32\,\mu C/cm^2 = 0.32\,\mu C/mm^2$, $L_{SW} = 4\,mm$, $U_s = 3.9\,mm/\mu s$) into Eq. (10.2) and Eq. (10.3). The calculated transverse shock wave transit time, $\tau_{TSW} = 1.03\,\mu s$, and the calculated stress-induced current density, $j_{TSW} = 31\,A/cm^2$, are very close to those obtained in the experiments.

However, the substitution of parameters for a longitudinally shock-compressed PZT 95/5 film into Eq. (10.2) and Eq. (10.3) gives us a completely different picture. The calculated current density, $j_{LSW} = 3900\,A/cm^2$, is an order of magnitude higher than that recorded in the experiments with longitudinally shocked PZT 95/5 films, $360\,A/cm^2$. The FWHM of the current pulse produced by longitudinally shock-compressed films, $0.1\,\mu s$ (see Table 10.1), is an order of magnitude longer than the calculated shock wave transit time in the films ($\tau_{LSW} = 0.008\,\mu s$).

Such a significant difference between the calculated and experimentally obtained pulse duration and stress-induced current density generated by longitudinally shocked films cannot be explained by the influence of the measuring circuit. The inductance of the FEG output terminals made as a strip transmission line (see Figure 10.2) was negligible. In two experiments with longitudinally compressed films we monitored stress-induced current with a Pearson 411 probe (10 ns transition time). In the other four experiments we monitored current with a Prodyn I-265 probe (0.2 ns transition time) or with a Pearson 411 and a Prodyn I-265 at the same time. We did not observe a significant difference between waveforms recorded with these probes. The rise time of current generated by longitudinally compressed films, 149 ± 13 ns, was more than an order of magnitude longer than the transition times of the two probes.

A comparison of the amplitude and pulse duration of stress-induced currents generated by transversely shocked single-layer PZT 95/5 films with calculated parameters allows one to reach the conclusion that the transverse shock depolarization of PZT 95/5 films can be described by the model assuming an immediate FE-to-AFE phase transition and depolarization of ferroelectric material in the shock wave front.

Opposite to transversely shocked films, parameters of the current pulses generated by longitudinally shocked single-layer PZT 95/5 films differ significantly from calculated ones. The experimental results indicate that the behavior of longitudinally shock-compressed PZT 95/5 films is more complex than that under transverse shock. This complex behavior of longitudinally shocked PZT 95/5 films is in agreement with the results obtained earlier with bulk PZT 95/5 ceramic specimens [10, 12, 13].

Halpin reported [12, 13] on detailed experimental studies of the depolarization of bulk PZT 95/5 ceramic specimens longitudinally shock-compressed by planar shock waves initiated by projectiles accelerated with gas gun systems. Experiments were conducted with PZT 95/5 ceramic disks having diameters of 12.7 mm and thicknesses of 0.76 mm subjected to longitudinal stress ranging from 0.4 to 3.3 GPa. Over the full range of stress investigated it was found that there was a wide variation in the waveforms of the stress-induced current (current peak magnitude and position, pulse duration) and the amount of stress-induced charge released by bulk ceramic specimens, which suggests the very complicated behavior of the longitudinally shocked PZT 95/5. The observation was made that the duration of stress-induced current pulses was greater than the stress wave transit time. It was shown that longitudinally shocked PZT 95/5 specimens were completely depolarized at stress levels ranging from 1.0 to 2.5 GPa. At stress levels below 1.0 GPa and above 2.5 GPa a fraction of the initial remanent polarization was retained.

Based on the obtained results Halpin drew the conclusion that the complete depolarization of PZT 95/5 under longitudinal stress is the result of FE-to-AFE phase transformation [12]. It was also concluded in [12] that the complex behavior of longitudinally shocked PZT 95/5 specimens reflects the presence of stress wave front tilt which influences the overall character of the current pulse. An unavoidable 0.1° tilt of the stress wave front in the ferroelectric specimens occurred as a result of angular misalignment of the accelerated projectile and targeted impacting surfaces in the gas gun systems [12].

Setchell reported in [10] on experimental studies of depolarization of PZT 95/5 ceramic disks (diameter 25.4 mm and thickness 4.0 mm) under longitudinal stress ranging from 0.9 to 2.4 GPa. The experiments were conducted on a 63.5-mm diameter, compressed-gas gun capable of achieving impact velocities from 0.03 to 1.3 km/s. Similar to the results obtained by Halpin [12], a wide variation in the stress-induced current magnitudes, pulse durations, and stress-induced electric charges were observed over the full range of stress [10]. It was shown that the PZT 95/5 specimens were completely depolarized at 0.9 GPa stress. Progressively smaller electric charge was recorded as stress was increased from 0.9 to 2.4 GPa. The magnitude of stress resulted in the complete depolarization of PZT 95/5 reported by Setchell [10] which is twice lower than that reported by Halpin [13]. The only difference in these two studies was the PZT 95/5 specimen thickness (0.76 and 4.0 mm) and diameter (12.7 and 25.4 mm). Based on these results [10, 12] one may conclude that the magnitude of longitudinal stress resulting in complete depolarization of PZT 95/5 depends on the geometric dimensions of the ferroelectric specimens.

To explain the complex behavior of longitudinally stressed PZT 95/5 specimens, Halpin analyzed current waveforms using a model that took into consideration the tilt of the stress wave front in the ferroelectric specimens [12, 13]. The ferroelectric disk specimen was divided into a few parts that were subjected to different kinds of stress and, correspondingly, possessed different physical and electrical properties. In accordance with this model, strong opposing electric fields were generated on either side of the shock wave during shock wave transit. It was assumed that a finite resistivity was induced in the shock-compressed zone of ferroelectric material. The waveforms of stress-induced current calculated with this model were very similar to those obtained in the experiments [13]. A detailed description of the model and the results of the analysis can be found in [12, 13].

The magnitude difference of more than an order between the experimental and calculated amplitude of stress-induced current produced by PZT 95/5 films under longitudinal stress in our experiments is not in agreement with the results [12] obtained with bulk PZT 95/5 ceramic specimens. The explosive-generated shock wave geometry in the films was not perfectly planar (see Figure 10.2); experimental results indicate that, likely due to the relatively long shock travel distance, the lack of planarity did not have a significant effect on the depolarization and generation of current by PZT 95/5 films under transverse stress. The lack of planarity, however, could have an effect on the depolarization and generation of current by PZT 95/5 films under longitudinal stress because of the short stress wave travel distance through the film that was comparable with the thickness of the stress wave front. Due to the slightly hemispherical stress wave geometry, the film can be divided into parts that were subjected to different stress wave directions and, as such, possessed different physical and electrical properties during stress wave transit. In turn, shock passage can cause the generation of opposing electric fields in the compressed and uncompressed zone of the film, inducing a finite resistivity in the compressed zone. These factors may result in the complex behavior of the film under longitudinal stress and the distortion of the stress-induced current waveform.

10.8 Transverse Shock Depolarization of Single-Layer PZT 95/5 Films with Different Thicknesses

The film thickness is one of the important parameters that have an effect on the mechanical and electrical properties of the film (film capacitance, voltage generated by films under stress, breakdown field, etc.). Experimental investigations were performed on transverse shock depolarization of PZT 95/5 films with thicknesses 32, 64 and 128 μm. The films were polarized across the thickness to their remanent polarization, 32 μC/cm^2. The electrode area (4×6.3 mm^2) and size of the encapsulating ceramic body ($10 \times 8 \times 4$ mm^3) were identical for films of all three types. The schematics of the transverse shock wave ferroelectric generator and measuring circuit used in these experiments are shown in Figure 10.1.

Typical waveforms of the current and the dynamics of the electric charge released by 64-μm-thick and 128-μm-thick PZT 95/5 films under transverse shock loading are shown in Figures 10.7 and 10.8. The waveforms of the current generated by the films were similar to those produced by 32-μm-thick PZT 95/5 films under transverse shock (see Figure 10.5). The amplitudes of the current generated by the films (Figures 10.7 and 10.8) were $I_{TSW} = 10.2$ A and 11.1 A,

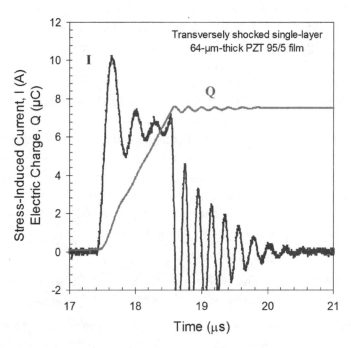

Fig. 10.7. Typical waveform of the current and the dynamics of the electric charge released by 64-μm-thick PZT 95/5 film under transverse stress. The film area was 25 mm^2.

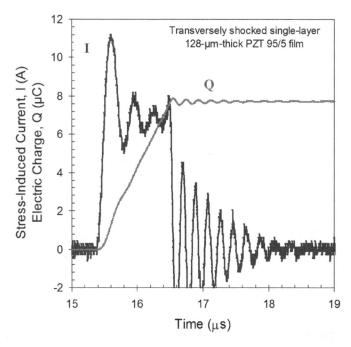

Fig. 10.8. Typical waveform of the current and the dynamics of the electric charge released by 128-μm-thick PZT 95/5 film under transverse stress. The film area was 25 mm^2.

respectively. The current pulse duration was 1.15 μs for films of both types.

The films with thickness 64 and 128 μm were constructed by stacking 32-μm-thick sheets (two and four 32 μm sheets per layer, respectively). Experimental results indicate that the design approach based on stacking ferroelectric tape can be successfully used to fabricate single-layer films with different thicknesses.

The results obtained with single-layer PZT 95/5 films having different thicknesses are summarized in Table 10.2. The stress-induced charges released by transversely shock-compressed 64-μm-thick and 128-μm-thick PZT 95/5 films were $Q_{TSW} = 7.7$ and 7.9 μC (corresponding charge density 30.8 and 31.6 μC/cm^2, respectively).

The films of both types were almost completely depolarized under transverse stress in a manner similar to the 32-μm-thick

Table 10.2.   Experimental results obtained with transversely shock-compressed
PZT 95/5 single-layer films with different thicknesses.

Film thickness (μm)	Film area (cm^2)	Current (A)	Current pulse duration (μs)	Electric charge (μC)	Charge density (μC/cm^2)
32	0.25	11.2 ± 1.4	1.0 ± 0.1	7.8 ± 0.2	31.1 ± 2.2
64	0.25	10.9 ± 1.4	1.1 ± 0.1	7.7 ± 0.2	30.8 ± 2.4
128	0.25	11.3 ± 1.4	1.1 ± 0.1	7.9 ± 0.2	31.6 ± 1.9

films (Table 10.2). The current pulse duration did not depend on the film thickness and was equal to the stress wave transit time. Based on these results, one may conclude that the thickness of ferroelectric films does not have a significant effect on the stress-induced depolarization and generation of current under transverse shock compression.

10.9   Transverse Shock Depolarization of Multi-Layer PZT 95/5 Film Modules

The amount of stress-induced electric charge released by ferro-electrics under shock compression determines the energy that can be delivered by ferroelectric generators into the load. The multi-layer ferroelectric film modules with layers electrically connected in parallel could provide the multiplication of stress-induced charges released by the films. However, this approach raises a question about possible mechanical or electrical interference between adjacent film layers during shock wave transit that could affect shock depolarization and the generation of stress-induced current and voltage.

Prior to the preparation and investigation of multi-layer PZT 95/5 film modules with layers connected in parallel, the multi-element approach was experimentally verified with specially designed and constructed multi-element PZT 95/5 ceramic modules containing from 2 to 8 bulk ceramic elements. The modules contained PZT 95/5 ceramic plates (each 25-mm long × 19-mm wide × 2-mm thick) electrically connected in parallel. Experimental results indicate that multi-element ceramic modules were completely depolarized under transverse stress. There was no indication of mechanical or electrical interference between the elements of the modules during depolarization. The results obtained with multi-element PZT 95/5 ceramic modules are described in detail in Chapter 7.

After successful experiments with multi-element PZT 95/5 ceramic modules, 10-layer and 20-layer PZT 95/5 film modules (layers connected in parallel) were prepared and investigated. The schematics of transverse and longitudinal shock wave ferroelectric generators used in these investigations are shown in Figures 10.1 and 10.2. The layer thickness in the multi-layer film modules was 64 μm. After construction, the multi-layer films were polarized across the thickness dimensions to their remanent polarization $P_0 = 32 \, \mu$C/cm^2. The layer area (4×6.3 mm^2) and the size of the encapsulating ceramic body ($10 \times 8 \times 4$ mm^3) of multi-layer film specimens were identical to those for single-layer films described above.

Typical waveforms of the stress-induced current and electric charge released by 10-layer and 20-layer PZT 95/5 film modules under transverse stress are shown in Figure 10.9. The amplitudes of the currents generated by 10-layer and 20-layer film modules were 116 and 244 A, respectively. The experimental results indicate that the amplitude of stress-induced current was directly proportional to the number of layers. The current pulse durations, $\tau_p = 1.1 \, \mu$s, did not change with changing the number of layers and was practically equal to the shock wave transit time for both types of films.

Comparing the waveforms of current produced by transversely compressed multi-layer film modules (Figure 10.9) and single-layer films (Figures 10.5, 10.7 and 10.8), one can see that they are not similar. The waveforms of current generated by multi-layer modules were not quasi-rectangular like those for single-layer films. Apparently, this difference was caused by electrical interference between ferroelectric layers of multi-layer film modules during stress wave transit and depolarization. We did not observe this effect in the experiments with multi-layer PZT 95/5 bulk ceramic modules having the ferroelectric layer thickness of 2 mm (see Chapter 7 of this book). The waveforms of current produced by multi-layer ceramic modules

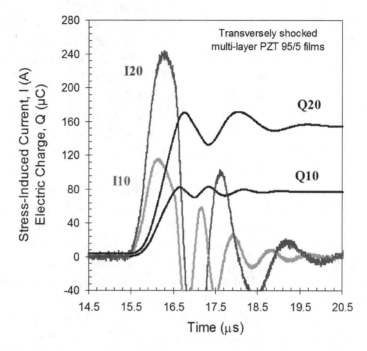

Fig. 10.9. Typical waveforms of the current and the dynamics of the electric charge released by 10-layer (I10 and Q10) and 20-layer (I20 and Q20) PZT 95/5 film modules (layers connected in parallel) under transverse stress.

were very similar to those for single-layer ceramic elements (see Figure 7.8). The only difference was the amplitude of current that was directly proportional to the number of ceramic layers. Probably, the interference between ferroelectric film layers appears when the layer thickness is less than a few hundreds of micrometers.

Despite the difference between waveforms of current generated by multi-layer film modules and single-layer films, the stress-induced electric charge released by 10-layer and 20-layer film modules was $Q_{TSW} = 80$ and $161\,\mu C$ respectively, with corresponding charge density 32.0 and $32.2\,\mu C/cm^2$. One may conclude that in a similar manner to single-layer films, the multi-layer PZT 95/5 film modules underwent the FE-to-AFE phase transition and became completely depolarized under transverse stress. The results obtained with 10-layer and 20-layer film modules indicate no mechanical or electrical interference between layers in the multi-layer film modules during stress wave transit and depolarization.

10.10   Longitudinal Shock Depolarization of Multi-Layer PZT 95/5 Film Modules

The 20-layer film modules were investigated in the longitudinal shock compression mode. Typical waveforms of the current and electric charge released by a 20-layer PZT 95/5 film module under longitudinal stress are shown in Figure 10.10. Opposite to the transverse mode where all the layers of multi-layer film modules were simultaneously subjected to shock compression, in the longitudinal mode the shock wave consequently passed layer after layer separated by inter-layer insulation. However, the waveform of the current produced by a longitudinally shock-compressed 20-layer film module was not a series of pulses, but a single pulse with no breaks

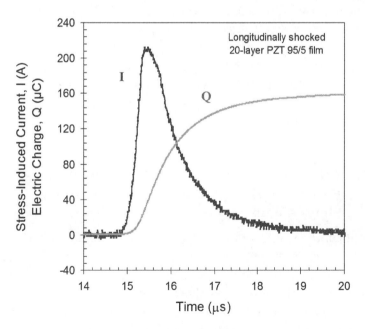

Fig. 10.10.   Typical waveform of the stress-induced current and the dynamics of electric charge released by a 20-layer PZT 95/5 film module (layers connected in parallel) under longitudinal stress.

or irregularities. The current waveform generated by the multi-layer film module (Figure 10.10) was similar to that obtained with longitudinally shocked single-layer film (see Figure 10.6).

The current generated by longitudinally shocked 20-layer film modules was significantly lower than the estimations based on the results obtained with longitudinally shocked single-layer films (see Table 10.1). Single-layer films produced current pulses with amplitude $I_{LSW} = 87$ A (Table 10.1), which was higher by a factor of eight than the current pulses produced by single-layer films under transverse stress. The current produced by longitudinally shocked 20-layer film modules, $I_{LSW} = 205 \pm 9$ A, was not higher, but lower than the current generated by identical 20-layer film modules under transverse shock compression, $I_{TSW} = 269 \pm 8$ A.

The stress-induced charge generated by longitudinally shocked 20-layer film modules was $Q_{LSW} = 142 \pm 17 \,\mu$C. The corresponding charge density, $28.4 \pm 0.3 \,\mu$C/cm^2, was about 15% lower than the remanent polarization. Therefore, 20-layer PZT 95/5 film modules were completely depolarized under longitudinal shock compression.

Low amplitudes of stress-induced current and incomplete depolarization of 20-layer PZT 95/5 film modules indicate that the behavior of longitudinally stressed multi-layer film modules was even more complicated than that of the single-layer films described in Section 10.5.2 above.

10.11 Ultrahigh-Current Generation by Multi-Layer PZT 95/5 Film Modules under Transverse Shock Compression

To confirm the results obtained with 10-layer and 20-layer PZT 95/5 films, a series of 40-, 60- and 80-layer films (layer thickness $32 \,\mu$m) with the layers connected in parallel was designed, constructed and investigated in the transverse shock depolarization mode. The layer area $(4 \times 6.3 \,\text{mm}^2)$ and the size of the encapsulating ceramic body $(10 \times 8 \times 4 \,\text{mm}^3)$ of the multi-layer film specimens were identical to those for single-layer films, and for 10-layer and 20-layer film modules described above.

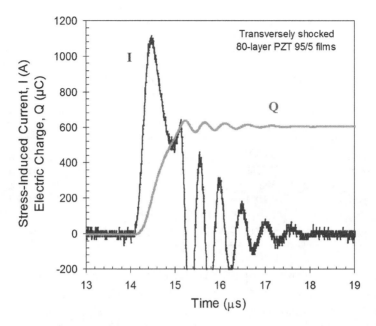

Fig. 10.11. Typical waveform of the current and the dynamics of the electric charge released by 80-layer PZT 95/5 film module (layers connected in parallel) under transverse stress.

A typical waveform of the stress-induced current and the dynamics of electric charge released by a transversely stressed 80-layer film module is shown in Figure 10.11. The stress-induced charge was $Q_{TSW} = 632\,\mu C$ with corresponding charge density $31.5\,\mu C/cm^2$. The 80-layer film was completely depolarized under transverse shock compression. The amplitude of the current was $I_{TSW} = 1.1\,kA$ and the pulse duration was $1.1\,\mu s$.

The experimental results obtained with transversely and longitudinally shock-compressed multi-layer PZT 95/5 film modules are summarized in Table 10.3. The 10- to 80-layer film modules were completely depolarized under transverse stress due to the FE-to-AFE phase transition, and they produced current pulses with amplitudes that were directly proportional to the number of layers.

The current pulse duration, however, did not depend on the number of layers and was equal to the stress wave transit time.

Table 10.3. Experimental results obtained with multi-layer PZT 95/5 film modules subjected to transverse and longitudinal shock loading.

Number of layers	Effective area (cm^2)	Current amplitude (A)	Current pulse duration (μs)	Electric charge (μC)	Charge density ($\mu C/cm^2$)
			Transverse stress		
10 layers	2.5	131 ± 3	1.1 ± 0.1	83.5 ± 3.2	33.4 ± 1.1
20 layers	5	269 ± 8	1.1 ± 0.1	170.2 ± 5.1	34.1 ± 1.2
40 layers	10	541 ± 12	1.1 ± 0.1	343.4 ± 9.2	34.3 ± 1.2
80 layers	20	1091 ± 34	1.1 ± 0.1	691 ± 11	34.5 ± 1.2
			Longitudinal stress		
20 layers	5	213 ± 7	2.2 ± 0.2	171.4 ± 6.2	34.2 ± 1.2

The obtained results indicate no mechanical or electrical interference between film layers in multi-layer film modules during stress wave transit and depolarization.

Figure 10.12 shows the experimentally obtained dependence of amplitudes of stress-induced current on the electric charge produced by multi-layer PZT 95/5 films under transverse stress. The current amplitude is directly proportional to the amount of electric charge released by multi-layer film modules. These experimental results make it possible to estimate the amplitude of the stress-induced current, I_{TSW}, for a given amount of electric charge released by PZT 95/5 films:

$$I_{TSW} = \alpha \cdot Q_{TSW}, = \alpha \cdot P_0 \cdot A, \qquad (10.4)$$

where α is the coefficient of proportionality ($1.6 \cdot 10^6$ A/C) determined from the current versus charge plot (Figure 10.12), Q_{TSW} is the electric charge released by films under transverse stress, P_0 is the remanent polarization of the films, and A is the total area of the films in the module.

In accordance with Eq. (10.4), it can be expected that a multi-layer PZT 95/5 film module with the total film electrode area of $195\,cm^2$ and the total volume of $1.25\,cm^3$ (film layer thickness $64\,\mu m$) can produce a $10\,kA$ current pulse. The polarization charge stored in this module is $6250\,\mu C$ (remanent polarization $P_0 = 32\,\mu C/cm^2$).

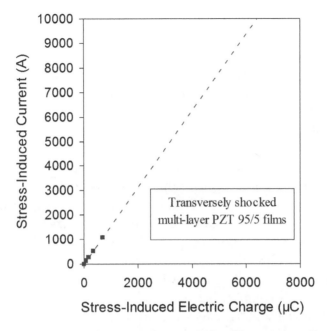

Fig. 10.12. Experimentally obtained stress-induced current amplitudes as a function of electric charge released by PZT 95/5 multi-layer film modules under transverse stress (diamonds) and an extrapolated curve.

10.12 Summary

- The shock depolarization of single-layer and multi-layer PZT 95/5 ferroelectric films has been investigated experimentally.

- It was found that single-layer PZT 95/5 films become completely depolarized under 2.4 GPa transverse stress and release electric charge with density equal to their remanent polarization.

- The important result is that the specific electric charge released by shock-compressed PZT 95/5 films, $10^4 \, \mu C/cm^3$, is an order of magnitude higher than that for PZT 95/5 bulk ceramic specimens.

- The experimental results indicate that the thickness of single-layer PZT 95/5 films, lying in a range from 32 to 128 μm, did not have a significant effect on either the transverse shock depolarization of those films or on the amplitude of stress-induced current. The duration of current pulses produced under transverse stress did not depend on the film thickness, but was equal to the stress wave transit time. Therefore, the film thickness can be used as a variable parameter to fit the physical and electrical properties of the films to certain ultrahigh-power applications.

- It was found that 32-μm-thick single-layer PZT 95/5 films become completely depolarized under 2.4 GPa longitudinal shock compression and generate stress-induced current with amplitude almost an order of magnitude higher than that generated by identical films under transverse stress.

- Based on the obtained results the conclusion was reached that under transverse and longitudinal shock loading the PZT 95/5 films underwent a pressure-induced ferroelectric-to-antiferroelectric phase transition and became completely depolarized.

- The experimental results indicated that the behavior of single-layer PZT 95/5 films under longitudinal shock compression is complicated in comparison with that under transverse shock loading. This complex behavior can be caused by an appearance of electric fields in the shock-compressed and uncompressed zones of the films during longitudinal shock wave transit and depolarization.

- It was experimentally demonstrated that multi-layer (10 to 80 layers) PZT 95/5 film modules with layers electrically connected in parallel become completely depolarized under transverse shock compression within miniature ferroelectric generators. At the same time, waveforms of current generated by multi-layer film modules were not similar to those for single-layer films. Apparently, this difference was caused by electrical interference between ferroelectric layers of multi-layer film modules during stress wave transit and depolarization.

- The behavior of multi-layer PZT 95/5 film modules under longitudinal shock compression was even more complicated than that for longitudinally shocked single-layer films. Multi-layer PZT 95/5 film modules were not completely depolarized under longitudinal shock loading and retained a fraction of their initial remanent polarization.

- The transversely shock-compressed 80-layer PZT 95/5 film modules were capable of producing kiloampere current pulses. The amplitude of current generated by shock-compressed multi-layer PZT 95/5 film modules was directly proportional to the number of layers connected in parallel. The duration of current pulses produced under transverse stress did not depend on the number of layers of multi-layer film modules and was equal to the shock wave transit time. The multi-layer film module approach can be used for the generation of high currents by miniature FEGs.

- The established relationship between the amplitude of stress-induced current and the electric charge released by multi-layer PZT 95/5 film modules under transverse stress makes it possible to predict the current pulse amplitude generated by multi-layer film modules up to tens of kiloamperes.

- The results of the study of the shock depolarization of ferroelectric films contribute to knowledge of the properties of ferroelectric materials and promise new ultrahigh-power applications for ferroelectric films.

Bibliography

1. D. Damjanovic, Ferroelectric, dielectric and piezoelectric properties of thin films and ceramics, *Rep. Prog. Phys.* **61**(9) (1998) pp. 1267–1410.

2. J.F. Scott, *Ferroelectric memories*, Vol. 3 of the Springer series on "Advanced Microelectronics" (Springer, Heidelberg, April 2000).

3. P. Tran-Huu-Hue, F. Levassort, F.V. Meulen, J. Holc, M. Kosec and M. Lethiecq, Preparation and electromechanical properties of PZT/PGO thick films on alumina substrates, *J. Eur. Ceram. Soc.* **21** (2001) pp. 1445–1449.

4. N. Setter and D. Damjanovic, Ferroelectric thin films: Review of materials, properties and applications, *J. Appl. Phys.* **100** (2006) p. 051606.

5. http://trstechnologies.com/

6. R.E. Tipton, A 2D Lagrange MHD code, in *Proc.4th Int. Conf. Megagauss Magnetic Field Generation and Related Topics* (Editors: C.M. Fowler, R.S. Caird and D.J. Erickson) (Plenum Press, New York, U.S.A. 1987) pp. 299–302.

7. S.I. Shkuratov, J. Baird, V.G. Antipov, E.F. Talantsev, H.R. Jo, J.C. Valadez and C.S. Lynch, Depolarization mechanisms of $PbZr_{0.52}Ti_{0.48}O_3$ and $PbZr_{0.95}Ti_{0.05}O_3$ poled ferroelectrics under high strain rate loading, *Appl. Phys. Lett.* **104** (2014) p. 212901.

8. P.C. Lysne, Kinetic effects in the electrical response of a shock-compressed ferroelectric ceramics, *J. Appl. Phys.* **46**(11) (1975) pp. 4078–4079.

9. P.C. Lysne and C.M. Percival, Electric energy generation by shock compression of ferroelectric ceramics: Normal-mode response of PZT 95/5, *J. Appl. Phys.* **46**(4) (1975) pp. 1519–1525.

10. R.E. Setchell, Shock wave compression of the ferroelectric ceramic Pb0.99(Zr0.95Ti0.05)0.98Nb0.02O3: Depoling currents, *J. Appl. Phys.* **97** (2005) p. 013507.

11. I.J. Fritz, Uniaxial stress effects in a 95/5 lead zirconate titanate ceramic, *J. Appl. Phys.* **49**(9) (1978) pp. 4922–4928.

12. W.J. Halpin, Current from a shock-loaded short-circuited ferroelectric ceramic disk, *J. Appl. Phys.* **37**(1) (1966) pp. 153–163.

13. W.J. Halpin, Resistivity estimates for some shocked ferroelelectrics, *J. Appl. Phys.* **39**(8) (1968) pp. 3821–3826.

Chapter 11

Ultrahigh Energy Density Harvested from Shock-Compressed Domain-Engineered Relaxor Ferroelectric Single Crystals

11.1 Introduction

Lead zirconate titanate solid solutions are widely used in modern low-power ferroelectric transducers, sensors, and actuators and in ultrahigh-power ferroelectric generators. However, imperfections of crystal structure, voids and inclusions, and the uncontrollable crystallographic orientation of grains of PZT polycrystalline ceramics create problems for some engineering applications. Attempts to grow large-size PZT single crystals have not been successful [1].

Extensive studies of alternative MPB systems other than PZT led to the discovery of a new class of ferroelectric materials, so-called relaxor ferroelectrics [2–8]. These ferroelectric materials, such as $(1\text{-}x)Pb(Mg_{1/3}Nb_{2/3})O_3\text{-}(x)PbTiO_3$ (PMN-PT) and $(1\text{-}y\text{-}x)Pb$ $(In_{1/2}Nb_{1/2})O_3\text{-}(y)Pb(Mg_{1/3}Nb_{2/3})O_3\text{-}(x)PbTiO_3$ (PIN-PMN-PT) received special attention because of their intriguing and extraordinary dielectric, ferroelectric and piezoelectric properties [2–8].

The successful growth of large-size relaxor single crystals led to a breakthrough in ferroelectric research [2–9]. The ferroelectric properties of single crystals can be tailored through domain engineering (creating domains using a specific geometric structure relative to crystallographic orientation, creating dipole alignment that results in optimal properties).

271

There is significant interest in expanding the usage of ferroelectric single crystals to high-power and ultrahigh-power systems, such as high-intensity focused ultrasound therapy [10], resonance based transducers [10], and explosive ferroelectric generators [11]. High-power systems require power that is from four to six orders of magnitude higher than that associated with low-power transducers.

It was reported [12] that an increase in the applied stress of up to a few tens of megapascals resulted in the harvesting of $0.75\,\mathrm{kJ/m^3}$ of energy density from PIN-PMN-PT crystals; that is, significantly higher than that produced by other ferroelectrics. A study of ferroelectric crystals under gigapascal mechanical stress provides information about the ultimate energy density that can be harvested from these materials. This information can also be used to identify possible limitations on operation in the ultrahigh-power mode.

In this chapter, the results are presented on experimental investigations of energy harvested from domain-engineered PIN-PMN-PT single crystals under high strain rate loading. The obtained results are directly compared to those for PZT 52/48 and PZT 95/5 ferroelectrics. It is shown that rhombohedral $[111]_\mathrm{C}$ cut and poled PIN-PMN-PT crystals possessing the highest possible remanent polarization become completely depolarized under transverse shock compression and release electric charge density significantly higher than that for PZT 52/48 and 95/5. The remarkable observation is that the energy density generated by single crystals in the high-voltage mode, $0.3\,\mathrm{MJ/m^3}$, is four times higher than that for PZT 52/48 and 95/5 [13]. These results provide the basis to the successful development of a new class of ferroelectric materials for ultrahigh-power ferroelectric generators and demonstrate a unique ability to control the ferroelectric properties of single crystals through the size and crystallographic orientation of domains to fit certain applications.

11.2 Preparation of PIN-PMN-PT Single Crystal Specimens

Lead indium niobate-lead magnesium niobate-lead titanate (25-29%) $\mathrm{Pb(In_{1/2}\,Nb_{1/2})\,O_3}$-(47-35%) $\mathrm{Pb(Mg_{1/3}\,Nb_{2/3})\,O_3}$-(28-36\%) $\mathrm{PbTiO_3}$ single crystal specimens were prepared by TRS Technologies [14].

Single crystals were grown using the modified Bridgman technique [4, 15, 16]. Ternary PIN-PMN-PT single crystals at room temperature can be in either the ferroelectric rhombohedral phase or the ferroelectric tetragonal phase. These two phases are separated by a morphotropic phase boundary that can contain a monoclinic phase. When the composition is close to the MPB the ferroelectric properties of the crystals are enhanced. One of the goals of this work was to identify PIN-PMN-PT single crystals with the highest possible remanent polarization. To reach this goal the composition of the crystals, crystallographic orientation and poling conditions were varied to create domains with a specific geometric structure. The as-grown PIN-PMN-PT crystals exhibit different phases along the growth direction due to the segregation of Ti. The studied specimens were selected to be in the ferroelectric rhombohedral phase and compositionally in very close proximity to the MPB, $0.27Pb(In_{1/2}Nb_{1/2})O_3$–$0.47Pb(Mg_{1/3}Nb_{2/3})O_3$–$0.3PbTiO_3$. All specimens were oriented using a real-time Laue X-ray orientation system with an accuracy of $0.5°$.

The effective macroscopic symmetry associated with the domain structure is a key issue for the domain engineering of single crystals. The rhombohedral phase has eight possible spontaneous polarization directions. The studied PIN-PMN-PT single crystals were cut and poled along the $[111]_C$ crystallographic direction. For rhombohedral crystals, poling along the $[111]_C$ direction induces $3\,m$ symmetry of the single-domain state.

The size of the PIN-PMN-PT single crystal specimens was 5.0-mm long × 5.0-mm wide × 5.0-mm thick. Vacuum sputtered Cr/Au films were deposited on the desired surfaces as electrodes. The specimens were cut and poled along the $[111]_C$ direction ($3\,m$ symmetry) at a DC electric field of $7.5\,kV/cm$ at $50°C$. The complete set of material constants was determined by combined resonance and ultrasonic methods. Experimental details are described elsewhere [16].

Figure 11.1 shows a typical polarization hysteresis loop for a $[111]_C$-oriented PIN-PMN-PT crystal with $3\,m$ symmetry measured at $1\,Hz$. The remanent polarization of PIN-PMN-PT crystals was found to be $48\,\mu C/cm^2$. It is significantly higher (from 30 to 90%) than that for rhombohedral and tetragonal PIN-PMN-PT crystals

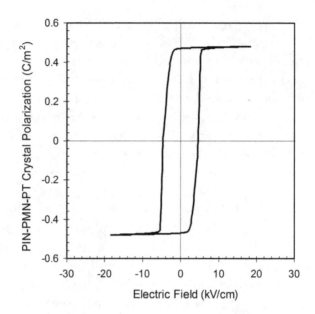

Fig. 11.1.   Polarization hysteresis loops of PIN-PMN-PT single crystal cut and poled along the $[111]_C$ crystallographic direction measured at an AC electric field of 15 kV/cm at 1 Hz.

cut and poled along $[001]_C$ and $[011]_C$ crystallographic directions [17]. The higher remanent polarization of $[111]_C$ cut and poled PIN-PMN-PT crystals is attributed to the single-domain state and polarization vector being aligned normal to the electrodes in the crystals. The physical properties of PIN-PMN-PT crystals cut and poled along the $[111]_C$ direction are listed in Table 11.1. The tolerance of all parameters does not exceed ±5%.

11.3   High Strain Rate Loading Techniques

PIN-PMN-PT single crystals were studied in the short-circuit depolarization mode and the high-voltage depolarization mode. Schematics of the transverse FEG with single crystal specimen and the measuring circuit used in the short-circuit depolarization experiments are shown in Figure 11.2.

   Ferroelectric generators based on spherically expanding shock waves were used in these studies (see Chapter 5 of this book

Table 11.1. Physical properties of rhombohedral PIN-PMN-PT single crystals cut and poled along the [111]$_C$ crystallographic direction, and properties of PZT 52/48 and PZT 95/5 ceramic specimens.

Property	PIN-PMN-PT	PZT 52/48	PZT 95/5
Density (10^3 kg/m^3)	8.1	7.5	7.7
Curie point (°C)	167	320	230
Dielectric constant at 1 kHz (poled)	1180	1300	295
Dielectric constant at 1 kHz (depoled)	940	1140	225
Piezoelectric constant d$_{33}$ (10^{-12} m/V)	1800	295	68
Elastic constant s$_{11}^E$ (10^{-12} m^2/N)	16.4	12.8	7.7
Remanent polarization (μC/cm^2)	48	29	32

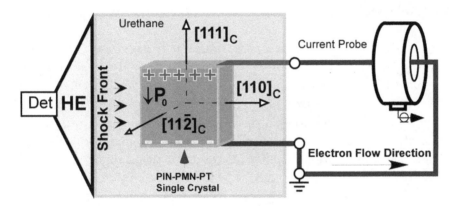

Fig. 11.2. Schematics of the transverse ferroelectric generator and measuring circuit used in the short-circuit shock depolarization studies of PN-PMN-PT single crystals. P_0 is the remanent polarization vector. The polarity of the surface charge is shown by (+) and (−) signs.

for details). They contain two parts, a detonation chamber and a PIN-PMN-PT single crystal encapsulated within a plastic body. After initiation of the detonator, the detonation wave propagates through HE. The HE detonation shock propagates through a ure-thane potting material and into the single crystal specimen.

The ferroelectric generator was placed in the blast chamber. The electrodes of the PIN-PMN-PT crystal were short-circuited. The stress-induced current was monitored with a Pearson 411 current probe placed outside the blast chamber. Additional experimental details are described elsewhere [18–20].

Schematics of the transverse FEG with single crystal specimen and the measuring circuit used in the high-voltage depolarization experiments are shown in Figure 11.3. The design of the FEG was identical to that used in the short-circuit depolarization experiments (Figure 11.2). The electrodes of the PIN-PMN-PT crystal were connected to a Tektronix P6015A high-voltage probe (resistance 100 MΩ, capacitance 3 pF, transition time 4 ns) that was placed outside the blast chamber. It was practically an open circuit operation of the FEG. The stress-induced charge was not transferred from the FEG into the external load, but was utilized for charging the single crystal specimen itself. Additional experimental details are described elsewhere [21–24].

To determine the uniaxial stress amplitude and stress distribution in the ferroelectric crystals, a simulation of high strain rate loading using the CALE computer code was performed. The CALE is a two-dimensional Arbitrary Lagrange/Eulerian second-order accurate hydrodynamics program [25]. The simulation was run in cylindrical coordinates, assuming rotational symmetry around the axis and zero material rotation. The CALE results indicate that the uniaxial

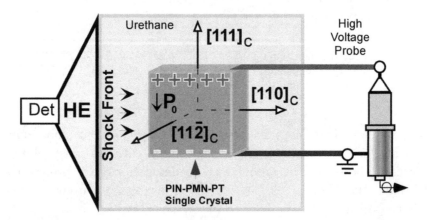

Fig. 11.3. Schematics of the transverse ferroelectric generator and measuring circuit used in the high-voltage shock depolarization studies of PN-PMN-PT single crystals.

stress in the middle cross-section of the single crystal specimen was 3.9 ± 0.1 GPa.

11.4 Shock Depolarization of Domain-Engineered PIN-PMN-PT Single Crystals

The shock depolarization results obtained with PIN-PMN-PT crystal specimens were compared to those for the PZT 52/48 and PZT 95/5 ferroelectric ceramics. Polycrystalline PZT 52/48 and PZT 95/5 ferroelectrics were chosen for direct comparison with PIN-PMN-PT crystals in this study because they are widely used in explosive ferroelectric generators. The ceramic specimens were poled along the thickness by the manufacturers to their remanent polarization $29\,\mu C/cm^2$ (PZT 52/48) and $32\,\mu C/cm^2$ (PZT 95/5). The sizes and geometric shapes of the specimens affect the stress-induced behavior [18]. The specimen geometry of PZT 52/48 and PZT 95/5 was identical to that for PIN-PMN-PT crystals, 5.0-mm long $\times$ 5.0-mm wide $\times$ 5.0-mm thick, to enable a direct comparison of experimental results obtained from single crystals and ceramic materials. The properties of PZT 52/48 and PZT 95/5 ceramics are listed in Table 11.1.

When the electrodes of the ferroelectric crystals were short-circuited (see diagram in Figure 11.1), the high strain rate loading produced current versus time profiles. The stress-induced electric charge is the time integral of the stress-induced current.

Figures 11.4 through 11.6 show typical waveforms of the current and dynamics of electric charge for the single crystal and ceramic specimens under transverse shock compression. The direction of the stress-induced current flow was identical for PIN-PMN-PT crystals and PZT 52/48 and PZT 95/5 ceramic specimens.

The $[110]_C$ transverse direction for rhombohedral PIN-PMN-PT $[111]_C$-oriented crystals has a negative d_{31} piezoelectric coefficient. Under transverse shock compression the linear piezoelectric effect should increase the material polarization, but this would cause a current to flow in the direction opposite to what was observed (Figure 11.4). The direction of stress-induced current flow indicates

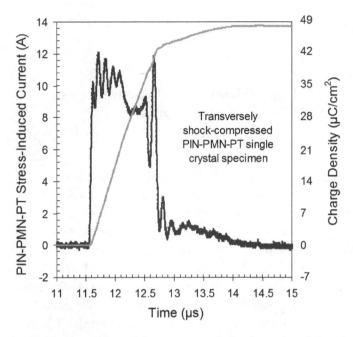

Fig. 11.4. Typical waveform of the current and the dynamics of electric charge released by PIN-PMN-PT single crystal specimen under transverse shock compression in the short circuit depolarization mode. Specimen size was $5 \times 5 \times 5 \, mm^3$.

that the charge released by PIN-PMN-PT crystals under high strain rate loading was not caused by the piezoelectric effect.

Before high strain rate loading, the initial remanent polarization of the ferroelectrics is balanced by the surface charge density. When the initial remanent polarization is decreased or completely lost due to the stress-induced domain reorientation, the surface charge is released at the electrodes of the specimens. The polarity of the stress-induced charge released by PIN-PMN-PT crystals and PZT 52/48 and 95/5 ceramics under transverse stress (Figure 11.4 to 11.6) was identical to the polarity of the surface charge that balanced the remanent polarization. These results indicate that, similarly to PZT 52/48 and PZT 95/5 ceramics, the PIN-PMN-PT crystals were depolarized under transverse shock compression.

The experimental results for stress-induced charge density for PIN-PMN-PT crystals and for PZT 52/48 and PZT 95/5 ceramics

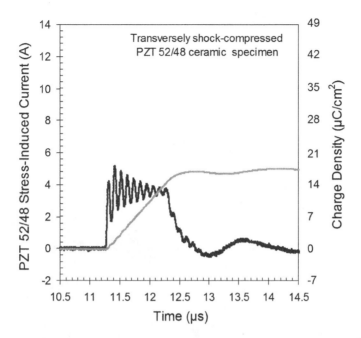

Fig. 11.5. Typical waveform of current and dynamics of electric charge released by PZT 52/48 ceramic specimen under transverse shock compression in the short circuit depolarization mode. Specimen size was $5 \times 5 \times 5 \, \mathrm{mm}^3$.

are summarized in Table 11.2. The stress-induced charge density released by PIN-PMN-PT crystals, $\omega_{PIN} = 48 \, \mu\mathrm{C/cm}^2$, is significantly higher than that for PZT 52/48 and PZT 95/5 ceramic specimens ($\omega_{52/48} = 15 \, \mu\mathrm{C/cm}^2$ and $\omega_{95/5} = 32 \, \mu\mathrm{C/cm}^2$, respectively).

A comparison of the stress-induced charge density released by PZT 52/48 and PZT 95/5 ceramics with their remanent polarization (Table 11.2) shows a partial (52%) depolarization of PZT 52/48 and a complete depolarization of PZT 95/5. These PZT 52/48 and 95/5 results are in good agreement with previous results of transverse depolarization of the two ferroelectrics (see Chapter 6 of this book).

PZT 52/48 specimens were in the tetragonal ferroelectric phase. This composition lies near the MPB that separates the FE tetragonal from the FE rhombohedral structures. PZT 52/48 does not undergo a phase transition under high strain rate loading, it remains in the

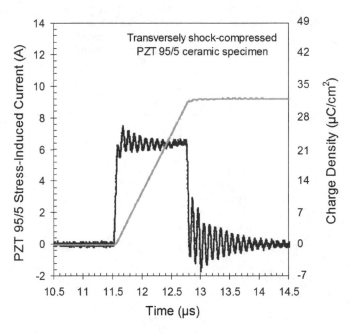

Fig. 11.6. Typical waveform of the current and the dynamics of electric charge released by PZT 95/5 ceramic specimen under transverse shock compression in the short circuit depolarization mode. Specimen size was $5 \times 5 \times 5\,\text{mm}^3$.

Table 11.2. Experimental results obtained for rhombohedral PIN-PMN-PT $[111]_\text{C}$-oriented crystals, and for PZT 52/48 and PZT 95/5 ceramic specimens in the short-circuit depolarization mode. The size of all investigated specimens was $5 \times 5 \times 5\,\text{mm}^3$.

Property	PIN-PMN-PT	PZT 52/48	PZT 95/5
Stress-induced charge density, $\omega(\mu\text{C/cm}^2)$	48 ± 2	15 ± 2	32 ± 1
Remanent polarization*, $P_0(\mu\text{C/cm}^2)$	48	29	32

*The tolerance of parameters does not exceed $\pm 5\%$.

tetragonal phase under pressure at the experimental conditions used in this work. Partial depolarization of PZT 52/48 is the result of the reorientation of non-180° domains caused by release waves traveling behind the compressive front (see Chapter 6 of this book for more details).

The PZT 95/5 specimens were in the rhombohedral FE phase. The composition of PZT 95/5 lies very close to a boundary between ferroelectric rhombohedral and antiferroelecric orthorhombic phases in the phase diagram. The complete depolarization of PZT 95/5 subjected to high strain rate loading is the result of the stress-induced phase transformation from the FE rhombohedral to the AFE orthorhombic phase. This is consistent with the results of hydrostatic studies [26]. It has been demonstrated [24] that in hydrostatically loaded PZT 95/5 the FE (rhombohedral) to AFE (orthorhombic) phase transformation occurs abruptly at a pressure of 0.32 GPa.

A comparison of the stress-induced charge density and remanent polarization for PIN-PMN-PT crystals (Table 11.2) shows that the crystals compressed along the $[110]_C$ direction were completely depolarized under transverse shock. Polarization can be eliminated by a phase transition to a non-polar phase, or polarization reorientation can be induced by the stress difference between the components of an applied mechanical stress that is more compressive in the polarization direction than in the transverse directions. This stress difference is the driving force for non-180° domain wall motion and polarization reorientation that result in the minimization of the free energy of the system and the loss of the initial remanent polarization.

The complete depolarization of PIN-PMN-PT crystals cannot be explained by domain disappearance in a non-polar cubic phase. Uniaxial pressure tends to push the crystal structure to a lower symmetry [27, 28]. The non-polar cubic phase has higher symmetry than the uncompressed rhombohedral PIN-PMN-PT crystal. The rhombohedral-to-cubic phase transition can be reached in a PIN-PMN-PT crystal by thermally heating it above the Curie point or by very high uniaxial compressive stress exceeding 7 GPa [28].

High resolution X-ray diffraction was performed on PIN-PMN-PT, PZT 95/5, and PZT 52/48 specimens before and after high strain loading experiments using a Bragg-Brentano diffractometer (PANalytical X'Pert Pro). Specimens were crushed, ground and sieved using a 100 mesh to produce a powder. The XRD data were collected in a broad 2θ range (20° through 90°) using Cu $K\alpha$ radiation ($\lambda = 0.154060$ nm). A continuous scanning mode was used and the data were recorded with a scanning step 0.017° 2θ and a scanning time of 4 seconds per step. All diffraction patterns were subjected to the fitting algorithm provided by the PANalytical X'Pert Highscore Plus software to determine the position, intensity, broadening and shape of each peak. X'Pert HighScore Plus uses the Pseudo-Voigt profile function, which is the weighted mean between a Lorentz and a Gauss function. The d-spacing was performed using the X'Pert Highscore FWHM single peak fitting routine.

Figure 11.7 shows the X-ray diffraction patterns for PIN-PMN-PT, PZT 95/5, and PZT 52/48 specimens.

The obtained results (Figure 11.7(a)) indicate that the XRD patterns and, correspondingly, the atomic structure of PIN-PMN-PT and PZT 95/5 specimens are very similar. The peak splitting in the {111} reflections indicates a predominantly rhombohedral distortion (inset in Figure 11.7(a)). For PZT 52/48 the peak splitting was observed in the {100} and {200} reflections in the diffraction patterns and this indicates a predominantly tetragonal distortion (insets in Figure 11.7(b)).

The XRD patterns of PIN-PMN-PT specimens after high strain rate loading and in the as-received condition are shown in Figure 11.7(c). The two XRD patterns have identical interplanar reflections, but the peaks for the shocked specimen were shifted to the higher 2θ. This could be caused by a smaller volume of the unit cell in the specimen subjected to high strain loading.

The {111} reflections of two specimens are compared in the inset in Figure 11.7(c). The double peaks were observed in both {111} reflections. This indicates that the specimen was in the rhombohedral phase after high strain rate loading. A similar result was observed for the PZT 95/5 specimens. The X-ray diffraction indicates that PZT 95/5 depolarized by the high strain rate loading is in the rhombohedral phase, i.e. it returns from the orthorhombic to the rhombohedral state after loading. This observation is consistent with the results of hydrostatic studies of PZT 95/5 [26]. It was shown that when the pressure for hydrostatically loaded PZT 95/5 is decreased from 0.32 to 0.14 GPa, it undergoes a transformation from the orthorhombic state back to the rhombohedral state.

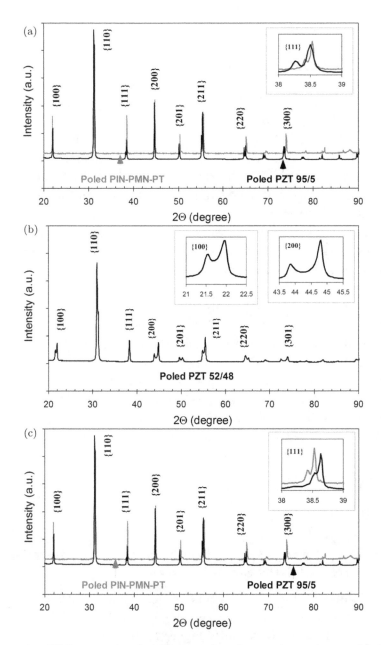

Fig. 11.7. XRD patterns of single crystal and ceramic specimens. (a) PIN-PMN-PT and PZT 95/5 specimens in the as-received condition. (b) PZT 52/48 specimens in the as-received condition. (c) PIN-PMN-PT specimens after high strain rate loading and in the as-received condition.

Based on the obtained results it appears possible that, similar to PZT 95/5, the PIN-PMN-PT is undergoing a ferroelectric rhombohedral to orthorhombic phase transition under high strain rate loading. It should be mentioned that the rhombohedral to orthorhombic phase transformation in PIN-PMN-PT crystals caused by fatigue (polarization degradation) was observed in [30, 31].

The PIN-PMN-PT orthorhombic phase is a multidomain polar phase with domains having polarization aligned along the $[011]_C$ directions. As a result of the rhombohedral to orthorhombic phase transition, the $[111]_C$ single domain state is transformed into a multi-domain structure that is compressed by the unloading waves propagating behind the shock front. This leads to domain wall motion, polarization reorientation and loss of the initial remanent polarization. To minimize energy, the resulting multi-domain state is reoriented to give net zero polarization.

This explanation is in agreement with the results of studies of PIN-PMN-PT and PMN-PT crystals under uniaxial static stress. It was demonstrated in [26] that rhombohedral $[011]_C$-oriented PIN-PMN-PT and PMN-PT single crystals can be driven to the orthorhombic phase by a small uniaxial compressive stress applied in the $[100]_C$ direction (transverse compression) or by an electric field applied in the $[011]_C$ direction. This phase transition was observed as a sharp decrease in the compliance, piezoelectric, and dielectric coefficients, and the polarization reorientation. This phase transformation is associated with the accommodation of the monoclinic phase that separates the two phases in compositions near the MPB [28].

11.5 High-Voltage Generation by Shock-Compressed PIN-PMN-PT Single Crystals

The obtained experimental results (Table 11.2) indicate that the density of stress-induced charge released by PIN-PMN-PT under transverse shock is significantly higher than that for PZT ceramics. This is the result of both the high remanent polarization of PIN-PMN-PT crystals and their ability to be completely depolarized under shock compression.

When the electrodes of a ferroelectric specimen are open, the stress-induced charge is utilized for charging the ferroelectric element itself, resulting in a high electric field and a high electric potential across the element. The unique ability of ferroelectric materials to generate high voltage under stress is used in a variety of modern engineering applications including ultrahigh-power ferroelectric generators [11]. Conventional pulsed power systems with operating voltage ranging from 100 to 500 kV are capable of producing high-power microwave radiation with peak power up to gigawatt level [11, 32]. The ability of ferroelectric crystals to produce high voltage under high stress is critical for their usage in ferroelectric generators.

The experiments with PIN-PMN-PT, PZT 52/48 and PZT 95/5 specimens in the high-voltage mode (see a diagram of the experimental setup in Figure 11.3) were conducted. Figures 11.8 through 11.10 show typical waveforms of the stress-induced voltage

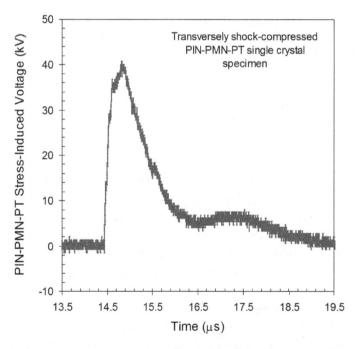

Fig. 11.8. Typical waveform of the stress-induced voltage produced by a PIN-PMN-PT single crystal specimen in the high-voltage mode.

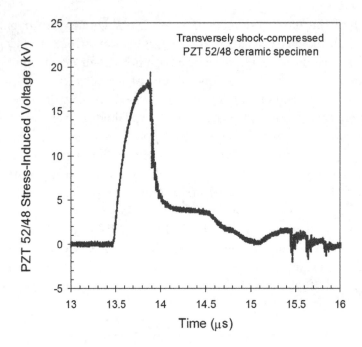

Fig. 11.9. Typical waveform of the stress-induced voltage produced by a PZT 52/48 ceramic specimen in the high-voltage mode.

produced by single crystal and ceramic specimens in the high-voltage mode. The stress-induced voltage waveforms were single pulses with identical polarity for the three ferroelectrics.

The voltage started rising at the moment of time when the compression front entered the front face of the single crystal or ceramic specimen. The voltage increased to its maximum for a few hundred nanoseconds and then decreased due to an internal breakdown within the specimen (Figures 11.8 to 11.10) [19, 33, 34]. The experimental results indicate that the peak voltage was achieved when only a part of the ferroelectric specimen was depolarized under stress. This is typical for ferroelectric specimens of different shapes and sizes (see Chapter 8 of this book for details).

The PIN-PMN-PT single crystal specimen produced voltage pulses with amplitudes 41.1 kV under transverse shock compression (Figure 11.8). The corresponding electric field across the crystal was

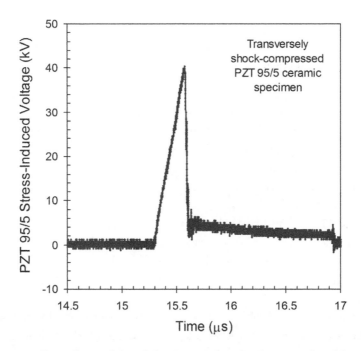

Fig. 11.10. Typical waveform of the stress-induced voltage produced by a PZT 95/5 ceramic specimen in the high-voltage mode.

8.22 MV/m. This is double the higher voltage and electric field than for PZT 52/48 (Figure 11.9) and it is practically equal to those for PZT 95/5 (Figure 11.10).

The PZT 52/48 and PZT 95/5 results (Figures 11.9 and 11.10) are in good agreement with previous results obtained with the two ferroelectrics in the high-voltage mode (see Chapters 8 and 9 of this book). The amplitude of the voltage pulses limited by the dielectric strength of the ferroelectric materials can be considered as the highest possible voltage that can be produced by the single crystal and ceramic specimens under stress (Figures 11.8 to 11.10). The results of the high-voltage experiments with the three ferroelectrics are summarized in Table 11.2.

The ferroelectric element of an FEG combines three stages of conventional microwave pulsed power systems in one, i.e. a prime power source, a high-voltage generator, and a capacitive energy storage device [11]. The energy density of a capacitive energy storage device is directly proportional to the amplitude of the voltage to the power of two and the capacitance of the specimen:

$$W = \frac{\varepsilon_0 \varepsilon_r A V^2}{2\,d \cdot vol} = \frac{\varepsilon_0 \varepsilon_r V^2}{2\,d^2} \tag{11.1}$$

where ε_r is the relative dielectric permittivity of the ferroelectric, ε_0 is the dielectric permittivity of free space ($8.85 \cdot 10^{-12}$ F/m), A is the area of the electrodes, d is the distance between the electrodes, V is the voltage across the ferroelectric specimen, and *vol* is the volume of the specimen.

Table 11.3 summarizes the energy density generated by the three ferroelectrics in the high-voltage mode. The obtained results indicate that the energy density of PIN-PMN-PT, $W_{PIN} = 0.33\,\text{MJ/m}^3$ is more than four times higher than that for PZT 52/48 and PZT 95/5 ceramics.

It should be mentioned that the energy densities for the three ferroelectrics listed in Table 11.3 are the lower bounds of these parameters because they were determined with the permittivities of the ferroelectrics obtained from standard low electric field measurements (see Chapter 8 of this book for details).

The amplitude of high voltage produced by ferroelectric specimens under high stress is increasing with increasing thickness according to the power law $V_{FEG}(d) = \chi \cdot d^{1-\xi}$ (where d is the thickness of the specimen, χ is the amplitude of voltage generated by a 1.0 mm-thick ferroelectric specimen, and ξ is a coefficient that is justified by the mechanism of electric breakdown) [17, 34, 35]. It gives grounds

Table 11.3. Experimental results obtained for rhombohedral PIN-PMN-PT [111]$_C$-oriented crystals, and for PZT 52/48 and PZT 95/5 ceramic specimens in the high-voltage mode, energy densities generated by the three ferroelectrics. The size of all the investigated specimens was $5 \times 5 \times 5\,\text{mm}^3$.

	PIN-PMN-PT	PZT 52/48	PZT 95/5
Stress-induced voltage (kV)	40.4 ± 0.18	17.8 ± 0.16	40.1 ± 0.19
Electric field (MV/m)	8.08 ± 0.36	3.56 ± 0.32	8.02 ± 0.38
Energy density, W (MJ/m^3)	0.33	0.082	0.072

for estimating the voltage and energy produced by PIN-PMN-PT: a single crystal with a thickness of 6 cm and a total volume of 70 cm³ could produce a 400 kV microsecond pulse with tens of joules of total energy. These voltage and energy levels are sufficient to provide the generation of a subnanosecond microwave pulse with hundreds of megawatts of peak power.

11.6 Summary

- PIN-PMN-PT ferroelectric single crystals possessing the record high remanent polarization, $48\,\mu C/cm^2$, (not achievable for all known ferroelectric materials) were developed.
- The developed PIN-PMN-PT single-domain crystals were subjected to transverse shock depolarization studies along with high-resolution X-ray diffraction. The obtained experimental results were directly compared to those for PZT 95/5 and PZT 52/48 ferroelectric ceramics under identical loading conditions.
- The experimental results indicate that the polarity of electric charge released by PIN-PMN-PT [111]$_C$-oriented crystals under transverse shock loading was opposite to the polarity of the charge generated due to the piezoelectric effect but it was identical to the polarity of the surface charge that balances the initial remanent polarization. These results allow one to draw the conclusion that PIN-PMN-PT [111]$_C$-oriented crystals are depolarized under transverse shock compression.
- The remarkable result is that under 3.9 GPa transverse stress PIN-PMN-PT crystals release electric charge density equal to their remanent polarization $(48\,\mu C/cm^2)$ and become completely depolarized. This stress-induced charge can be utilized for the generation of current and high voltage in the FEG load circuit.
- The mechanism of shock depolarization of PIN-PMN-PT crystals is different from those for PZT 52/48 and PZT 95/5.
- In the high-voltage mode, the stress-induced electric charge results in the generation of high voltage across the PIN-PMN-PT single crystals. The important result is that the energy density generated

by crystals in the high-voltage mode exceeds $0.3\,\mathrm{MJ/m^3}$ and is four times higher than that for PZT 52/48 and PZT 95/5 ceramics. These results are a breakthrough in the research and development of ferroelectric materials for applications in explosive ferroelectric generators.

• Based on the obtained results the conclusion can be reached that ferroelectric single crystal technology is very promising for explosive pulsed power applications. This is a demonstration of the unique ability of domain engineering to control precisely the ferroelectric properties of single crystals such that they fit certain applications, something that is not achievable with polycrystalline ceramics.

Bibliography

1. B. Jaffe, W.R. Cook, Jr., and H. Jaffe, *Piezoelectric Ceramics* (Academic Press, London, 1971).
2. S.E. Park and T.R. Shrout, Ultrahigh strain and piezoelectric behavior in relaxor based ferroelectric single crystals, *J. Appl. Phys.* **82**(4) (1997) pp. 1804–1811.
3. Z.W. Yin, H.S. Luo, P.C. Wang and G.S. Xu, Growth, characterization and properties of relaxor ferroelectric PMN–PT single crystals, *Ferroelectrics* **229** (1999) pp. 207–216.
4. H. Luo, G. Xu, H. Xu, P. Wang, and W. Yin, Compositional homogeneity and electrical properties of lead magnesium niobate titanate single crystals grown by a modified Bridgman technique, *Jpn. J. Appl. Phys.* **39** (2000) pp. 5581–5585.
5. Y. Guo, H. Luo, T. He, and Z. Yin, Peculiar properties of a high Curie temperature Pb(In1/2Nb1/2)O3–PbTiO3 single crystal grown by the modified Bridgman technique, *Solid State Comm.* **123** (2002) pp. 417–420.
6. G. Xu, K. Chen, D. Yang, and J. Li, Growth and electrical properties of large size Pb..In1/2Nb1/2...O3–Pb..Mg1/3Nb2/3...O3–PbTiO3 crystals prepared by the vertical Bridgman technique, *Appl. Phys. Lett.* **90** (2007) p. 032901.
7. J. Tian, P. Han, X. Huang, and H. Pan, Improved stability for piezoelectric crystals grown in the lead indium niobate-lead magnesium niobate-lead titanate system, *Appl. Phys. Lett.* **91** (2007) p. 222903.
8. S. Zhang, L. Jun, W. Hackenberger, T.R. Shrout, Characterization of Pb..In1/2Nb1/2... O3–Pb..Mg1/3Nb2/3...O3–PbTiO3 ferroelectric crystal with enhanced phase transition temperatures, *J. Appl. Phys.* **104** (2008) p. 064106.

9. E. Sun and W. Cao, Relaxor-based ferroelectric single crystals: Growth, domain engineering, characterization and applications, *Progress in Materials Science* **65** (2014) pp. 124–210.

10. Q. Zhou, K.H. Lam, H. Zheng, W. Qiu and K.K. Shung, Piezoelectric single crystal ultrasonic transducers for biomedical applications, *Progress in Materials Science* **66** (2014) pp. 87–111.

11. L.L. Altgilbers, J. Baird, B. Freeman, C.S. Lynch and S.I. Shkuratov, *Explosive Pulsed Power* (Imperial College Press, London, 2010).

12. W.D. Dong, P.F. Finkel, A. Amin and C.S. Lynch, Giant electro-mechanical energy conversion in [011] cut ferroelectric single crystals, *Appl. Phys. Lett.* **100** (2012) p. 042903.

13. S.I. Shkuratov, J. Baird, V.G. Antipov, E.F. Talantsev, J.B. Chase, W.S. Hackenberger, J. Luo, H.R. Jo and C.S. Lynch, Ultrahigh energy density harvested from domain-engineered relaxor ferroelectric single crystals under high strain rate loading, *Sci. Rep.* **7** (2017) p. 46758.

14. http://trstechnologies.com/

15. S. Zhang, F. Li, J. Luo, R. Xia, W. Hackenberger and T. Shrout, Investigation of single and multidomain Pb..In0.5Nb0.5...O3–Pb..Mg1/3Nb2/3...O3–PbTiO3 crystals with *mm*2 symmetry, *Appl. Phys. Lett.* **97** (2010) p. 132903.

16. S. Zhang, F. Li, N. Sherlock, J. Luo, H. Lee, R. Xia, R. Meyer, W. Hackenberger and T. Shrout, Recent developments on high Curie temperature PIN–PMN–PT ferroelectric crystals, *J. Crystal Growth* **318** (2011) pp. 846–850.

17. X. Huo, S. Zhang, G. Liu, R. Zhang, J. Luo, R. Sahul, W. Cao and T. Shrout, Complete set of elastic, dielectric, and piezoelectric constants of [011]$_C$ poled rhombohedral Pb(In0.5Nb0.5)O3-Pb(Mg1/3Nb2/3)O3-PbTiO3:Mn single crystals. *J. Appl. Phys.* **113** (2013) p. 074106.

18. S.I. Shkuratov, J. Baird, V.G. Antipov and E.F. Talantsev, Depolarization mechanisms of PbZr0.52Ti0.48O3 and PbZr0.95Ti0.05O3 poled ferroelectrics under high strain rate loading, *Appl. Phys. Lett.* **104** (2014) p. 212901.

19. S.I. Shkuratov, J. Baird and E.F. Talantsev, Extension of thickness-dependent dielectric breakdown law on adiabatically compressed ferroelectric materials, *Appl. Phys. Lett.* **102** (2013) p. 052906.

20. S. I. Shkuratov, J. Baird and E. F. Talantsev, Miniature 120-kV generator based on transverse shock depolarization of Pb($Zr_{0.52}Ti_{0.48}$)O3 ferroelectrics, *Rev. Sci. Instrum.* **82**(8) (2011) p. 086107.

21. S.I. Shkuratov, E.F. Talantsev, L. Menon, H. Temkin, J. Baird, and L.L. Altgilbers, Compact high-voltage generator of primary power based on shock wave depolarization of lead zirconate titanate piezoelectric ceramics, *Rev. Sci. Instrum.* **75**(8) (2004) pp. 2766–2769.

22. S.I. Shkuratov, E.F. Talantsev and J. Baird, Electric breakdown of longitudinally shocked Pb($Zr_{0.52}Ti_{0.48}$)O3 ceramics, *J. Appl. Phys.* **110**(2) (2011) p. 024113

23. S.I. Shkuratov, J. Baird, V.G. Antipov and E.F. Talantsev, Autonomous pulsed power generator based on transverse shock wave depolarization of ferroelectric ceramics, *Rev. Sci. Instrum.* **81**(12) (2010) p. 126102.

24. S.I. Shkuratov, J. Baird, V.G. Antipov, E.F. Talantsev, C.S. Lynch and L.L. Altgilbers, PZT 52/48 depolarization: quasi-static thermal heating versus longitudinal explosive shock, *IEEE Trans. Plasma Sci.* **38**(8) (2010) pp. 1856–1863.

25. R.E. Tipton, A 2D Lagrange MHD code, in *Proc.4th Int. Conf. Megagauss Magnetic Field Generation and Related Topics* (Editors: C.M. Fowler, R.S. Caird and D.J. Erickson) (Plenum Press, New York, U.S.A. 1987) pp. 299–302.

26. I.J. Fritz, Uniaxial stress effects in a 95/5 lead zirconate titanate ceramic. *J. Appl. Phys.*, **49**(9) (1978) pp. 4922–4928.

27. S. Zhang, S. Taylor, F. Li, J. Luo and R.J. Meyer, Jr., Piezoelectric property of relaxor-PbTiO3 crystals under uniaxial transverse stress, *Appl. Phys. Lett.* **102** (2013) p. 172902.

28. J.A. Gallagher, J. Tian and C.S. Lynch, Effects of composition and temperature on the large field behavior of [011]$_C$ relaxor ferroelectric single crystals, *Appl. Phys. Lett.* **105** (2014) p. 052909.

29. M. Ahart, S. Sinogeikin, O. Shebanova, D. Ikuta, Z.-G. Ye, Ho.-k. Mao, R.E. Cohen and R.J. Hemley, Pressure dependence of the monoclinic phase in $(1-x)$Pb(Mg1/3Nb2/3)O3-xPbTiO3 solid solutions. *Phys. Rev. B* **86** (2012) p. 224111.

30. X.J. Lou, Polarization fatigue in ferroelectric thin films and related materials, *J. Appl. Phys.* **105** (2009) p. 024101.

31. S. Zhang, J. Luo, F. Li and R.J. Meyer, Polarization fatigue in Pb(In0.5Nb0.5)O3–Pb(Mg1/3Nb2/3)O3–PbTiO3 single crystals, *Acta Materialia* **58** (2010) pp. 3773–3780.

32. G.A. Mesyats, *Pulsed Power* (Kluwer Academic/Plenum Publishers, New York, 2005).

33. P. C. Lysne, P.C. Prediction of dielectric breakdown in shock-loaded ferroelectric ceramics, *J. Appl. Phys.* **46**(1) (1975) pp. 230–232.

34. S.I. Shkuratov, E.F. Talantsev and J. Baird, Electric breakdown of longi-tudinally shocked Pb(Zr$_{0.52}$Ti$_{0.48}$)O3 ceramics, *J. Appl. Phys.* **110** (2011) p. 024113.

35. S.I. Shkuratov, J. Baird, V.G. Antipov, E.F. Talantsev, W.S. Hackenberger, A.H. Stults and L.L. Altgilbers, High voltage generation with transversely shock-compressed ferroelectrics: breakdown field on thickness dependence, *IEEE Trans. Plasma Sci.* **44**(10) (2016) pp. 1919–1927.

Chapter 12

Mechanism of Complete Stress-Induced Depolarization of Relaxor Ferroelectric Single Crystals without Transition through a Non-Polar Phase

12.1 Introduction

Relaxor ferroelectric single crystals such as PMN-PT and PMN-PIN-PT have triggered a revolution in electromechanical systems due to their superior piezoelectric properties [1–7]. Ferroelectric crystals have been extensively employed in the development of a new generation of sensors and low-power transducers with improved performance. Recent research has focused on achieving an increase in power from the hundred milliwatt level to the kilowatt level for high-power resonant transducers used in therapeutic applications [8, 9] and to the megawatt level for ultrahigh-power ferroelectric generators [10]. The operation of ferroelectric crystals in the low-strain mode is reliable, while their operation in the high-power mode is limited by stress and electric field-induced phase transitions, loss of remanent polarization, and damage to the crystals. Although this degradation of piezoelectric properties can limit their application in high-power systems requiring high frequency, rapid depolarization under shock loading can result in the generation of megawatt power levels for a brief interval of time.

The ferroelectric and piezoelectric properties of crystals can be enhanced through domain engineering, i.e. creating different domain configurations by polarizing crystals in specific directions relative to crystallographic orientation. Specific orientations can be selected where, once poled, there is no driving force for domain wall motion under a positive electric field. This results in enhanced electro-mechanical coupling with a corresponding low dielectric loss and extraordinarily large piezoelectric coefficients. The resulting macroscopic (volume average) polarization symmetry is associated with the engineered dipole alignment of the rhombohedral domains.

In this chapter, results are presented from systematic studies of the remanent polarization change in PIN-PMN-PT crystals with different macroscopic domain pattern symmetries subjected to high strain rate loading. The important finding is that the domain-engineered state plays an important role in both the remanent polarization and the charge density released under shock compression. The experimental results indicate that domain configuration has a significant effect on stress-induced depolarization.

12.2 Single Crystal Specimen Preparation

PIN-PMN-PT single crystals were grown by TRS Technologies using the modified Bridgman technique [3, 11, 12]. Ternary PIN-PMN-PT single crystals at room temperature can be in either the ferroelectric rhombohedral phase or the ferroelectric tetragonal phase. These two phases are separated by a morphotropic phase boundary that can contain a monoclinic phase. When the composition is close to the MPB the ferroelectric properties of the crystals are enhanced. The as-grown PIN-PMN-PT crystals exhibit different phases along the growth direction due to the segregation of Ti. All crystal specimens were selected to be in the ferroelectric rhombohedral phase and compositionally in close proximity to the morphotropic phase boundary, $0.27\%Pb(In_{1/2}Nb_{1/2})O_3-0.41\%Pb(Mg_{1/3}Nb_{2/3})O_3-32\%\ 0.3PbTiO_3$. The specimens were oriented using a real-time Laue X-ray orientation system with an accuracy of $0.5°$.

The effective macroscopic symmetry associated with the domain structure is a key issue for the domain engineering of single crystals.

The rhombohedral phase has eight possible spontaneous polarization directions. The studied PIN-PMN-PT single crystals were cut and poled along $[001]_C$, $[011]_C$ and $[111]_C$ crystallographic directions. For rhombohedral crystals, poling along the $[001]_C$ direction induces $4\,mm$ macroscopic multidomain pattern symmetry, poling along the $[011]_C$ direction induces $mm2$ macroscopic multidomain pattern configuration, and poling along the $[111]_C$ direction induces $3m$ symmetry of the single-domain state.

Each PIN-PMN-PT crystal specimen was measured as 5.0 mm long × 5.0 mm wide × 5.0 mm thick. Vacuum-sputtered Cr/Au films were deposited on the desired surfaces as electrodes. Crystal specimens were poled at a DC electric field of 7.5 kV/cm in an oil bath at 50°C. The complete set of material constants was determined by combined resonance and ultrasonic methods.

Typical polarization hysteresis loops of PIN-PMN-PT single crystal specimens cut and poled along the $[001]_C$, $[011]_C$ and $[111]_C$ crystallographic directions are presented in Figure 12.1. Remanent polarization of PIN-PMN-PT crystals cut along $[001]_C$, $[011]_C$ and $[111]_C$ directions was found to be 25, 38 and 48 $\mu C/cm^2$, respectively. The physical properties of PIN-PMN-PT crystals are listed in Table 12.1. The tolerance of parameters does not exceed ± 5%.

12.3 Shock Loading of Ferroelectric Single Crystals with Different Domain Pattern Symmetries

Poled rhombohedral PIN-PMN-PT crystals were shock-compressed perpendicular to the direction of polarization (transverse shock). Schematics of the transverse shock wave generators containing single crystals with different domain pattern symmetries and the measuring circuit are shown in Figures 12.2 through 12.4. In PIN-PMN-PT $[001]_C$- and $[011]_C$-oriented crystals, the direction of shock wave propagation was parallel to the $[100]_C$ crystallographic direction (Y-axis in Figures 12.2 and 12.3).

The FEG design was based on spherically expanding shock waves (see Chapter 5 of this book for details). The FEG contained two parts, a detonation chamber and a PIN-PMN-PT single crystal

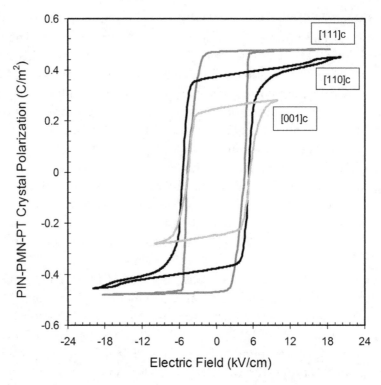

Fig. 12.1. Typical polarization hysteresis loops (AC electric field at 1 Hz) for rhombohedral PIN-PMN-PT $[001]_C$-oriented crystal with $4mm$ symmetry, $[011]_C$-oriented crystal with $mm2$ symmetry and $[111]_C$-oriented crystal with $3m$ symmetry.

Table 12.1. Physical properties of rhombohedral PIN-PMN-PT $[001]_C$-, $[011]_C$- and $[111]_C$-oriented crystals and PZT 52/48 and PZT 95/5 ceramic specimens.

| Property | PIN-PMN-PT single crystals | | | PZT ceramics | |
	$[001]_C$ cut	$[011]_C$ cut	$[111]_C$ cut	52/48	95/5
Domain pattern symmetry	$4mm$	$mm2$	$3m$	∞m	∞m
Density (10^3 kg/m^3)	8.1	8.1	8.1	7.5	7.7
Dielectric constant	6700	5100	1180	1300	295
Elastic constant, s_{11}^E (10^{-12} m^2/N)	45.4	24.8	16.4	12.8	7.7
Remanent polarization (μC/cm^2)	25	38	48	29	32

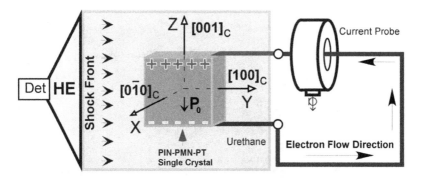

Fig. 12.2. Schematics of the transverse ferroelectric generator containing a multidomain PIN-PMN-PT $[001]_C$-oriented crystal with $4mm$ symmetry and the measuring circuit used in the shock depolarization studies. P_0 is the remanent polarization vector. The polarity of the surface charge is shown by $(+)$ and $(-)$ signs.

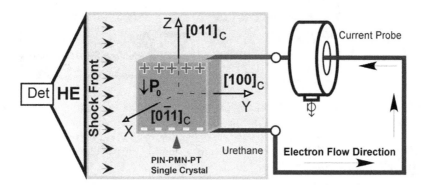

Fig. 12.3. Schematics of the transverse ferroelectric generator containing a multidomain PIN-PMN-PT $[011]_C$-oriented crystal with $mm2$ symmetry.

encapsulated within a plastic body. After the initiation of the detonator, the detonation wave propagated through HE. The HE detonation shock propagated through the urethane potting material and into the single crystal specimen.

The diameter of the ferroelectric generators was 38 mm. The FEG was placed in the blast chamber. The electrodes of the PIN-PMN-PT crystal were short-circuited. The stress-induced current

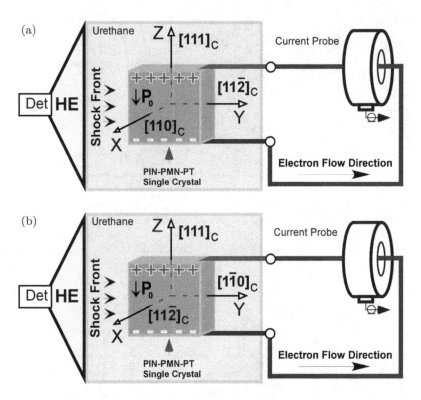

Fig. 12.4. Schematics of the transverse ferroelectric generator containing a single-domain PIN-PMN-PT $[111]_C$-oriented crystal with $3m$ symmetry. (a) Direction of shock wave propagation is parallel to the $[11\bar{2}]_C$ crystallographic direction. (b) Direction of shock wave propagation is parallel to the $[1\bar{1}0]_C$ crystallographic direction.

was monitored with a Pearson 411 current transformer (transition time 10 nanoseconds) placed outside the blast chamber.

PIN-PMN-PT $[111]_C$-oriented crystals were studied in two modes of transverse shock compression. In the first mode, the direction of shock wave propagation was parallel to the $[11\bar{2}]_C$ crystallographic direction (Y-axis in Figure 12.4(a)). In the second mode, the direction of shock wave propagation was parallel to the $[1\bar{1}0]_C$ crystallographic direction (Y-axis in Figure 12.4(b)).

12.4   Uniaxial Stress Distribution in Shock-Compressed Ferroelectric Single Crystals

In these studies the stress waves in the ferroelectric crystals originated from the detonating HE charge (see Figures 12.2 to 12.4) and the stress wave parameters affecting the ferroelectric crystals were determined by the geometry of the HE chamber, the initiation point, the geometry of the experimental device, the shock impedance matching among the parts of the experimental system and the ferroelectric crystal specimen.

To determine the uniaxial stress amplitude and stress distribution in the ferroelectric crystals we performed a simulation of high strain rate loading by using the CALE computer code. The CALE is a two-dimensional Arbitrary Lagrange/Eulerian second-order accurate hydrodynamics program [13].

The simulation was run in cylindrical coordinates, assuming rotational symmetry around the axis and zero material rotation. In order for the numerical integration to proceed, the space was divided into "zones" by a mesh of coaxial cylindrical surfaces and a set of planes. Each zone was washer-shaped, with a nearly square section in the R-Z plane, with the coordinates R (cylindrical radius from the symmetry axis), Z (axial position), and angle (around the symmetry axis).

The simulation was run in cylindrical coordinates, assuming rotational symmetry around the axis and zero material rotation. In order for the numerical integration to proceed, the space was divided into "zones" by a mesh of coaxial cylindrical surfaces and a set of planes. Each zone was washer-shaped, with a nearly square section in the R-Y plane, with the coordinates R (cylindrical radius from the symmetry axis), Y (axial position), and angle (around the symmetry axis). Figure 12.5 shows the CALE results for the uniaxial stress distribution in a single crystal specimen at $0.6\,\mu$s after the compression front entered the specimen.

The CALE results for the uniaxial stress at different distances from the axis in the middle cross-section of the PIN-PMN-PT crystal are shown in Figure 12.6. The uniaxial stress is practically uniform in the cross-section, 3.9 ± 0.1 GPa.

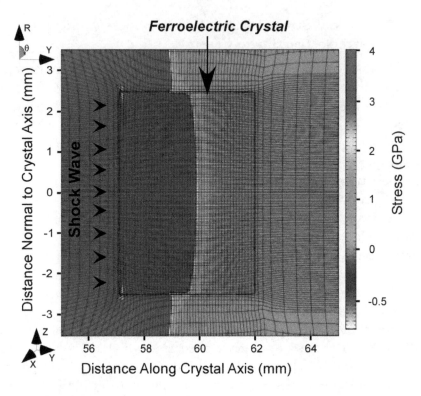

Fig. 12.5. CALE results for the uniaxial stress distribution in a single crystal specimen at $0.6\,\mu s$ after the compression front entered the specimen.

12.5 Shock Depolarization of PIN-PMN-PT Single Crystals with Different Domain Pattern Symmetries

When shock loaded, the high-amplitude stress wave that propagates across the ferroelectric crystals produces a current versus time profile. The stress-induced electric charge released is the time integral of the stress-induced current.

A typical waveform of the stress-induced current and the dynamics of electric charge released by a multidomain PIN-PMN-PT [001]$_C$-oriented crystal with $4mm$ symmetry is shown in Figure 12.7. The crystal was shock-compressed along the [100]$_C$ direction (see the FEG diagram in Figure 12.2). The inset in Figure 12.7 represents

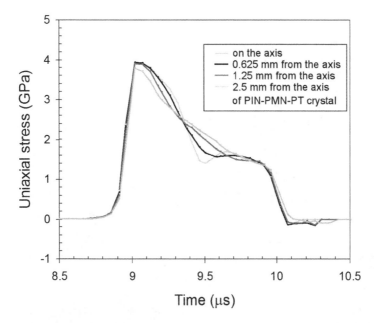

Fig. 12.6.   CALE results for the uniaxial stress at different distances from the axis in the middle cross-section of a PIN-PMN-PT crystal.

typical stress-induced current waveforms for PZT 95/5 ceramic specimens.

The PZT 95/5 and PZT 52/48 ceramic specimens were used as references in this study. The specimen geometry for PZT 52/48 and PZT 95/5 was identical to that for PIN-PMN-PT crystals, 5.0 mm long × 5.0 mm wide × 5.0 mm thick, to enable a direct comparison of the experimental results obtained from single crystals and those obtained from ceramic materials. The properties of PZT 52/48 and 95/5 ceramics are listed in Table 12.1.

The direction of stress-induced current provides information about the origin of electric charge generated by ferroelectrics under high stress. The direction of the stress-induced current produced by [001]$_C$-oriented crystals was identical to that for PZT 95/5 ceramics (Figure 12.7).

Under transverse shock compression, the linear piezoelectric effect would cause a current to flow in the direction opposite to what

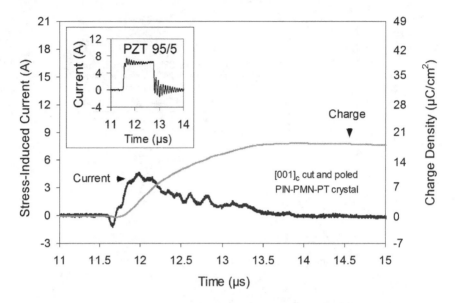

Fig. 12.7. Typical waveform of the current and the dynamics of electric charge released by multidomain PIN-PMN-PT $[001]_C$-oriented crystals with $4mm$ symmetry under transverse shock compression (see the FEG diagram in Figure 12.2). The inset represents typical stress-induced current waveforms for PZT 95/5 ceramic specimens used as a reference.

was observed in the experiments. This indicates that the charge released by the $[001]_C$-oriented crystals with $4mm$ symmetry was caused not by the piezoelectric effect, but by other factors that led to depolarization.

A typical waveform of the stress-induced current and the dynamics of electric charge released by multidomain $[011]_C$-oriented crystals with $mm2$ symmetry are shown in Figure 12.8. The inset in Figure 12.8 represents typical stress-induced current waveforms for PZT 52/48 ceramic specimens.

The waveform of stress-induced current produced by the $[011]_C$-oriented crystal under transverse stress (Figure 12.8) is more complicated than that for the $[001]_C$-oriented crystal (Figure 12.7). It is not a single pulse, but a series of pulses. However, the direction of the stress-induced current produced by the $[011]_C$-oriented crystals was identical to that for $[001]_C$-oriented crystals and PZT ceramics.

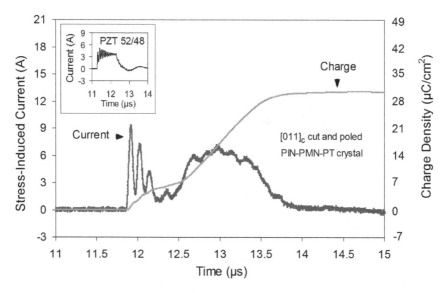

Fig. 12.8. Typical waveform of the current and the dynamics of electric charge released by multidomain PIN-PMN-PT $[011]_C$-oriented crystals with $mm2$ symmetry under transverse shock compression (see FEG diagram in Figure 12.3). The inset represents typical stress-induced current waveforms for PZT 52/48 ceramic specimens used as a reference.

This indicates that the charge released by multidomain crystals with $mm2$ domain pattern symmetry was caused by factors that led to depolarization. The densities of stress-induced charge released by $[011]_C$- and $[001]_C$-oriented crystals were $\omega_{[001]} = 19\,\mu\mathrm{C/cm}^2$ and $\omega_{[011]} = 32\,\mu\mathrm{C/cm}^2$, respectively. Crystals of both types were only partially depolarized under transverse shock compression.

Figure 12.9 presents typical waveforms of the stress-induced current and the dynamics of electric charge density released by the $[111]_C$-oriented single-domain crystals with $3m$ symmetry. The waveforms of stress-induced current generated by single-domain crystals shock-compressed along two transverse directions ($[1\bar{1}0]_C$ and $[11\bar{2}]_C$) were not identical (Figures 12.9(a) and 12.9(b)). However, experimental results indicate that there is no difference in stress-induced charge density released by these crystals $\omega_{[111]} = 48\,\mu\mathrm{C/cm}^2$, i.e. the crystals were completely depolarized.

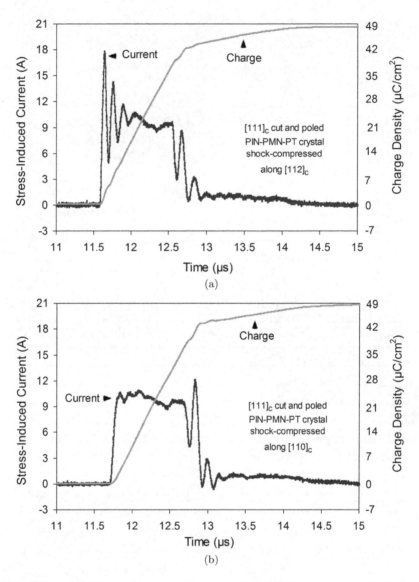

Fig. 12.9. Typical waveform of the current and the dynamics of electric charge released by single-domain PIN-PMN-PT [111]$_C$-oriented crystals with $3m$ symmetry under transverse shock compression. (a) Direction of shock wave propagation was parallel to the [11$\bar{2}$]$_C$ crystallographic direction (see the FEG diagram in Figure 12.4(a)). (b) Direction of shock wave propagation was parallel to the [1$\bar{1}$0]$_C$ crystallographic direction (see the FEG diagram in Figure 12.4(b)).

Table 12.2. Experimentally obtained stress-induced charge densities for rhombohedral PIN-PMN-PT [001]$_C$-, [011]$_C$- and [111]$_C$-oriented crystals and PZT ceramic specimens.

Property	PIN-PMN-PT single crystals			PZT ceramics	
	[001]$_C$	[011]$_C$	[111]$_C$	52/48	95/5
Stress-induced charge density, $\omega(\mu C/cm^2)$	19 ± 1	32 ± 2	48 ± 2	15 ± 2	32 ± 1
Remanent polarization*, $P_0(\mu C/cm^2)$	25	38	48	29	32

*The tolerance of parameters does not exceed ±5%.

The direction of the stress-induced current produced by crystals with the three domain configurations (Figures 12.7 to 12.9) was identical to that for PZT ceramics and opposite to the direction of current that would be caused by the linear piezoelectric effect. This indicates that the charge released by the crystals with the three domain configurations was caused by stress-induced depolarization. Similar to PZT ceramics, the rhombohedral PIN-PMN-PT crystals were depolarized under high strain rate loading.

The experimental results for PIN-PMN-PT crystals along with PZT ceramics are summarized in Table 12.2. The obtained results (Figures 12.7 to 12.9 and Table 12.2) indicate that the initial engineered domain state has a significant effect on the waveforms of the stress-induced current and the charge densities generated by the crystals. The single-domain crystals with $3m$ symmetry were completely depolarized under high stress and released the charge density 48 $\mu C/cm^2$, which is 250% and 150% higher than that for multidomain crystals with $4mm$ and $mm2$ symmetries, respectively. Understanding the mechanism responsible for the stress-induced complete depolarization of single-domain crystals is important for the development of a new class of materials for ultrahigh-power ferroelectric generators.

12.6 Mechanism of Complete Stress-Induced Depolarization of PIN-PMN-PT Single Crystals without Transition through a Non-Polar Phase

12.6.1 *Mechanism of stress-induced depolarization of ferroelectrics*

Polarization reorientation can occur when the component of stress parallel to the polarization direction is more compressive than that perpendicular to the polarization direction. Thus, uniaxial transverse adiabatic compression would not be expected to induce immediate rotation of the polarization.

The hydrostatic compression component of the stress tensor, however, can induce structural transformations with an associated volume reduction, resulting in the depolarization of the ferroelectrics. The complete depolarization of PZT 95/5 (see Table 12.2) is the result of a stress-induced phase transformation from a ferroelectric rhombohedral phase to a non-polar anti-ferroelectric orthorhombic phase. This transformation is also observed in hydrostatic studies where in PZT 95/5 the FE-to-AFE phase transformation occurs abruptly at a pressure of 0.32 GPa.

The results of a simulation of the high strain rate loading of single crystals using the CALE computer code (see Figures 12.5 and 12.6) indicate that the uniaxial stress is practically uniform in the cross-section of the crystal, 3.9 ± 0.1 GPa. Based on these results one can conclude that the complete depolarization of single-domain crystals with $3m$ symmetry (Table 12.2) cannot be explained by domain disappearance through a phase transformation to the higher symmetry non-polar cubic phase because the crystals do not undergo the rhombohedral-to-cubic phase transition under 3.9 GPa uniaxial adiabatic compression.

12.6.2 *X-ray diffraction*

High-resolution X-ray diffraction was performed on ferroelectric specimens before and after high strain rate loading experiments using a Bragg-Brentano diffractometer PANalytical X'Pert Pro in order to confirm the structure of the specimens, to determine the lattice parameters, and to calculate the volume of the unit cell. All specimens were crushed into powder with a mortar and pestle. All powder batches were passed through a 100-μm sieve to collect the fine particles.

The XRD data were collected in a broad 2θ range (20° through 90°) using Cu $K\alpha$ radiation ($\lambda = 0.154060$ nm). The continuous scanning mode was used and the data were recorded with a scanning step 0.017°2θ and a scanning time of 4 seconds per step. All diffraction patterns were subjected to the fitting algorithm provided by the PANalytical X'Pert Highscore Plus software to determine the position, intensity, broadening and shape of each peak. X'Pert HighScore Plus uses the Pseudo-Voigt profile function, which is the weighted mean between a Lorentz and a Gauss function. The d-spacing was performed using the X'Pert Highscore FWHM single peak fitting routine.

Figure 12.10(a) shows the XRD patterns for PIN-PMN-PT crystals in the as-received condition. The dominant {001}, {011} and {111} reflections indicate that the crystal specimens were [001]$_C$-, [011]$_C$- and [111]$_C$-oriented in the rhombohedral phase.

Figure 12.10(b) presents the XRD patterns for the [111]$_C$-oriented PIN-PMN-PT crystal and PZT ceramic specimens before loading. The XRD patterns indicate that the atomic structures of PIN-PMN-PT and PZT 95/5 specimens are very similar. The {111} reflections show a predominantly rhombohedral distortion for PIN-PMN-PT and PZT 95/5. For PZT 52/48, the peak splitting that was observed in the {100} and {200} reflections indicates a predominantly tetragonal distortion.

Figure 12.10(c) shows XRD patterns for the PIN-PMN-PT [111]$_C$-oriented crystals before and after loading. The two XRD patterns have identical interplanar reflections. The {111} reflections (see inset) indicate that the specimen was in the rhombohedral phase after loading. Similar results were observed for PZT 95/5 specimens. The X-ray diffraction results also indicate that PZT 95/5 depolarized by high strain rate loading is in the rhombohedral phase, i.e. it returns from the orthorhombic to the rhombohedral state after unloading. This observation is consistent with the results of the hydrostatic

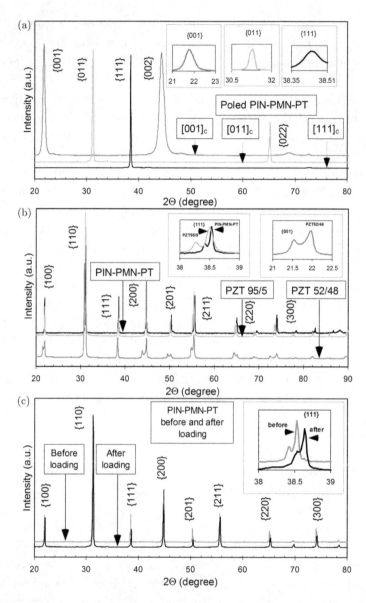

Fig. 12.10. XRD patterns for PIN-PMN-PT single crystals and PZT 95/5 and PZT 52/48 ceramic specimens in the as-received condition and after high strain rate loading experiments: (a) $[001]_C$-, $[011]_C$- and $[111]_C$-oriented crystals in the as-received condition; (b) $[111]_C$-oriented crystals, PZT 95/5 and PZT 52/48 specimens in the as-received condition; (c) $[111]_C$-oriented crystals before and after high strain rate loading.

studies of PZT 95/5 [14]. It was shown that when the pressure for hydrostatically-loaded PZT 95/5 is decreased from 0.32 to 0.14 GPa, it undergoes a transformation from the orthorhombic state back to the rhombohedral state.

Based on the obtained results, one could arrive at the conclusion that, similarly to PZT 95/5, the PIN-PMN-PT crystals pass through a rhombohedral to *non-polar* orthorhombic (R-O) phase transition induced by uniaxial strain adiabatic compression, but if this were the case it should occur in all three types of domain-engineered crystals. There is a well understood stress-induced FE_R to FE_O phase transformation occurring in the domain-engineered single crystals that can help to explain the observed depolarization behavior.

12.6.3 *Release waves*

Complete depolarization of the $[111]_C$-oriented crystal cannot be explained by a shock-induced R-O phase transition because, unlike PZT 95/5, this is a polar-to-polar phase transformation. A possible explanation is provided by considering the multiaxial time-dependent state of stress in the crystals and its interaction with a known stress-driven phase transformation. The $[111]_C$-oriented crystals transform from the single-domain rhombohedral phase to a lower-symmetry multi-domain orthorhombic phase, while the $[011]_C$- and $[001]_C$-oriented crystals transform from the multi-domain rhombohedral phase to the multi-domain orthorhombic phase under certain states of stress. The stress-induced R-O phase transformation has been observed in $[011]_C$- and $[001]_C$-oriented PIN-PMN-PT crystals under quasi-static compressive stress not exceeding 100 Mpa [14, 15]. The R-O phase transformation in $[011]_C$-oriented PIN-PMN-PT crystals associated with high-cycle fatigue (polarization degradation) has also been observed [16].

There is one more factor that can have an effect on the depolarization of ferroelectrics under shock loading, i.e. the release waves traveling behind the shock front. They decompress the material in the directions perpendicular to shock front propagation.

The release waves traveling behind the shock front can have an effect on the depolarization of ferroelectrics under shock loading.

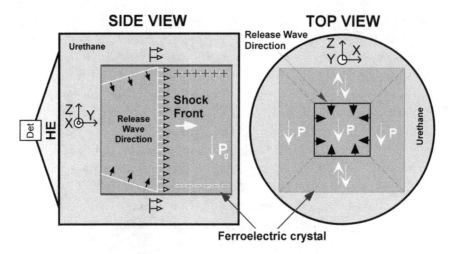

Fig. 12.11. Snapshot in time of a ferroelectric crystal under high strain rate loading, where the combination of the stresses creates an inhomogeneous stress field. The shock wave moves in the Y-direction. P_0 is the remanent polarization vector (Z-direction). The polarity of the surface charge in the crystal is shown by (+) and (−) signs.

They decompress the material in the directions perpendicular to the shock front propagation. Consider the cross-sections of a ferroelectric crystal in Figure 12.11. The adiabatic uniaxial compression of the crystal in the Y-direction occurs with a corresponding smaller level of compression in the X- and Z-directions. Using an elastic isotropic approximation, the X and Z stress components immediately behind the shock front are given by:

$$\sigma_{xx} = \sigma_{yy} = \frac{\sigma_{zz}\,\nu}{1 - \nu} \tag{12.1}$$

where σ_{xx}, σ_{zz}, and σ_{yy} are components of stress in the X-, Z-, and Y-directions respectively, and ν is the Poisson's ratio. As the shock wave moves into the specimen in the Y-direction, the stress field changes along the specimen behind the shock location. The acoustic impedance of the urethane is lower than that of the ferroelectric and it results in reducing the lateral stress components σ_{xx} and σ_{zz}.

In each of the wedge-shaped regions surrounding the square center area in Figure 12.11, the stress component normal to the side surface is decreased due to the interface boundary conditions. In the right and left wedges this has little effect on the polarization, because σ_{xx} is more compressive than σ_{zz}. Note that σ_{xx} provides a driving force for maintaining the polarization in the Z-direction, and it will also have an effect on the X-component of the polarization. In the top and bottom wedges, however, σ_{zz} is more compressive than σ_{xx}. This provides a driving force for rotating the polarization from the Z-direction to the X- and Y-directions through a ferroelastic effect (non-180° domain reorientation).

The partial depolarization of transverse-shock-compressed PZT 52/48 (Table 12.2) is the result of the effect of the release waves described above. PZT 52/48 is predominantly in the tetragonal FE phase at ambient conditions, and it remains predominantly in the tetragonal FE phase under the experimental conditions used in this work. As the transverse shock propagates through the entire specimen, the effect of the release wave rotates part of the original polarization [17].

12.6.4 *Mechanism of complete stress-induced depolarization of PIN-PMN-PT single crystals*

Neither the polar-to-polar phase transformation in the $[111]_C$-oriented crystals nor the release wave effect makes the mechanism of the complete depolarization of the crystals immediately apparent. However, the results described above give the grounds to propose a new mechanism responsible for the complete stress-induced depolarization of the $[111]_C$-oriented crystals. This mechanism is based on a combination of the shock-induced phase transformation and the reorientation of domains caused by release waves traveling behind the shock compressive front. The PIN-PMN-PT crystal polarization variants in the rhombohedral and orthorhombic phases are shown in Figure 12.12.

As adiabatic compression is applied in the Y-direction, a mechanism was proposed [18] wherein a transformation from the rhombohedral to orthorhombic phase takes place in the crystals. The four variants of $[001]_C$-oriented crystal being populated in the R phase are transformed into four possible O-phase variants (Figure 12.12(a)). The two O-phase variants lying in the X-Z plane become populated.

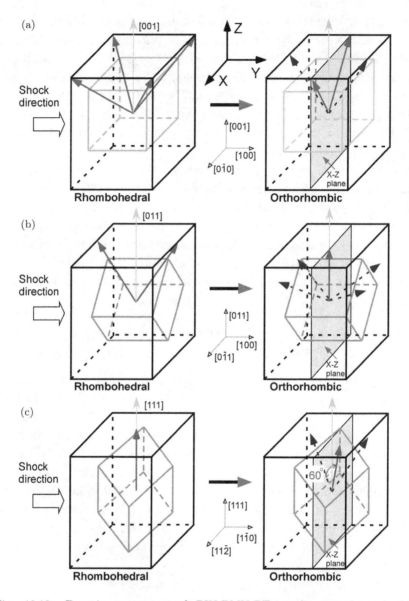

Fig. 12.12. Domain structures of PIN-PMN-PT single crystals and the rhombohedral–orthorhombic phase change process under high strain rate loading: (a) multi-domain $[001]_C$-oriented crystal with $4mm$ symmetry; (b) multi-domain $[011]_C$-oriented crystal with $mm2$ symmetry; (c) single-domain $[111]_C$-oriented crystal with $3m$ symmetry. The cubic phase is shown by gray in each crystal cut. The orthorhombic phase variants eliminated due to the shock compression are marked on the figure.

The other two variants lying in the Y-Z plane are eliminated due to the shock compression. There could be either a continuous rotation of the polarization through an intermediate monoclinic phase or a continuous evolution of the volume fraction of the O phase under shock compression.

There is no external electric field applied to the crystals along the Z-direction. Because of this the two R-phase variants in the $[011]_C$-oriented crystal are transformed into five O-phase variants (Figure 12.12(b)). The four of five O-phase variants having the Y-direction components are eliminated due to the shock compression and one variant becomes populated.

The $[001]_C$- and $[011]_C$-oriented crystals being in the orthorhombic phase are compressed in the X-Z plane perpendicular to the shock wave direction by the release waves propagating behind the shock front. The time of the mechanical destruction of the crystals behind the shock front is two orders of magnitude longer than the release wave propagation time. Upon passing the release waves, the crystal structure is transformed back into the rhombohedral phase with a different domain state and reduced polarization.

Opposite to the $[001]_C$- and $[011]_C$-oriented crystals that undergo a multidomain-to-multidomain phase transition under shock compression, the polarization in the $[111]_C$-oriented crystal begins in a single-domain rhombohedral state, which is no longer supported by the structure, and which transforms to a multi-domain orthorhombic state. There are no selection criteria (i.e. there is no priority in this process) in the multi-domain state which is textured by the shock compression and release waves propagating behind the shock front. In the multi-domain state the polarization is not oriented in one direction as it was in the single-domain state. The one R-phase variant in the $[111]_C$-oriented crystal is transformed into three O-phase variants (Figure 12.12(c)). The angle between the O-phase variants is 60° [19]. Two of the three O-phase variants lying in the Y-Z plane are eliminated due to shock compression and only one variant lying in the X-Z plane becomes populated. The crystal has anisotropic elastic properties, resulting in a different release wave velocity perpendicular to the polarization direction than that parallel to the polarization direction, making the stress pattern more complicated. The release wave provides a driving force for rotating the polarization in the X-Z plane through the ferroelastic effect. A multi-domain structure would develop to minimize the field and therefore, the shocked crystal becomes completely depolarized when it is transformed back into the rhombohedral phase.

There is one more important factor contributing to the complete depolarization of the single-domain crystals. In multi-domain crystals, the domain walls provide a resistance to depolarization, and defects in the crystals can reinforce this through a memory effect. Both effects are expected to be greatly reduced in single-domain $[111]_C$-oriented crystals. Based on all these, there is evidence that the mechanism of the complete depolarization of the $[111]_C$-oriented crystals is not related to a non-polar phase transition but is caused by a shock-induced transformation from the single-domain rhombohedral phase to the multi-domain orthorhombic phase compressed by the release waves, leading to domain wall motion, polarization reorientation, and loss of the initial remanent polarization. To minimize energy, the resulting multidomain state is reoriented to give net zero polarization.

The mechanisms of stress-induced depolarization are similar for PIN-PMN-PT crystals with different domain configurations. However, the multidomain-engineered PIN-PMN-PT crystals are more resistant to stress-induced depolarization and are not completely depolarized under high stress. The differences in the depolarization of $[001]_C$- and $[011]_C$-oriented crystals could be caused by different phase change processes and different domain wall motion and domain reorientation induced by complex time-dependent multi-directional stress.

A single-domain state makes the behavior of the crystals under high stress different from that of the multidomain crystals. The complete depolarization of single-domain $[111]_C$-oriented crystals with high stress-induced electric charge density is very important for applications in ultrahigh-power ferroelectric generators harvesting the ultimate energy density from ferroelectric materials.

The mechanism of complete stress-induced depolarization without transition through the non-polar phase described above opens the way for the development of a new class of high energy density ferroelectric materials for ultrahigh-power applications. It is possible that this mechanism provides complete depolarization for not only rhombohedral crystals, but also single-domain crystals based on other structures.

12.7 Summary

- It was experimentally demonstrated that a domain-engineered state plays an important role in the remanent polarization of PIN-PMN-PT ferroelectric crystals.
- The remanent polarization of rhombohedral PIN-PMN-PT single-domain $[111]_C$-oriented crystals with $3m$ symmetry was found to be $48\,\mu C/cm^2$. It is significantly higher than the remanent polarization for all known ferroelectric materials (ceramics, single crystals, films, polymers, composites).
- The remanent polarizations of rhombohedral PIN-PMN-PT multi-domain $[001]_C$-oriented crystals with $4mm$ symmetry and multi-domain $[011]_C$-oriented crystals with $mm2$ symmetry were found to be 25 and $38\,\mu C/cm^2$, respectively.
- The PIN-PMN-PT single crystals with different domain symmetries were subjected to transverse shock depolarization studies along with high-resolution X-ray diffraction. The obtained experimental results were directly compared to those for PZT 95/5 and PZT 52/48 ferroelectric ceramics under identical loading conditions.
- The experimental results indicate that domain configuration has a significant effect on the stress-induced depolarization. The rhombohedral PIN-PMN-PT single-domain $[111]_C$-oriented crystals were completely depolarized under $3.9\,GPa$ transverse shock compression and released stress-induced charge with density equal to their remanent polarization, $48\,\mu C/cm^2$. On the other hand the PIN-PMN-PT $[001]_C$- and $[011]_C$-oriented crystals were just partially depolarized under transverse stress.
- Since the 1960s, there has been a belief that a complete stress-induced depolarization of ferroelectric materials can only be caused

by phase transitions from a polar (ferroelectric) phase to a non-polar (antiferroelectric or cubic) phase. For instance, a complete depolarization of shock-compressed PZT 95/5 ferroelectric ceramics (widely used in explosive ferroelectric generators) is the result of a pressure induced FE-to-AFE phase transition driven by the hydrostatic compression component of the stress tensor.

- However, the study of the shock depolarization of domain-engineered ferroelectric single crystals revealed the new mechanism of the complete stress-induced depolarization of ferroelectric materials.

- The mechanism of complete stress-induced depolarization of the PIN-PMN-PT single-domain $[111]_C$-oriented crystals is unique without transition through a non-polar phase. The single crystals are undergoing a ferroelectric rhombohedral to orthorhombic phase transition under high strain rate loading. The PIN-PMN-PT orthorhombic phase is a multi-domain polar phase with domains having polarization aligned along the $[011]_C$ crystallographic direction. As a result of the rhombohedral to orthorhombic phase transition the $[111]_C$ single-domain state is transformed into a multi-domain structure that is compressed by the unloading waves propagating behind the shock front. This leads to domain wall motion, polarization reorientation and loss of the initial remanent polarization. To minimize energy, the resulting multi-domain state is reoriented to give net zero polarization.

- A new mechanism of complete stress-induced depolarization without transition through a non-polar phase opens the way for the development of a new class of high energy density ferroelectric materials for ultrahigh-power ferroelectric generators. It is possible that this new mechanism provides complete depolarization for not only rhombohedral crystals, but also single-domain crystals based on other structures.

Bibliography

1. S.E. Park and T.R. Shrout, Ultrahigh strain and piezoelectric behavior in relaxor based ferroelectric single crystals, *J. Appl. Phys.* **82**(4) (1997) pp. 1804–1811.

2. S. Zhang and F. Li, High performance ferroelectric relaxor-PbTiO3 single crystals: Status and perspective, *J. Appl. Phys.* **111**(3) (2012) p. 031301.

3. H. Luo, G. Xu, H. Xu, P. Wang and W. Yin, Compositional homogeneity and electrical properties of lead magnesium niobate titanate single crystals grown by a modified Bridgman technique, *Jpn. J. Appl. Phys.* **39** (2000) pp. 5581–5585.

4. Y. Guo, H. Luo, T. He and Z. Yin, Peculiar properties of a high Curie temperature Pb(In1/2Nb1/2)O3–PbTiO3 single crystal grown by the modified Bridgman technique, *Solid State Comm.* **123** (2002) pp. 417–420.

5. G. Xu, K. Chen, D. Yang and J. Li, Growth and electrical properties of large size Pb.. In1/2Nb1/2... O3–Pb.. Mg1/3Nb2/3... O3–PbTiO3 crystals prepared by the vertical Bridgman technique, *Appl. Phys. Lett.* **90** (2007) p. 032901.

6. J. Tian, P. Han, X. Huang and H. Pan, Improved stability for piezoelectric crystals grown in the lead indium niobate-lead magnesium niobate-lead titanate system, *Appl. Phys. Lett.* **91** (2007) p. 222903.

7. S. Zhang, L. Jun, W. Hackenberger and T.R. Shrout, Characterization of Pb,, In1/2Nb1/2... O3–Pb, Mg1/3Nb2/3... O3–PbTiO3 ferroelectric crystal with enhanced phase transition temperatures, *J. Appl. Phys.* **104** (2008) p. 064106.

8. E. Sun and W. Cao, Relaxor-based ferroelectric single crystals: Growth, domain engineering, characterization and applications, *Progress in Materials Science* **65** (2014) pp. 124–210.

9. S. Zhang, F. Li, X. Jiang, J. Kim, J. Luo and X. Geng, Advantages and challenges of relaxor-PbTiO3 ferroelectric crystals for electroacoustic transducers — A review, *Prog. Mater. Sci.* **68**(205) pp. 1–66.

10. S.I. Shkuratov, J. Baird, V.G. Antipov, E.F. Talantsev, J.B. Chase, W.S. Hackenberger, J. Luo, H.R. Jo and C.S. Lynch, Ultrahigh energy density harvested from domain-engineered relaxor ferroelectric single crystals under high strain rate loading, *Sci. Rep.* **7** (2017) p. 46758.

11. S. Zhang, F. Li, J. Luo, R. Xia, W. Hackenberger and T. Shrout, Investigation of single and multidomain Pb,, In0.5Nb0.5... O3–Pb,, Mg1/3Nb2/3... O3–PbTiO3 crystals with *mm*2 symmetry, *Appl. Phys. Lett.* **97** (2010) p. 132903.

12. S. Zhang, F. Li, N. Sherlock, J. Luo, H. Lee, R. Xia, R. Meyer, W. Hackenberger and T. Shrout, Recent developments on high Curie temperature PIN–PMN–PT ferroelectric crystals, *J. Crystal Growth* **318** (2011) pp. 846–850.

13. R.E. Tipton, A 2D Lagrange MHD code, in *Proc.4ᵗʰ Int. Conf. Megagauss Magnetic Field Generation and Related Topics* (Editors: C.M. Fowler, R.S. Caird and D.J. Erickson) (Plenum Press, New York, USA. 1987) pp. 299–302.

14. S. Zhang, S. Taylor, F. Li, J. Luo and R.J. Meyer, Jr., Piezoelectric property of relaxor-PbTiO3 crystals under uniaxial transverse stress, *Appl. Phys. Lett.* **102** (2013) p. 172902.

15. J.A. Gallagher, J. Tian and C.S. Lynch, Effects of composition and temperature on the large field behavior of $[011]_C$ relaxor ferroelectric single crystals, *Appl. Phys. Lett.* **105** (2014) p. 052909.

16. S. Zhang, J. Luo, F. Li and R.J. Meyer, Polarization fatigue in $Pb(In0.5 Nb0.5)O3–Pb(Mg1/3Nb2/3)O3–PbTiO3$ single crystals, *Acta Mater.* **58** (2010) pp. 3773–3780.

17. S.I. Shkuratov, J. Baird, V.G. Antipov, E.F. Talantsev, H.R. Jo, J.C. Valadez and C.S. Lynch, Depolarization mechanisms of $PbZr_{0.52}Ti_{0.48}O_3$ and $PbZr_{0.95}Ti_{0.05}O_3$ poled ferroelectrics under high strain rate loading, *Appl. Phys. Lett.* **104** (2014) p. 212901.

18. S.I. Shkuratov, J. Baird, V.G. Antipov, E.F. Talantsev, J.B. Chase, W.S. Hackenberger, J. Luo, H.R. Jo and C.S. Lynch, Mechanism of complete stress-induced depolarization of relaxor ferroelectric single crystals without transition through non-polar phase, *Appl. Phys. Lett.* **112** (2018) p. 122903.

19. M. Davis, Phase transitions, anisotropy and domain engineering: the piezoelectric properties of relaxor-ferroelectric single crystals. *Ecole Polytechnique Federale De Lausanne, These No 3513* (May 2006).

Chapter 13

Transversely Shock-Compressed Ferroelectrics: Electric Charge and Energy Transfer into Capacitive Load

13.1 Introduction

A variety of conventional pulsed power systems are based on capacitive energy storage devices [1, 2]. The Arkadiev-Marx generator is a typical example of a capacitor-based pulsed power system [1]. In conventional pulsed power systems, electric energy is provided by high-voltage sources powered from 110/220 V – 50/60 Hz supply lines. The operational theory of these generators and their uses is well developed [1, 2].

Certain modern applications, however, require that the pulsed power system be autonomous, i.e. that it use no external power supply line. Another important requirement is that these autonomous pulsed power systems be compact [3]. Based on these requirements, explosive ferroelectric generators are ideal sources for charging capacitor-based pulsed power systems, because of their ability to produce high-voltage pulses with amplitudes of up to a few hundreds of kilovolts (see Chapters 7 and 8).

In this chapter, the results of systematic studies of the operation of transversely shock-compressed PZT 52/48 and PZT 95/5 ferroelectrics with capacitive loads in a wide range of load capacitances are presented. To enable a direct comparison of the results obtained with PZT 52/48 and PZT 95/5, an identical specimen geometry for the two ferroelectrics was used in the experiments described in this chapter.

The experimental results indicate that electric breakdown within shocked ferroelectric elements is one of the limiting factors for the electric charge and energy transfer from transverse shock wave FEG into capacitive load. The limitations on the operation of transverse FEG in the charging mode caused by the electric conductivity of the mechanically fragmented zone of shocked ferroelectric elements are discussed.

13.2   Transverse Shock Wave FEG with Capacitive Load

The charging mode experiments were conducted with transverse shock wave FEGs containing the PZT 52/48 and PZT 95/5 ferroelectric elements. To enable a direct comparison of the results obtained with PZT 52/48 and PZT 95/5, an identical specimen geometry for the two ferroelectrics was used in the experiments (12.7 mm thick × 12.7 mm wide × 50.8 mm long).

Schematics of the transverse shock wave ferroelectric generator with a ferroelectric element and the measuring circuit used in this study are shown in Figure 13.1. The diameter of the generators was 38 mm. The FEG was based on spherically expanding shock waves. More details on this type of FEG can be found in Chapter 5 of this book. The shock propagation direction was perpendicular to the remanent polarization vector (transverse shock).

The FEG was placed in the blast chamber. The output terminals of the FEG were connected to the capacitor bank. The voltage across the capacitive load was monitored with either a Tektronix P6015A high-voltage probe (resistance 100 MΩ, capacitance 3 pF, transition time 4 ns) or with a North Star PVM-5 high-voltage probe (resistance 400 MΩ, capacitance 12 pF, transition time 8 ns). The current in the load circuit was monitored with a Pearson Electronics current probe (model 411).

The capacitor bank (C_{Load} in Figure 13.1) contained low inductive TDK ceramic high-voltage capacitors with a nominal voltage of twice the charging voltage amplitude in the experiments. To suppress

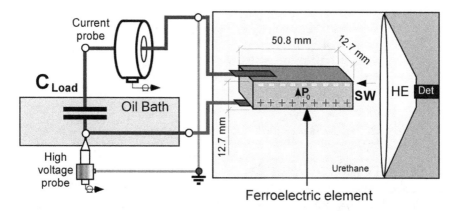

Fig. 13.1. Schematics of the transverse shock wave FEG with a ferroelectric element and the measuring circuit used in investigations of FEG–capacitive load systems. P_0 is the remanent polarization vector. SW is the shock wave vector. The polarity of the surface charge in the ferroelectric element is shown by $(+)$ and $(-)$ signs.

corona discharge and electric breakdown between the electrodes of the load the capacitors were placed in an oil bath. The capacitor bank, high-voltage and high-current probes were placed outside the blast chamber.

The electrical operation of shock-compressed ferroelectrics connected to an external capacitive load is different from that of conventional capacitive storage devices. When the shock wave travels across the ferroelectric element, the electric charge is released on the electrodes of the element. This charge is utilized for charging the ferroelectric element itself and an external capacitive load. Shock compression results in both the shock-induced depolarization of ferroelectric material and a continuous changing of its electrical and mechanical properties during the charging process. These changes might have an effect on the electric charge and energy transfer from the shocked ferroelectrics into external capacitive loads.

The equivalent circuit of transverse shock wave FEG connected to a capacitive load is shown in Figure 13.2. The Q_G represents the electric charge released due to the shock depolarization of the ferroelectric element. The C_{Load} and C_{FEG} represent capacitances of the load and ferroelectric element. The C_1 and C_2 represent the

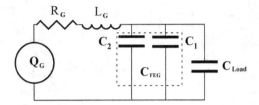

Fig. 13.2.   Equivalent circuit of transverse FEG with a capacitive load.

capacitance of the uncompressed and compressed zones, respectively. The R_G and L_G represent the resistance and inductance of the system. The C_{FEG} and C_{Load} are connected in parallel and they form the capacitance of the ferroelectric element-capacitive load system:

$$C_{tot} = C_{FEG} + C_{Load}, \tag{13.1}$$

where C_{tot} is the total capacitance of the FEG-capacitive load system.

Under shock compression the ferroelectric element is divided into two zones, the uncompressed zone and the compressed zone, as is shown in Figure 13.3. The capacitance of the shocked specimen is:

$$C_{FEG} = C_1 + C_1 = \frac{\varepsilon_0 \varepsilon_{r1} A_1}{d} + \frac{\varepsilon_0 \varepsilon_{r2} A_2}{d} \tag{13.2}$$

where C_1 and C_2 are the capacitances of the uncompressed and compressed zones, ε_0 is the permittivity of free space ($\varepsilon_0 = 8.85 \cdot 10^{-12}\,\mathrm{F \cdot m}$), ε_{r1} and ε_{r2} are the relative permittivities of polarized and depolarized ferroelectrics ($\varepsilon_{r1} = 1300$ and $\varepsilon_{r2} = 1140$ for PZT 52/48; $\varepsilon_{r1} = 295$ and $\varepsilon_{r2} = 225$ for PZT 95/5), d is the ferroelectric element thickness ($d = 12.7\,\mathrm{mm}$). A_1 and A_2 are the areas of the electrodes of the uncompressed and compressed zones. The areas of the electrodes of the uncompressed and compressed zones are as follows:

$$A_1 = w(l - L_{SW}) \tag{13.3}$$

$$A_2 = wL_{SW} \tag{13.4}$$

where w is the width of the ferroelectric element ($w = 12.7\,\mathrm{mm}$), l is the length of the ferroelectric element ($l = 50.8\,\mathrm{mm}$), and L_{SW} is the

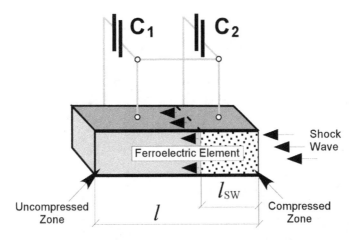

Fig. 13.3. Schematics of transversely shock-compressed ferroelectric element and capacitances of the uncompressed zone, C_1, and the compressed zone, C_2.

shock wave transit time in the element:

$$L_{SW} = \tau_{SW} U_s \qquad (13.5)$$

where τ_{SW} is the shock wave transit time, and U_s is the shock wave velocity.

13.3 Operation of Transversely Shock-Compressed PZT 52/48 with Capacitive Loads

13.3.1 *Energy generated by PZT 52/48 without capacitive load*

Before the experiments with capacitive loads were conducted, a few PZT 52/48 specimens obtained from ITT Corp. for this study were investigated in the short-circuit depolarization mode and the open circuit mode. These data were used for direct comparison with the results obtained in the charging mode and for analysis of the experimental results.

Figure 13.4 presents a typical waveform of high voltage produced by the PZT 52/48 element transversely shock-compressed within an FEG in the open circuit mode. The voltage increased to its maximum and then rapidly decreased due to the electric breakdown within the

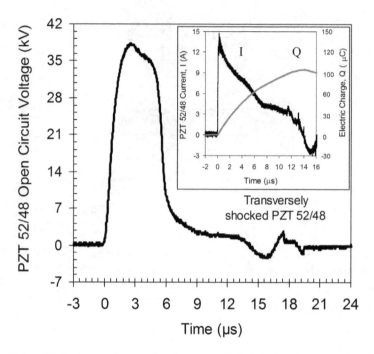

Fig. 13.4. Typical waveform of voltage produced by a transversely shock-compressed PZT 52/48 element ($12.7 \times 12.7 \times 50.8\,\text{mm}^3$) in the open circuit mode. The inset shows a typical waveform of the current and the dynamics of electric charge released by a transversely shocked PZT 52/48 specimen ($12.7 \times 12.7 \times 50.8\,\text{mm}^3$) in the short-circuit depolarization mode.

PZT 52/48 element (see Chapters 7 and 8 of this book for details). The voltage amplitude, $V_{OC52/48} = 38.2 \pm 1.7kV$, and voltage rise time, $2.5 \pm 0.2\,\mu s$, were in good agreement with those obtained earlier for elements of this type under identical loading conditions [5, 6].

The inset in Figure 13.4 shows a typical waveform of the short-circuit depolarization current and the dynamics of stress-induced charge released by the PZT 52/48 element under transverse shock. The current amplitude was $I_{SW52/48} = 14.9 \pm 2.1$ A and the electric charge was $Q_{SW} = 97 \pm 6\mu C$. It is about 50% of the total charge stored in the specimen (remanent polarization 32 $\mu C/cm^2$, total charge 206 μC). The stress-induced current amplitude, pulse duration and amount of charge released by transversely shocked PZT 52/48 were in good agreement with the results obtained earlier with PZT 52/48 specimens of this type [4]. This indicates good reproducibility of the transverse shock depolarization of PZT 52/48.

In the open circuit mode, the stress-induced charge is not transferred into the external circuit but utilized for charging the PZT 52/48 element of the FEG itself. The energy generated by the transversely shocked PZT 52/48 element in the open circuit mode was determined. When the voltage reached the maximum $(V_{OC52/48}(2.5\mu s) = 38\,kV$ in Figure 13.4) the shock wave passed through 9 mm of the PZT 52/48 element (total length 50.8 mm) and depolarized it. The capacitance of the element, $C_{SW52/48}$, at this moment of time can be determined with Eq. (13.2).

The substitution of the parameters of the PZT 52/48 element into Eq. (13.2) ($\tau_{SW} = 2.5\ \mu s$, $U_{SW} = 3.8\,mm/\mu s$, $A_1 = 530\,mm^2$, $A_2 = 114\,mm^2$, $d = 12.7\,mm$, $\varepsilon_{r1} = 1300$ and $\varepsilon_{r2} = 1140$) gives us the capacitance of the element at the moment of time when the voltage reached its maximum, $C_{FEG}(2.5\,\mu s) = 0.52\,nF$.

The energy generated by the PZT 52/48 element in the open circuit mode can be determined from the following equation:

$$W_{OC} = \frac{1}{2} C_{FEG} V_{OC}{}^2 \qquad (13.6)$$

The substitution of $C_{FEG}(2.5\,\mu s) = 0.52\,nF$ and $V_{OC52/48}(2.5\,\mu s) = 38\,kV$ into Eq. (13.6) gives us $W_{OC52/48} = 0.4\,J$. It is the lower bound of the generated energy because the capacitance of the PZT 52/48 element, C_{FEG}, was determined using the relative dielectric permittivity of PZT 52/48 obtained from low electric field measurements. The actual permittivity of the ferroelectric material under high electric field can be significantly higher than that measured at low electric field (see Chapter 8 of this book for details).

There is another way to determine the energy generated by PZT 52/48 in the open circuit mode. Assuming that there is no leakage current in the PZT 52/48 element during the first few microseconds of shock wave transit, one can determine the capacitance of the element using data obtained from shock depolarization experiments (Figures 13.4 and 13.5). The capacitance of the element, C_{FEGexp}, can be determined as follows:

$$C_{FEGexp} = \frac{Q_{exp}}{V_{OC}} \tag{13.7}$$

where V_{OC} is the amplitude of high voltage ($V_{OC52/48}(2.5\,\mu s) = 38\,\text{kV}$ in Figure 13.4), Q_{exp} is the experimentally obtained stress-induced charge released by the PZT 52/48 element (see the inset in Figure 13.4) at the same moment of time when the voltage reached the maximum, $Q_{exp52/48}(2.5\,\mu s) = 33\,\mu\text{C}$. It gives us $C_{FEGexp52/48}(2.5\,\mu s) = 0.9\,\text{nF}$.

The capacitance of the PZT 52/48 element determined from the experimental data $C_{FEGexp52/48}(2.5\,\mu s) = 0.9\,\text{nF}$ is a factor of 1.6 higher than that calculated with the low field dielectric permittivity of PZT 52/48 (0.52 nF). The substitution of $C_{FEGexp52/48}(2.5\,\mu s) = 0.9\,\text{nF}$ and $V_{OC52/48}(2.5\,\mu s) = 38\,\text{kV}$ into Eq. (13.6) gives us the upper bound of the generated energy, $W_{OCexp52/48} = 0.65\,\text{J}$.

The difference between the capacitance of the PZT 52/48 element, $C_{FEGexp52/48}$, when determined from the experimental data and when calculated with Eq. (13.2) using the low field permittivity of PZT 52/48, indicates that high electric fields have a significant effect on the dielectric properties of PZT 52/48, in particular on its permittivity.

As is discussed in Section 8.7 of Chapter 8 of this book, an application of high electric field to ferroelectric materials results in an increase of their dielectric permittivity. This phenomenon is experimentally proved for PZT 95/5 in a narrow range of electric field strength [7–10]. Because of the lack of experimental data for the dielectric permittivities of PZT 52/48 and PZT 95/5 in a wide range of electric field strengths, the capacitances based on the low field dielectric permittivities of PZT 52/48 and PZT 95/5 will be used for analysis of the experimental results obtained in the charging mode.

13.3.2 *Energy generated by PZT 52/48-capacitive load systems*

Experiments with PZT 52/48-capacitive load systems were conducted with a wide range of load capacitances from 1 to 9 nF.

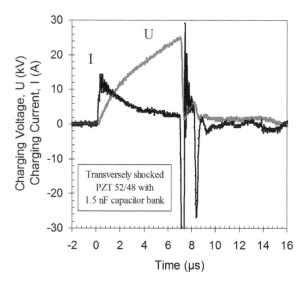

Fig. 13.5. Typical waveforms of the charging current and the charging voltage generated by a transversely shock-compressed PZT 52/48 element ($12.7 \times 12.7 \times 50.8 \, \text{mm}^3$) across a 1.5 nF capacitor bank.

Figure 13.5 presents typical waveforms of the charging voltage and the charging current generated in the PZT 52/48 – 1.5 nF system.

There are clear signs of breakdown in the charging voltage waveform in Figure 13.5. After reaching its maximum at 24 kV, the charging voltage rapidly decreased to zero. At this moment in time the shock wave front was in the middle of the ferroelectric element ($t = 7 \, \mu s$) and shock depolarization was in progress. The breakdown was accompanied by high negative current spikes.

It is obvious that the breakdown occurred not in the heavily insulated capacitor bank but within the ferroelectric specimen. The PZT 52/48 breakdown field in this experiment ($E_{break52/48} = 1.9 \, \text{kV/mm}$ in Figure 13.5) was 38% lower than that for identical PZT 52/48 specimens operating in the open circuit mode ($E_{break52/48} = 3.1 \, \text{kV/mm}$ in Figure 13.4).

Similar signs of electric breakdown were observed in the full range of load capacitances from 1 to 9 nF. In all experiments the charging process was interrupted by an electric breakdown within PZT 52/48 that occurred during shock wave transit when only a part of a ferroelectric element volume was depolarized under transverse shock compression. These experimental results indicate that breakdown in shocked-compressed ferroelectric elements plays a fundamental part in the operation of FEG-capacitive load systems.

Figure 13.6 presents experimentally obtained amplitudes of the charging voltage (i.e. PZT 52/48 breakdown voltage) as a function of the load capacitance. The PZT 52/48 breakdown voltage (charging voltage) strongly depends on the load capacitance. An increase in the load capacitance from 1 to 9 nF resulted in a decrease in the breakdown (charging) voltage from 26 to 6 kV. It should be mentioned that for all load capacitances, PZT 52/48 breakdown voltages and, correspondingly, breakdown fields were lower than those recorded for PZT 52/48 elements in the open circuit mode ($V_{OC52/48} = 38$ kV and $E_{break52/48} = 3.1$ kV/mm in Figure 13.4).

Figure 13.7 shows electric charge transfer from shock-compressed PZT 52/48 elements into capacitive loads. The total charge utilized

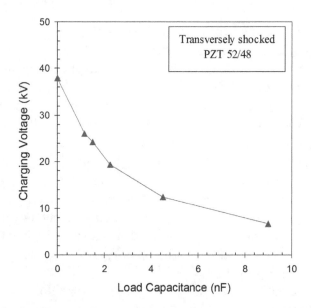

Fig. 13.6. Experimentally obtained amplitudes of the charging voltage as a function of the load capacitance for the PZT 52/48-capacitive load system. The PZT 52/48 specimen size was $12.7 \times 12.7 \times 50.8$ mm^3.

in the PZT 52/48-load system was significantly higher than that utilized in the open circuit mode.

Consider the electric charge transferred from shocked PZT 52/48 into the ferroelectric element itself and into a 1.5 nF capacitor bank. The charge transferred into a 1.5 nF capacitor bank was $Q_{CLoad} = 37\,\mu C$ (see Figure 13.6) and, correspondingly, the total charge utilized in the PZT 52/48-1.5 nF system, was $53\,\mu C$. It was a greater charge than the charge utilized in the open circuit mode ($33\,\mu C$).

The experimental results indicate (Figure 13.7) that an increase in the load capacitance from 1 to 9 nF resulted in an increase in the charge transfer from PZT 52/48 into the load. At the same time, charge utilized for charging the PZT 52/48 specimen itself decreased significantly with the increase of the load capacitance. The electric

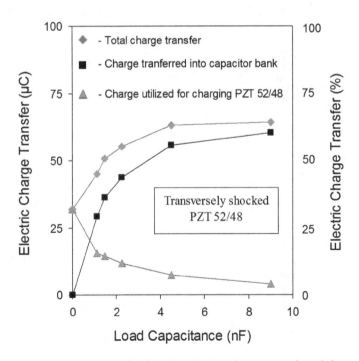

Fig. 13.7. Experimental results for the electric charge transferred from shock-compressed PZT 52/48 elements into capacitive loads as a function of the load capacitance. The PZT 52/48 element size was $12.7 \times 12.7 \times 50.8\text{mm}^3$.

charge transfer into the 9 nF load was 70% of the shock-induced charge released by the PZT 52/48 element.

The energy generated in the FEG-capacitive load system, W_{CMode}, can be determined as follows:

$$W_{CMode} = \frac{1}{2} C_{tot} V_{CLoad}{}^2 \qquad (13.8)$$

where V_{CLoad} is the amplitude of the charging voltage and C_{tot} is the total capacitance of the ferroelectric-capacitive load system.

Figure 13.8 summarizes the results for the energy generated in PZT 52/48-capacitive load systems with different load capacitances. The experimental results indicate that the PZT 52/48-capacitive load systems were capable of producing higher energy than PZT 52/48 in the open circuit mode.

In spite of the interruption of the charging process by electric breakdown within PZT 52/48, the energy generated in the system

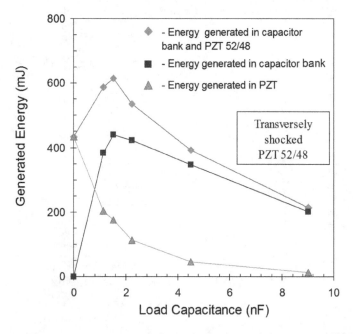

Fig. 13.8. Experimental results for the energy generated in the PZT 52/48-capacitive load system. The PZT 52/48 specimen size was $12.7 \times 12.7 \times 50.8 \text{mm}^3$.

was 35% higher than the lower bound of the energy generated by PZT 52/48 operating in the open circuit mode (see previous section).

There is a clear maximum in the generated energy plot (Figure 13.8). The maximum energy, 620 mJ, was generated with a 1.5 nF load. Apparently, the optimum load capacitance corresponding to the maximum energy generated in the PZT 52/48-capacitive load system depends on the geometric dimensions of a ferroelectric specimen. The specimen thickness determines the amplitude of high voltage generated under transverse stress. The specimen electrode area determines the amount of stress-induced charge.

In conclusion, it is experimentally demonstrated that transversely shock-compressed PZT 52/48 elements with volume of $8\,cm^3$ produce hundreds of kilowatts of pulsed power and are capable of pulse-charging capacitor banks of 1 to 9 nF to high voltages. The voltage generated in the charging mode in the full range of load capacitances used in these experiments was lower than that generated by PZT 52/48 in the open circuit mode. However, the energy generated in the PZT 52/48-capacitive load systems was higher than that generated by PZT 52/48 in the open circuit mode when the stress-induced charge is utilized only for charging the ferroelectric element itself. The experimental results indicate that the process of electric charge and energy transfer from shock-compressed PZT 52/48 into the capacitive load was interrupted by electric breakdown within PZT 52/48 elements. An increase in the PZT 52/48 breakdown field could reduce energy losses and provide significant increases in system efficiency.

13.4   Shock-Compressed PZT 52/48 Breakdown Field on Charging Time Dependence

The experimental results obtained in the charging mode raise an important question: why is the breakdown field of shock-compressed PZT 52/48 operating with a capacitive load different from that in the open circuit mode? The only difference between the two modes of operation is a higher capacitance of the system in the charging mode. This difference leads to some consequences.

A higher capacitance of the FEG-capacitive load system changes the charging process significantly in comparison with that for the open circuit mode. In the charging mode, the source of electric charge utilized for charging the PZT 52/48 element itself and the capacitive load is the same as in the open circuit mode, i.e. it is the shock depolarization of a PZT 52/48 element. Correspondingly, an increase in the systems capacitance results in a longer charging time. An exposure of the ferroelectric element to a high electric field for a longer time could result in a lower PZT 52/48 breakdown field.

The other difference between the two modes of FEG operation is a significantly higher energy generated by PZT 52/48-capacitive load systems in comparison with that for the open circuit mode (Figure 13.9). This higher energy can be utilized for the development of electrical instability at the initial stage of electric breakdown

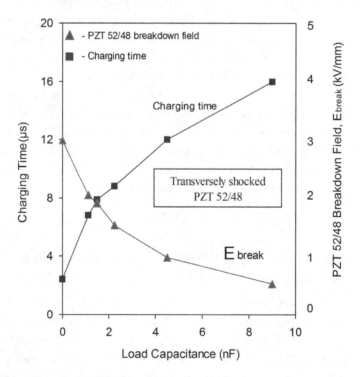

Fig. 13.9.   Experimentally obtained PZT 52/48 breakdown (charging) fields and charging times as a function of the load capacitance. The PZT 52/48 element size was $12.7 \times 12.7 \times 50.8\,\mathrm{mm}^3$.

within shocked ferroelectrics and can result in the breakdown of PZT 52/48 elements at a significantly lower electric field in comparison with that for the open circuit mode.

Figure 13.9 shows experimentally obtained PZT 52/48 breakdown fields and charging times as a function of the load capacitance. An increase in the load capacitance results in a significant increase in the charging time and, at the same time, in a decrease in the PZT 52/48 breakdown field.

Experimental results (Figure 13.9) indicate that there is a strong correlation between the PZT 52/48 breakdown field and the charging time: a lower PZT 52/48 breakdown field corresponds to a longer charging time.

Figure 13.10 shows the experimentally obtained PZT 52/48 breakdown fields as a function of the charging time. The PZT 52/48 breakdown field is inversely proportional to the charging time.

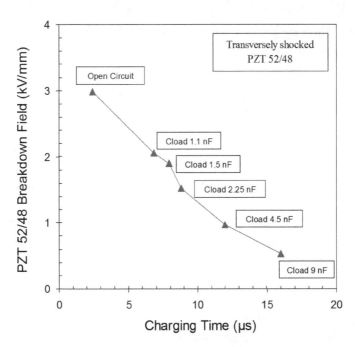

Fig. 13.10. PZT 52/48 breakdown fields as a function of the charging time experimentally obtained for PZT 52/48 elements operating with capacitive loads. The PZT 52/48 specimen size was $12.7 \times 12.7 \times 50.8\,\mathrm{mm}^3$.

In conclusion, the experimental results indicate that the charging time is one of the parameters that have a significant effect on the magnitude of the breakdown field of shock-compressed ferroelectrics and, correspondingly, on the energy generated in ferroelectric-capacitive load systems. There is a strong correlation between the breakdown field and charging time: a lower breakdown field corresponds to a longer charging time and vice versa, i.e. the breakdown field is inversely proportional to the charging time. The breakdown field with charging time dependence can be used for an estimation of the breakdown field of shocked PZT 52/48 for a given charging time. This experimentally obtained relationship should be taken into consideration during the theoretical analysis of the electric breakdown of shocked ferroelectrics.

13.5 Operation of Transversely Shock-Compressed PZT 95/5 with Capacitive Loads

13.5.1 *Energy generated by PZT 95/5 without capacitive load*

The experimental setup, the design of the ferroelectric generators, and the size of the ferroelectric elements and the measuring circuits used in the experiments with PZT 95/5 operating in the charging mode were the same as those for PZT 52/48 (see Figure 13.1). The size of the PZT 95/5 specimens obtained from TRS Technologies was identical to that for PZT 52/48, $12.7 \times 12.7 \times 50.8 \, \text{mm}^3$. The specimens were polarized by the manufacturer across their thickness to the remanent polarization $32 \, \mu\text{C/cm}^2$.

Before the experiments with capacitive loads, a few PZT 95/5 specimens obtained from TRS technologies for this study were subjected to transverse shock depolarization measurements to get data for actual amounts of electric charge and amplitudes of high voltage produced by the specimens in the short circuit and open circuit modes. These data were used for direct comparison with the results obtained in the charging mode and for analysis of the experimental results.

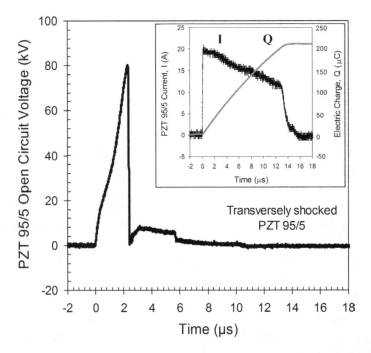

Fig. 13.11.   Typical waveform of high voltage produced by a transversely shock-compressed PZT 95/5 element ($12.7 \times 12.7 \times 50.8$ mm^3) in the open circuit mode. The inset shows a typical waveform of the current and the dynamics of electric charge released by a transversely shocked PZT 95/5 element ($12.7 \times 12.7 \times 50.8$ mm^3) in the short-circuit depolarization mode.

Figure 13.11 shows a typical waveform of high voltage produced by the PZT 95/5 element transversely shock-compressed within an FEG in the open circuit mode. In this mode of operation, stress-induced charge was not transferred into the external capacitive load, but was utilized for charging the PZT 95/5 element itself. The voltage increased to its maximum and then rapidly reduced to zero due to the electric breakdown within the PZT 95/5 element. The voltage amplitude, $V_{OC95/5} = 80.6 \pm 1.7\,kV$, the voltage rise time, $2.3 \pm 0.4\,\mu$s, and the PZT 95/5 breakdown field, 6.4 kV/mm, were in good agreement with those obtained earlier for PZT 95/5 under identical loading conditions [5–11].

The inset in Figure 13.11 shows a typical waveform of the short-circuit depolarization current and the evolution of electric charge released by PZT 95/5 under transverse shock. The current amplitude was $I_{SW95/5} = 21$ A and the electric charge released by PZT 95/5 was $Q_{SW95/5} = 208\,\mu$C. The PZT 95/5 element was completely depolarized under transverse shock. The stress-induced current pulse amplitude, pulse duration and amount of charge released by transversely shocked PZT 95/5 were in good agreement with the results obtained earlier with PZT 95/5 elements under identical loading conditions [4].

Consider energy generated by PZT 95/5 in the open circuit mode. The relative dielectric permittivities of polarized and depolarized PZT 95/5 obtained from low electric field measurements are $\varepsilon_{r1} = 295$ and $\varepsilon_{r2} = 225$, respectively. The capacitance of the PZT 95/5 specimen at the maximum voltage determined with Eq. (13.2) was $C_{FEG}(2.3\,\mu\text{s}) = 129\,$nF. The substitution of $C_{FEG} = 129\,$nF and $V_{OC95/5} = 81\,$kV into Eq. (13.6) gives us the lower bound of the energy generated in the PZT 95/5 element in the open circuit mode, $W_{OC95/5} = 0.42\,$J.

Assuming that there is no leakage current in the PZT 95/5 element during the first few microseconds of shock wave transit one can determine the upper bound of the energy generated by PZT 95/5 (similar to that for PZT 52/48 in Section 13.3.1 above).

The actual capacitance of the PZT 95/5 element at the moment of the maximum voltage, $C_{FEGexp95/5}$, can be determined with Eq. (13.7) using the data obtained from shock depolarization experiments (Figures 13.12 and 13.13). Substitution of the stress-induced charge released by the PZT 95/5 element at the same moment of time when the voltage reached the maximum ($Q_{exp95/5}(2.3\,\mu\text{s}) = 40\,\mu$C in Figure 13.12) and the amplitude of high voltage produced by PZT 95/5 ($V_{OC95/5}(2.3\,\mu\text{s}) = 81\,$kV in Figure 13.13) into Eq. (13.7) gives us $C_{FEGexp95/5}(2.3\,\mu\text{s}) = 0.49\,$nF. Correspondingly, the upper bound of the generated energy is $W_{OCexp95/5} = 1.6\,$J.

The capacitance of the PZT 95/5 element based on the experimental data $C_{FEGexp95/5}(2.3\,\mu\text{s}) = 0.49\,$nF is more than three times

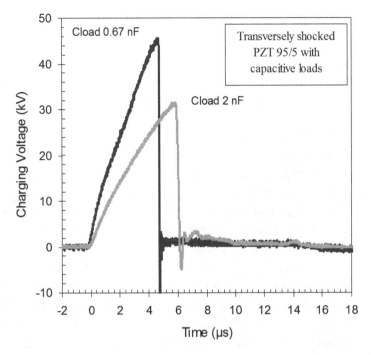

Fig. 13.12. Typical waveforms of charging voltage produced by transversely shock-compressed PZT 95/5 elements ($12.7 \times 12.7 \times 50.8\,\mathrm{mm}^3$) with the 0.67 and 2 nF capacitor banks.

higher than that determined with the low field dielectric permittivity of PZT 95/5. This is an indication of the significant effect of high electric field on the dielectric properties of PZT 95/5, particularly on its permittivity. An application of high electric field to PZT 95/5 results in an increase of the dielectric permittivity of the ferroelectric material (see Section 8.7 of Chapter 8 of this book for more details). This phenomenon was observed earlier in [7–10]. However, the experimental results were obtained for PZT 95/5 for a limited range of the electric field strength. Because of the lack of experimental data, the lower bounds of the permittivity and capacitance of PZT 95/5 elements will be used for an analysis of the results obtained in the charging mode.

13.5.2 *Energy generated by PZT 95/5-capacitive load systems*

The PZT 95/5-capacitive load experiments were conducted in a wide range of load capacitances from 0.5 to 9 nF. Typical waveforms of the charging voltage generated across 0.67 and 2 nF capacitor banks are shown in Figure 13.12. There are clear signs of breakdown within PZT 95/5 elements in both waveforms similar to those observed in the open circuit mode (see Figure 13.11). After reaching the maximum the voltage rapidly decreased to zero. The electric charge and energy transfer from shock-compressed PZT 95/5 into the capacitive load was interrupted due to the breakdown within the PZT 95/5 element similar to that observed in the experiments with PZT 52/48-capacitive load systems.

Experimental results (Figure 13.12) indicate that an increase of load capacitance resulted in a decrease of charging voltage (i.e. PZT 95/5 breakdown voltage) and, at the same time, in an increase of the charging time (i.e. PZT 95/5 breakdown time). Similar results were observed in the experiments with PZT 52/48 (see previous section).

Figure 13.13 shows the amplitudes of the voltage generated by the PZT 95/5-capacitive load system as a function of load capacitance. The experimental results indicate that the amplitudes of voltage generated by PZT 95/5 across capacitive loads were always lower than that for the open circuit mode. Similar results were obtained with PZT 52/48. It should be mentioned that in a full range of the load capacitance the amplitudes of charging voltage generated by PZT 95/5 were higher than those for PZT 52/48-capacitive load systems (see Figure 13.6).

Clear signs of electric breakdown within PZT 95/5 elements were observed in the charging voltage waveforms in a full range of the load capacitance. The electric charge and energy transfer from PZT 95/5 into the capacitive load was interrupted due to the breakdown within shock-compressed PZT 95/5 specimens. Similar results were obtained with PZT 52/48-capacitive load systems.

Figure 13.14 shows electric charge transfer from shock-compressed PZT 95/5 elements into capacitive loads as a function of the load

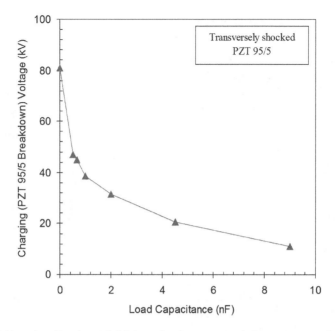

Fig. 13.13. Amplitudes of high voltage generated by transversely shock-compressed PZT 95/5 elements ($12.7 \times 12.7 \times 50.8 \, \text{mm}^3$) across capacitive loads as a function of the load capacitance.

capacitance. An increase in the load capacitance resulted in a progressively higher charge transferred from PZT 95/5 into the load. The charge transferred into the 9 nF load was 50% of the stress-induced charge produced by PZT 95/5 in the short-circuit depolarization mode (see Figure 13.11).

The experimental results indicate that PZT 95/5 is more efficient in the charging mode in comparison with PZT 52/48. The maximum charge transferred from PZT 95/5 into the capacitive load ($100 \, \mu\text{C}$ in Figure 13.14) was higher than that for PZT 52/48 ($70 \, \mu\text{C}$ in Figure 13.7).

Figure 13.15 shows energy generated in the PZT 95/5-capacitive load systems. Experimental results indicate that the energy generated in the PZT 95/5-capacitive load system was up to two times higher than for the open circuit mode. There is a clearly

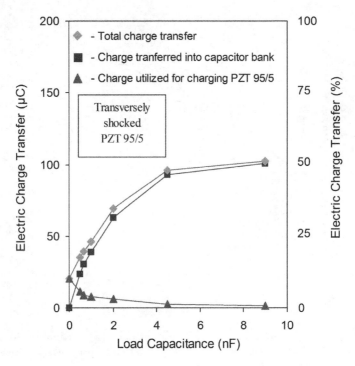

Fig. 13.14. Experimental results for the electric charge transferred from shock-compressed PZT 95/5 elements ($12.7 \times 12.7 \times 50.8 \, \text{mm}^3$) into capacitive loads as a function of the load capacitance.

visible maximum in the generated energy plot. The maximum energy was generated in the PZT 95/5-capacitive load system with load capacitance of 2 nF.

In conclusion, the operation of transversely shock-compressed PZT 95/5 in the charging mode is similar to that for PZT 52/48. PZT 95/5 elements with a volume of $8 \, \text{cm}^3$ produce and are capable of the pulse charging of capacitor banks with capacitance ranging from 0.5 to 9 nF to high voltages. The voltage generated by PZT 95/5 in the charging mode was always lower than that generated in the open circuit mode. At the same time, the energy generated in the PZT 95/5-capacitive load systems was always higher than that generated by PZT 95/5 in the open circuit mode. PZT 95/5 is more effective in the charging mode in comparison with PZT

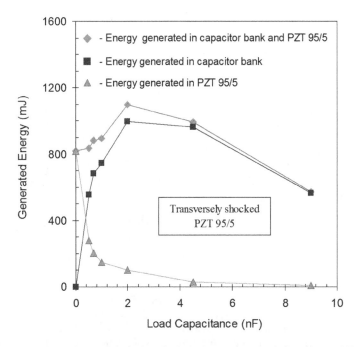

Fig. 13.15. Experimental results for energy generated by the PZT 95/5-capacitive load systems. The size of the PZT 95/5 elements was $12.7 \times 12.7 \times 50.8 \, \text{mm}^3$.

52/48. PZT 95/5-capacitive load systems generate energy 40% higher than that for PZT 52/48-based systems. Similarly to PZT 52/48, charge and energy transfer from transversely shocked PZT 95/5 into a capacitive load was interrupted by electric breakdown within PZT 95/5 elements. Electric breakdown within shocked ferroelectrics plays a fundamental role in the operation of ferroelectric-capacitive load systems.

13.6 Shock-Compressed PZT 95/5 Breakdown Field-Charging Time Relationship

Figure 13.16 shows the PZT 95/5 breakdown field and charging time as a function of the load capacitance. An increase of the load capacitance results in a significant increase in the charging time and, at the same time, in a decrease in the PZT 95/5 breakdown field.

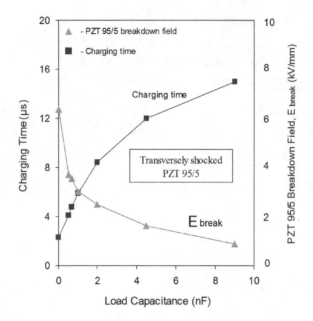

Fig. 13.16. Experimentally obtained PZT 95/5 breakdown field and charging time as a function of the load capacitance for PZT 95/5 elements operating in the charging mode. The PZT 95/5 element size was $12.7 \times 12.7 \times 50.8 \, \text{mm}^3$.

There is a strong correlation between the PZT 95/5 breakdown field and charging time: a lower breakdown field corresponds to a longer charging time. Figure 13.17 shows the PZT 95/5 breakdown field as a function of the charging time. The breakdown field is inversely proportional to the charging time. Similar results were obtained for transversely shock-compressed PZT 52/48 specimens operating in the charging mode (see Section 13.4 above).

In conclusion, the experimental results obtained with PZT 95/5 and PZT 52/48 indicate that the charging (breakdown) time is a fundamental parameter having a significant effect on the process of electric breakdown and the magnitude of the breakdown field in shock-compressed ferroelectrics. The experimentally obtained breakdown field-charging time relationship could help to identify the mechanism of the breakdown of shock-compressed ferroelectrics.

This breakdown field-charging time relationship is also important for practical applications of ferroelectric materials, in particular, for

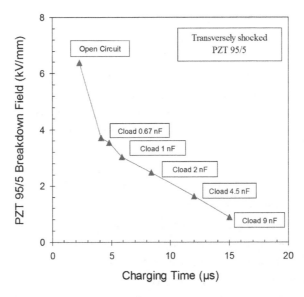

Fig. 13.17. Experimentally obtained PZT 95/5 breakdown fields as a function of the charging (breakdown) time for PZT 95/5 elements operating in the charging mode. The PZT 95/5 element size was $12.7 \times 12.7 \times 50.8 \, \text{mm}^3$.

the optimization of the operation of ferroelectric-capacitive load systems. The results of the optimization based on the abovementioned relationship are presented in the next section.

13.7 Effect of the Geometry of the Ferroelectric Element on Energy Transferred into Capacitive Load

In accordance with the dependence of the experimentally obtained breakdown field on charging time (Figures 13.10 and 13.17), a decrease in the charging time should result in an increase in the ferroelectric breakdown field and, correspondingly, an increase in the efficiency of the ferroelectric-capacitive load system.

One of the possible ways to charge the capacitor bank faster is to provide a higher charging current generated by a ferroelectric element. It can be done through an increase in the remanent polarization and/or an increase in the electrode area.

However, a significant increase in the remanent polarization of ferroelectric elements is not possible due to fundamental limits caused by the nature of ferroelectric materials. An increase in the ferroelectric element electrode area is not always desirable because it could lead to an increase in the size of miniature generators.

The other possible way to decrease the charging time is to provide higher dynamics of shock depolarization through changing the geometric dimensions of the ferroelectric element and its positioning within the FEG without changing the remanent polarization, or element electrode area, or element volume. The results of this approach are described below.

A schematic of transverse FEG with a redesigned PZT 95/5 element is shown in Figure 13.18. The general design of this FEG was identical to that shown in Figure 13.1. The diameter of the FEG was 38 mm. The direction of propagation of the shock wave was perpendicular to the polarization direction. The only difference was the geometrical dimensions of the PZT 95/5 element.

The PZT 95/5 element was redesigned to maximize the dynamics of the stress-induced charge. The size of this element was 12.7 mm thick $\times$ 25.4 mm wide $\times$ 25.4 mm long. The width of the element

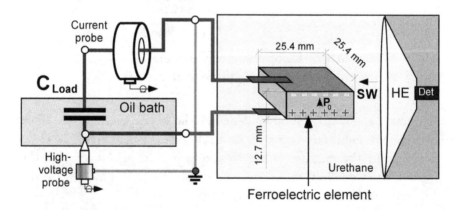

Fig. 13.18.   Schematics of transverse FEG with a redesigned ferroelectric element and the measuring circuit used in the experiments. P_0 is the remanent polarization vector. SW is the shock wave propagation direction. The polarity of the surface charge in the ferroelectric element is shown by (+) and (−) signs.

(where shock waves entered the ferroelectric element) was doubled (from 12.7 to 25.4 mm) in comparison to the old design (see Figure 13.1).

To enable a direct comparison of the results obtained for the redesigned PZT 95/5 elements with results obtained for the previous design, the volume of the redesigned PZT 95/5 element and the area of its electrodes were made equal to those for the rectangular PZT 95/5 element used in the experiments described above (see Figure 13.1).

Figure 13.19 presents a typical waveform of short-circuit depolarization current and an evolution of electric charge released by the redesigned PZT 95/5 element under transverse shock. The current amplitude was $I_{SW95/5} = 36.1$ A. The electric charge released by a square PZT 95/5 element, 200 μC, was practically equal to that for the rectangular PZT 95/5 element (see Figure 13.11).

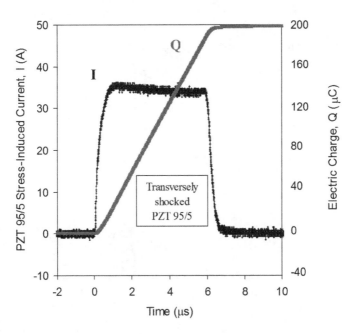

Fig. 13.19. Typical waveform of the current and the evolution of electric charge produced by the redesigned PZT 95/5 element ($12.7 \times 25.4 \times 25.4$ mm^3) under transverse stress in the short-circuit depolarization mode.

The charging mode experiments with redesigned PZT 95/5 elements which were conducted with a 2 nF capacitor bank. A typical waveform of the charging voltage generated by a redesigned PZT 95/5 element across the 2 nF capacitor bank is shown in Figure 13.20. There are clear signs of breakdown within the PZT 95/5 element in the charging voltage waveform (Figure 13.20) similar to those observed in the experiments with rectangular PZT 95/5 elements (see Figure 13.12). In this experiment (Figure 13.20) the PZT 95/5 breakdown (charging) voltage amplitude was 45.4 kV and the corresponding breakdown field was 3.6 kV/mm. It is a significantly higher breakdown field than that for rectangular PZT 95/5 elements of the old design with the same capacitive load (see Figure 13.12 and 13.17).

The experimental results (Figure 13.20) indicate that the PZT 95/5 breakdown field-charging time relationship was in action in the experiment with the redesigned PZT 95/5 element. The shorter PZT

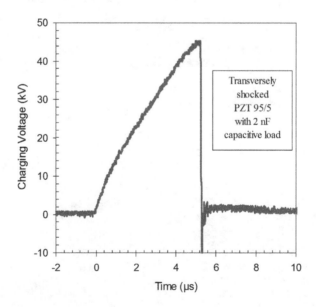

Fig. 13.20. Typical waveform of high voltage produced by a redesigned PZT 95/5 element ($12.7 \times 25.4 \times 25.4$ mm^3, see FEG schematic in Figure 13.18) across a 2 nF capacitor bank.

95/5 charging time corresponded to the higher breakdown field. The charging time for the redesigned PZT 95/5 element, 5.1 μs, was 60% shorter than that for the rectangular elements and the breakdown field for the redesigned PZT 9/5 element, 3.6 kV/mm, was 60% higher than that for the rectangular elements (see Figures 13.12 and 13.17).

The results obtained with redesigned PZT 95/5 elements (Figure 13.20) are in good agreement with the results of Mock and Holt [12]. They reported on the investigation of energy transfer from transversely shocked PZT 95/5 elements into capacitive loads. The size of the PZT 95/5 elements was 25.5 mm thick × 24.9 mm wide × 18.6 mm long. The direction of shock wave propagation was parallel to the 18.6 mm side of the element. It was demonstrated in [12] that transversely shocked PZT 95/5 was capable of charging a 1.2 nF capacitor bank to 81 kV.

The newly-designed PZT 95/5 element made it possible to double the energy generated by the PZT 95/5-capacitive load system (2.3 J) without any changes of the ferroelectric element volume, the remanent polarization, the size of the FEG or the mass of an HE charge.

Based on the obtained experimental results the conclusion can be reached that the dynamics of shock depolarization is one of the important parameters that have a significant effect on the operation of a ferroelectric-capacitive load system. An increase of the dynamics of stress-induced charge through the optimization of the geometrical dimensions of ferroelectric specimens or other means results in an increase in the energy generated by ferroelectrics in the charging mode.

13.8   Conductivity of Mechanically Fragmented Ferroelectrics as the Limiting Factor for the Operation of FEG-Capacitive Load Systems

High-pressure gases and the products of detonation of high explosives eventually destroy the body of the ferroelectric generator, the ferroelectric specimen and other parts of the system. The time required for the mechanical destruction of the ceramics behind the shock front can be estimated as [13]:

$$t_{dest} = \frac{L_{SW}}{u_p} \qquad (13.9)$$

where t_{dest} is the time required for mechanical destruction, L_{SW} is the shock wave travel distance in the ferroelectric specimen, and u_p is the particle velocity. The latter can be determined as [13]:

$$u_p = \frac{p_{SW}}{\rho_0 U_s} \qquad (13.10)$$

where p_{SW} is the shock pressure, ρ_0 is the density of ceramics before the shock action, and U_s is the shock front velocity. In our experiments, $U_s = 3.8\,\text{mm}/\mu\text{s}$, and, correspondingly, $u_p = 0.05\,\text{mm}/\mu\text{s}$. When the shock wave front reached the middle of the element, the destruction time would be $t_{dest} = 500\,\mu\text{s}$.

However, the analysis of the waveforms of the current and the voltage recorded with ferroelectrics charging the capacitive loads along with high-speed photographs of the explosive operation of ferroelectric generators revealed a mechanism of energy losses and a time limiting factor for the operation of ferroelectric-capacitive load systems that is related to the mechanical destruction of the ferroelectric element and that occurred during a relatively short (about $9\,\mu\text{s}$) interval of time.

Figure 13.21 shows a typical waveform of the charging current and the dynamics of electric charge transferred from a transversely shocked PZT 52/48 element into the 9 nF capacitor bank.

Figure 13.21 also shows a typical waveform of the stress-induced current generated by identical PZT 52/48 elements and the dynamics of stress-induced charge released in the short-circuit depolarization mode.

The charging current and short-circuit depolarization current were practically identical for the first nine microseconds of the shock wave transit through the element. There were no signs of energy losses related to a leakage current.

However, after $t = 9\,\mu\text{s}$ (Figure 13.21), the charging current and the short-circuit current behaved differently. The short-circuit depolarization current lasted until $t = 14.7\,\mu\text{s}$ (when the shock wave passed the PZT 52/48 specimen), while the charging current started

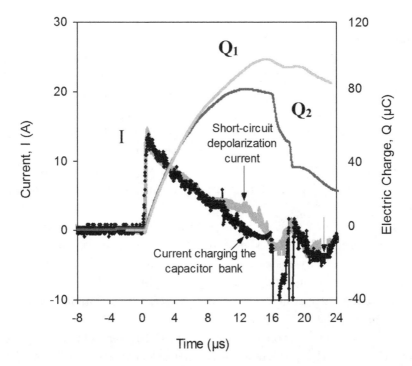

Fig. 13.21. Typical waveform of the charging current and the dynamics of electric charge transferred from a transversely shocked PZT 52/48 element ($12.7 \times 12.7 \times 50.8 \, \mathrm{mm}^3$) into a 9 nF capacitor bank (Q_2); typical waveform of the stress-induced current generated by identical PZT 52/48 element and the dynamics of stress-induced charge (Q_1) released in the short-circuit depolarization mode.

its rapid decrease much earlier, at $t = 9 \, \mu\mathrm{s}$. It reduced to zero and started to flow in the opposite (negative) direction.

At $t = 9 \, \mu\mathrm{s}$, the shock wave was still propagating through the ferroelectric element and depolarization was in progress. However, since this moment of time, the charge transferred into capacitive load (Q_2) was not rising, but was decreasing. Similar waveforms were obtained with PZT 52/48 operating with a 4.5 nF load.

These charge and energy losses cannot be explained by a leakage current in the shock-compressed part of the ferroelectric element because the electric field was relatively low (1 kV/mm or less). It was experimentally shown earlier [14, 15] that there is no significant electric conductivity in shocked PZT ceramic materials under a 4.4 kV/mm electric field.

To understand the observed phenomenon, we performed high-speed photography of the operation of the ferroelectric generator. A Cordin 10A framing camera at half a million frames per second was used to photograph the experiments. The plastic body was made with a transparent urethane compound so that processes within it could be observed. More details can be found elsewhere [16].

Figure 13.22 presents a series of high-speed photographs of the explosive operations of the generator. After the EBW detonator was fired and the HE charge detonated, subsequently, the detonation wave propagated in the HE charge to the top of the urethane body of the FEG.

In Figure 13.22, at $t = -8\,\mu$s, there is a clearly visible bright light corresponding to the detonation-generated plasma at the HE/urethane body interface when the detonation front reached the top of the urethane body. This is the beginning of the shock front propagation within the body, at a velocity in the urethane compound of $2.8\,\text{mm}/\mu$s.

At $t = 0\ \mu$s, the shock front reached the ferroelectric element and the shock depolarization of the ferroelectric element began (see the current and voltage waveforms in Figure 13.21). There is a visible radial expansion of the very top of the FEG body at the joint between the body and the explosive chamber. This mechanical deformation is the result of high-pressure gases coming from the HE chamber.

At $t = 8\,\mu$s, there is a visible mechanical fragmentation of the PZT/urethane interface (facing the HE chamber). The destruction wave is following the shock wave front.

At $t = 12\ \mu$s, the mechanical fragmentation of the PZT/urethane interface has undergone significant progress. There are numerous cracks clearly visible in the urethane surrounding the ferroelectric element.

There is a high probability that the mechanical fragmentation of the ferroelectric/urethane interface (Figure 13.22) could cause limited conductivity at the part of the ferroelectric element facing the high-explosive charge, resulting in the stress-induced energy losses observed in the experiments with long charging time (Figure 13.21). The destruction wave travels behind the shock front.

Based on these results the conclusion can be reached that the electrical operation of a ferroelectric-capacitive load system should not exceed 8 μs. Otherwise, significant energy losses could occur in the charging mode.

In conclusion, the analysis of the charging voltage and the current waveforms, along with the high-speed photography of the operation of explosive ferroelectric generators, allows one to conclude that the mechanical fragmentation of the ferroelectric/urethane interface can cause limited conductivity at the shocked part of the ferroelectric element. This phenomenon can cause energy loss and can be the time-limiting factor for the operation of ferroelectric-capacitive load systems. The high-voltage operation of explosive ferroelectric generators should not exceed 8 μs because of the mechanical fragmentation

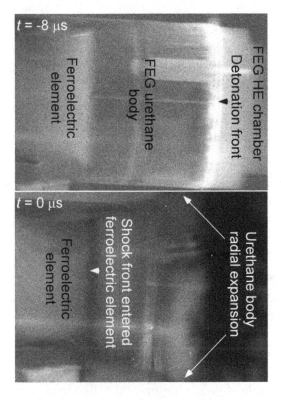

Fig. 13.22.   High-speed photographs of the operation of explosive FEG.

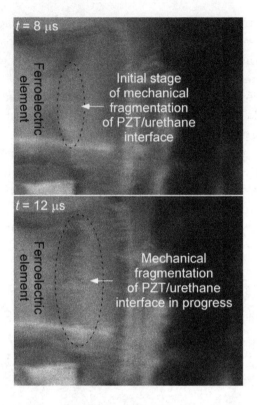

Fig. 13.22. (*Continued*)

of the shocked part of the ferroelectric element due to the action of the destruction wave following the shock wave front.

13.9 Summary

- The electric charge and energy transfer from transversely shock-compressed PZT 95/5 and PZT 52/48 ferroelectrics into capacitive loads have been investigated experimentally.
- It was experimentally demonstrated that ferroelectric elements with the volume of 8 cm^3 transversely shock-compressed within miniature FEGs produce hundreds of kilowatts of pulsed power and are capable of pulse charging capacitor banks to a high voltage with capacitances ranging from 0.5 to 9 nF.

- The experimental results indicate that the voltage generated by transversely shock-compressed ferroelectrics in the charging mode is always lower than that generated in the open circuit mode.

- However, the energy generated by transverse FEG-capacitive load systems is always higher than that generated by PZT 95/5 and PZT 52/48 elements in the open circuit mode when stress-induced charge is utilized only for charging the ferroelectric elements itself. PZT 95/5-capacitive loads systems are more effective than PZT 52/48-based systems. PZT 95/5-capacitive loads systems generate energy 40% higher than PZT 52/48-based systems.

- The experimental results indicate that in a full range of load capacitances, the electric charge and energy transfer from PZT 95/5 and PZT 52/48 into capacitive loads is interrupted by electric breakdown within the shock-compressed ferroelectric elements. The breakdown within shocked ferroelectrics plays a fundamental role in the operation of ferroelectric-capacitive load systems. It has a significant effect on the energy losses and the efficiency of the systems.

- The experimental results indicate that the charging time is one of the most important parameters affecting the magnitude of the breakdown field of shock-compressed ferroelectric elements and, correspondingly, the energy generated by ferroelectric-capacitive load systems.

- There is a strong correlation between the shocked ferroelectric breakdown field and the charging time: a lower breakdown field corresponds to a longer charging time and vice versa, i.e. the breakdown field is inversely proportional to the charging time.

- The experimentally obtained breakdown field on charging time dependences can be used to estimate the breakdown field of shocked PZT 95/5 and PZT 52/48 for a given charging time.

- The breakdown field-time relationship is important for the practical applications of ferroelectrics, in particular, for optimization of the operation of ferroelectric-capacitive load systems. A decrease in the charging time through a change in the geometrical dimensions of the ferroelectric element resulted in an increase in the energy produced by the PZT 95/5-capacitive load system by a factor of two without a change in the PZT 95/5 element volume or the FEG design.

- The breakdown field-charging time relationships should be taken into consideration in a theoretical analysis of the process of the electric breakdown of shocked ferroelectrics. It could help to identify the mechanism of the breakdown of shock-compressed ferroelectrics.

- The experimental results indicate that the dynamics of stress-induced charge is one of the important parameters of the operation of ferroelectric-capacitive load systems. An increase in the stress-induced charge dynamics through optimization of the geometrical dimensions of ferroelectric specimens, the position of the specimens within the FEGs, remanent polarization or other means can result in a significant increase in the energy generated by ferroelectrics-capacitive load systems.

- The analysis of the charging voltage and the current waveforms along with the high-speed photography of the operation of explosive ferroelectric generators indicate that mechanical fragmentation of the ferroelectric/urethane interface can cause the electrical conductivity of the fragmented zone of the specimen. This phenomenon can be the time-limiting factor for the operation of ferroelectric-capacitive load systems.

Bibliography

1. G.A. Mesyats, *Pulsed Power* (Kluwer/Plenum, New York, 2005).
2. H. Bluhm, Pulsed Power Systems: Principles and Applications (Springer, New York, 2006)
3. L.L. Altgilbers, J. Baird, B. Freeman, C.S. Lynch and S.I. Shkuratov, *Explosive Pulsed Power* (Imperial College Press, London, 2010).
4. S.I. Shkuratov, J. Baird, V.G. Antipov and E.F. Talantsev, Depolarization mechanisms of PbZr0.52Ti0.48O3 and PbZr0.95Ti0.05O3 poled ferroelectrics under high strain rate loading, *Appl. Phys. Lett.* **104** (2014) p. 212901.
5. Shkuratov, J. Baird and E.F. Talantsev, Miniature 120-kV autonomous generator based on transverse shock-wave depolarization of Pb(Zr0.52Ti0.48)O3 ferroelectric, *Rev. Sci. Instrum.* **82** (2011) p. 086107.
6. S.I. Shkuratov, J. Baird, V.G. Antipov, E.F. Talantsev, W.S. Hackenberger, A.H. Stults and L.L. Altgilbers, High voltage generation with transversely

shock-compressed ferroelectrics: Breakdown field on thickness dependence, *IEEE Trans. Plasma Sci.* **44**(10) (2016) pp. 1919–1927.

7. R.E. Setchell, S.T. Montgomery, D.E. Cox and M.U. Anderson, Delectric Properties of PZT 95/5 during shock compression under high electric filed, *AIP Conf. Proc. 845, CP845, Shock Compression of Condensed Matter — 2005*, edited by M.D. Furnish, M. Elert, T.P. Russell, and C.T. White, ©2006 American Institute of Physics, (2006) pp. 278–281.

8. J.C. Valadez, R. Sahul, E. Alberta, W. Hackenberger and C.S. Lynch, The effect of a hydrostatic pressure induced phase transformation on the unipolar electrical response of Nb modified 95/5 lead zirconate titanate, *J. Appl. Phys.* **111** (2012) p. 024109.

9. H.R. Jo and C.S. Lynch, Effect of composition on the pressure-driven ferroelectric to antiferroelectric phase transformation behavior of (Pb0.97La0.02)(Zr1−x−ySnxTiy)O3 ceramics, *J. Appl. Phys.* **116** (2014) p. 074107.

10. Y. Wu, G. Liu, Z. Gao, H. He and J. Deng, Dynamic dielectric properties of the ferroelctric ceramic Pb(Zr0.95Ti0.05)O3 in shock compression under high electrical field, *J. Appl. Phys.* **123** (2018) p. 244102.

11. S.I. Shkuratov, J. Baird, V.G. Antipov and E.F. Talantsev, Utilizing Pb(Zr0.95Ti0.05)O3 ferroelectric ceramics to scale down autonomous explosive-driven shock-wave ferroelectric generators, *Rev. Sci. Instrum.* **83** (2012) p. 076104.

12. W. Mock, Jr. and W.H. Holt, Pulse charging of nanofarad capacitors from the shock depoling of PZT 56/44 and PZT 95/5 ferroelectric ceramics, *J. Appl. Phys.* **49** (1978) pp. 5846–5854.

13. L. Davison, Fundamentals of Shock Wave Propagation in Solids (Springer, Heidelberg, 2008).

14. V.A. Borisenok, V.A. Kruchinin, V.A. Bragunets, S.V. Borisenok, V.G. Simakov, and M.V. Zhernokletov, Measuring shock-induced electrical conductivity in piezoelectrics and ferroelectrics: Single-crystal quartz, *Combustion, Explosion, and Shock Waves* **43** (2007) pp. 96–103.

15. V.A. Bragunets, V.G. Simakov, V.A. Borisenok, S.V. Borisenok and V.A. Kruchinin, Shock-induced electrical conductivity in some ferroelectrics, *Combustion, Explosion, and Shock Waves* **46** (2010) pp. 231–236.

16. S.I. Shkuratov, J. Baird and E.F. Talantsev, Extension of thickness-dependent dielectric breakdown law on adiabatically compressed ferroelectric materials, *Appl. Phys. Lett.* **102** (2013) p. 052906.

Chapter 14

Operation of Longitudinally Shock-Compressed Ferroelectrics with Resistive Loads

14.1 Introduction

The operation of explosive ferroelectric generators with resistive loads is important for certain applications [1]. One of the specific features of this mode of operation is that the ferroelectric element is always a part of the load circuit during the explosive and electrical operation of the FEG. The electrical parameters of the ferroelectric element change significantly under shock loading and this affects the electrical parameters and operation of the FEG-resistive load system. In this chapter the results are presented on experimental investigations of the operation of longitudinally shock-compressed PZT 52/48 ferroelectrics with resistive loads. The experiments were conducted with ferroelectric elements having different sizes in a wide range of ohmic loads. The experimental results indicate that the output voltage, current, and power produced by FEGs depend on both the electrical parameters of the loads and the geometrical dimensions of the ferroelectric elements [2–7]. It was experimentally demonstrated that miniature FEGs are capable of delivering pulsed powers with peak amplitude up to 0.4 MW and hundreds of amperes of current into active load. The objective of these efforts was to investigate both experimentally and theoretically the properties of explosive ferroelectric generators. Both efforts were conducted synergistically in order to fully understand the characteristics of the

FEGs and to optimize their operation [3–5]. A description of the theoretical model of explosive FEGs and the results of numerical investigations of the operation of FEG-resistive load systems are presented in the next chapter of this book.

14.2 Longitudinal Shock Wave FEG-Resistive Load System

A schematic diagram of the longitudinal shock wave FEG and the measuring circuit used for an investigation of the operation of FEGs with resistive loads is shown in Figure 14.1. The design of the FEG was identical to that used in the investigations of the longitudinal shock depolarization of ferroelectrics in the short-circuit mode, for high-voltage generation in the open circuit mode, and for the investigation of the operation of the FEG combined with the vector inversion generator (VIG) (see Chapters 7, 8 and 16 of this book). The output terminals of the FEG were connected to the resistive load. The voltage across the load was monitored with a Tektronix P6015A high-voltage probe. The current in the load circuit was monitored with a Pearson Electronics current probe (model 411).

The load of the FEG was made of low-inductive bulk carbon composition TWR resistors. The resistors were encapsulated with

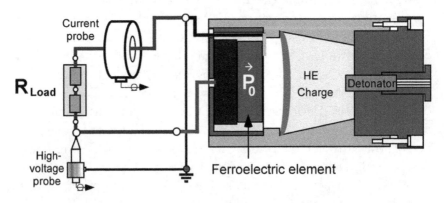

Fig. 14.1. Schematics of a longitudinal shock wave FEG and the measuring circuit used in the FEG-resistive load experiments. P_0 is the remanent polarization vector.

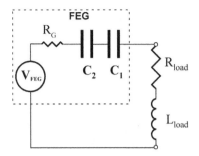

Fig. 14.2. Equivalent circuit of a longitudinal FEG operating with resistive load.

urethane compound to suppress the surface flashover. The inductance of the resistive loads did not exceed 250 nH. The FEG was placed in the blast chamber. The resistive loads, current and voltage probes were placed outside the blast chamber.

The equivalent circuit of an FEG operating with a resistive load is shown in Figure 14.2. The FEG-load system can be considered an *RC*-circuit. However, it is not a conventional *RC*-circuit. The ferroelectric element of the FEG is a dynamic capacitor that is charging due to the shock wave depolarization of the element itself and it is immediately discharging through the resistive load.

The dielectric properties and, correspondingly, capacitance of the ferroelectric element are continually changing during shock depolarization. The capacitance of the shock-compressed portion of the ferroelectric element is represented as C_2 (Figure 14.2). The capacitance of the uncompressed portion of the element is shown as C_1.

The electrical conductivity of the shock-compressed portion of the element can be higher than that for the uncompressed portion. The parameter R_G represents the resistance of the shock-compressed portion of the ferroelectric element. R_{Load} and L_{Load} represent the resistance and inductance of the load, respectively.

14.3 FEG Pulsed Power Generation with Resistive Loads

Parametric experimental investigations of the longitudinal shock depolarization of ferroelectrics operating with resistive loads were performed with PZT 52/48 ceramic elements. One of the adjusted parameters, which can facilitate maximizing the output of the FEG, is the load resistance. A parametric investigation of the load resistance was performed in the wide range from $1\,\Omega$ to $14\,\text{k}\Omega$.

The second important factor is the geometric size and shape of the ferroelectric elements. Experiments were conducted with PZT 52/48 ceramic disk elements having three different thicknesses: 0.65, 2.5 and 5.1 mm. The diameters of the ferroelectric elements varied in the narrow range from 25 to 26 mm.

In this section the experimental results for a few cases of the parametric investigations are presented. The experimental results obtained for all investigated cases are summarized in the next section (Section 14.4).

14.3.1 *FEG pulsed power generation with high ohmic loads*

Figure 14.3 shows typical waveforms of current, voltage and power generated by a longitudinally shocked 2.5-mm-thick PZT 52/48 element in the $40\,\Omega$ load. The amplitude of the load current was $I_{FEG}(13.3\,\mu s) = 86$ A and the current rise time was 1.0 μs.

The power delivered from the FEG into the $40\,\Omega$ load was determined as the product of the instantaneous FEG output voltage $V_{FEG}(t)$ and the instantaneous current in the load circuit, $I_{FEG}(t) : P_{FEG}(t) = I_{FEG}(t) \cdot V_{FEG}(t)$. The peak output power was $P_{FEG}(13.3\,\mu s) = 0.32$ MW (Figure 14.3).

The integration of the $P_{FEG}(t)$ waveform from 0 to t gives the momentary value of the energy delivered from the shock-compressed PZT 52/48 element in the resistive load. Figure 14.4 shows waveforms of the FEG output power and energy delivered in the $40\,\Omega$ load. The total energy delivered in the load was 226 mJ. The specific energy density of the ferroelectric element can be determined as follows:

$$W_{FEG} = \left(\int_0^{\tau_{SW}} P_{FEG}(t) \cdot dt \right) \Big/ vol, \qquad (14.1)$$

where W_{FEG} is the specific energy density of a shocked ferroelectric element, $P_{FEG}(t)$ is the instantaneous power generated by a shocked

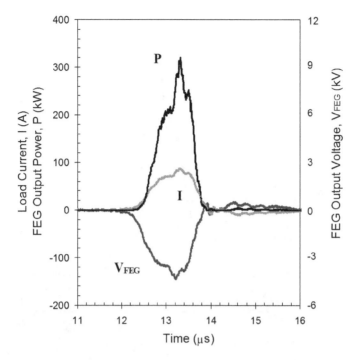

Fig. 14.3. Typical waveforms of current, voltage and power generated by a longitudinally shock-compressed 2.5-mm-thick PZT 52/48 element in the 40 Ω load.

ferroelectric, τ_{SW} is the shock wave transit time, and *vol* is the volume of the element. For the experiment shown in Figures 14.3 and 14.4 the energy density was $W_{FEG} = 185\,\mathrm{mJ/cm^3}$.

Figure 14.5 shows typical waveforms of current, voltage and power generated by a 5.1-mm-thick PZT 52/48 element in the 40 Ω load. The amplitude of the load current was $I_{FEG}(13.4\,\mu\mathrm{s}) = 72\,\mathrm{A}$ and the rise time of the current pulse was 1.2 μs. The peak power generated in the load was $P_{FEG}(13.4\,\mu\mathrm{s}) = 0.22\,\mathrm{MW}$.

The experimental results (Figures 14.3 to 14.5) indicate that miniature longitudinally shock-compressed PZT 52/48 elements are capable of producing pulsed powers exceeding 0.3 MW in a resistive load. Comparing the results obtained with 2.5-mm-thick and 5.1-mm-thick elements, it can be concluded that thinner PZT 52/48 elements are capable of generating higher current and higher

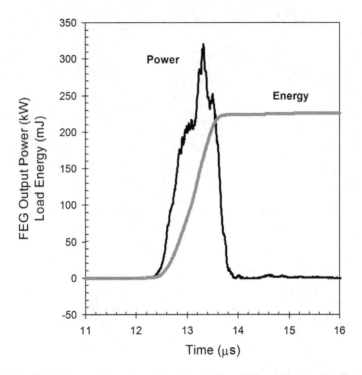

Fig. 14.4. Waveforms of power and energy delivered by a longitudinally shock-compressed 2.5-mm-thick PZT 52/48 element in the $40\,\Omega$ load.

power in an active load. Apparently, thin ferroelectric elements are optimal from the standpoint of generating higher currents and, correspondingly, higher energy densities in a resistive load.

14.3.2 *FEG pulsed power generation with low ohmic loads*

Figure 14.6 shows typical waveforms of current, voltage and power generated by a longitudinally shocked 0.65-mm-thick PZT 52/48 element in the $1\,\Omega$ load. The amplitude of the load current was $I_{FEG}(12.6\,\mu s) = 341$ A and the current rise time was $1.1\,\mu s$. The mean current amplitude generated in the experiments was 351 ± 32 A. The peak power generated in the load was $P_{FEG}(12.7\,\mu s) = 0.33$ MW.

Figure 14.7 shows typical waveforms of current, voltage and power generated by a 2.5-mm-thick PZT 52/48 element in the $1\,\Omega$ load.

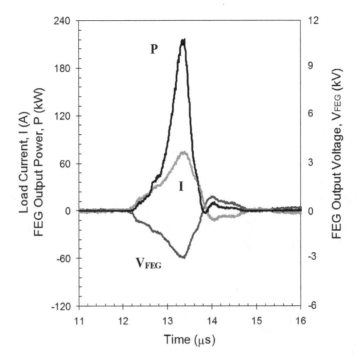

Fig. 14.5. Typical waveforms of current, voltage and power generated by a longitudinally shock-compressed 5.1-mm-thick PZT 52/48 element in the 40 Ω load.

The amplitude of the load current was $I_{FEG}(11.9\,\mu s) = 210\,A$ and the current rise time was $1.0\,\mu s$. The peak power generated in the load was $P_{FEG}(11.9\,\mu s) = 0.12\,MW$.

The experimental results indicate that miniature longitudinally shock-compressed PZT 52/48 elements with volume not exceeding $1\,cm^3$ are capable of producing current with amplitudes up to 400 A. Comparing the results obtained with two types of ferroelectric elements (Figures 14.6 to 14.7), the conclusion can be reached that the thinner PZT 52/48 element (0.65 mm) is capable of producing twice the load current as the thicker (2.5 mm) element in a low ohmic load. These results are in good agreement with the results obtained with the 40 Ω load (see the subsection above). In the longitudinal shock depolarization mode, thin ferroelectric elements are capable of generating higher currents and higher power.

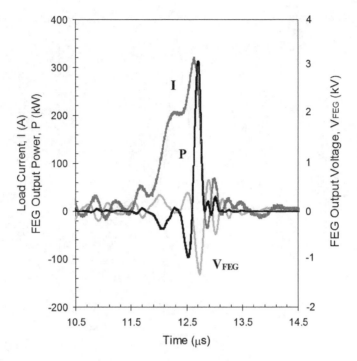

Fig. 14.6. Typical waveforms of current, voltage and power generated by a longitudinally shock-compressed 0.65-mm-thick PZT 52/48 element in the 1 Ω load.

14.4 Experimental Results Summary for FEG-Resistive Load Systems

Figure 14.8 shows the amplitudes of current generated by longitudinally shocked PZT 52/48 elements as a function of the load resistance. The experimental results indicate that the current in the load decreases as the load resistance increases. The reason for this can be the increase in the losses in the shock-compressed portion of the ferroelectric element due to bulk leaks stimulated by the shock wave. The experimental results suggest that the longitudinal FEG can be an effective current source until the load resistance begins to approach the resistance of the shock-compressed portion of the ferroelectric element.

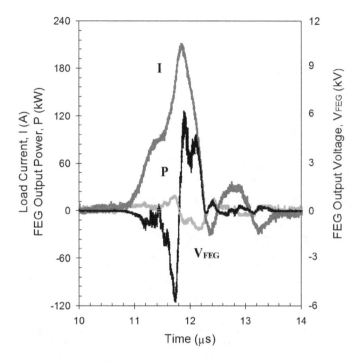

Fig. 14.7. Typical waveforms of current, voltage and power generated by a longitudinally shock-compressed 2.5-mm-thick PZT 52/48 element in the 1 Ω load.

Figure 14.9 shows amplitudes of voltage generated by longitudinal FEGs across resistive loads as a function of the load resistance. Increasing the load resistance leads to an increase in the amplitude of voltage produced by FEGs across active loads.

The voltage versus the resistance plots are practically linear on the logarithmic scale (Figure 14.9). This is an indication of an exponential relationship between the FEG output voltage amplitude and the resistance of the load.

The slopes of the voltage versus the resistance plots vary with PZT 52/48 element thickness. This means that the voltage generated by longitudinally shock-compressed ferroelectrics across active loads depends on both the resistance of the load and the geometrical dimensions of the ferroelectric elements.

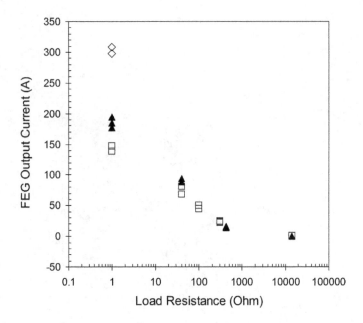

Fig. 14.8. Amplitudes of current generated by longitudinally shock-compressed PZT 52/48 elements in active loads as a function of the load resistance. Experimental data for 5.1-mm-thick PZT 52/48 elements (squares), for 2.5 mm-thick elements (triangles) and for 0.65-mm-thick elements (diamonds).

Figure 14.10 shows the output energy generated by longitudinally shock-compressed PZT 52/48 elements in an active load as a function of the load resistance. The plots for the energy indicate that this value passes through a maximum at a certain resistance, which indicates an optimal balance between the generated current and the leakage current.

The experimental results indicate that the optimum load resistance, corresponding to the FEG output energy maximum, depends on the element thickness. The maximum energy was generated by 2.5-mm-thick elements in the 40 Ω load, while 5.1-mm-thick elements generated the maximum energy in the 100 Ω load (Figure 14.10). The thickness of the PZT element is one of the important parameters of the FEG that can be adjusted to maximize the output energy delivered by the FEG to resistive loads. A detailed discussion of the experimental results is presented in the next chapter of this book.

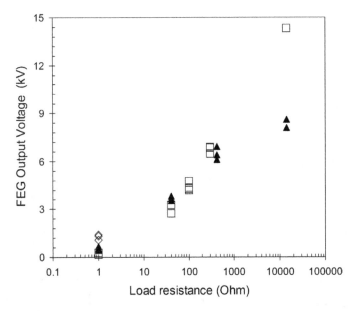

Fig. 14.9. Amplitudes of voltage pulses generated by longitudinally shock-compressed PZT 52/48 elements across active loads as a function of the load resistance. Experimental data for 5.1-mm-thick PZT 52/48 elements (squares), for 2.5 mm-thick elements (triangles) and for 0.65-mm-thick elements (diamonds).

14.5 Summary

- The longitudinal shock compression of ferroelectric elements connected to resistive loads has been investigated experimentally. The obtained results indicate that shock-compressed PZT 52/48 elements with volumes of 0.35 to 2.5 cm^3 are capable of producing pulsed powers exceeding 0.3 MW and hundreds of amperes of current in active loads. The energy density generated per unit volume of ferroelectric element ranges from 0.1–0.4 J/cm^3.

- The experimental results indicate that the current in the active load decreases as the load resistance increases. This can be caused by the losses in the shock-compressed portion of the ferroelectric element due to bulk leaks stimulated by the shock wave.

- The highest current (380 A) was generated by the thinnest PZT 52/48 element (0.65 mm). In the longitudinal shock depolarization mode, thin ferroelectric elements are optimal from the standpoint

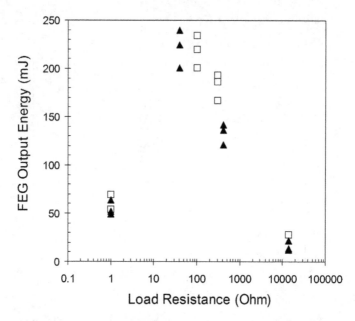

Fig. 14.10. Output energy generated in an active load by longitudinally shock-compressed 2.5-mm-thick (triangles) and 5.1-mm-thick (squares) PZT 52/48 elements as a function of the load resistance.

of generating higher currents and, correspondingly, higher energy densities in the load.

- It was experimentally demonstrated that load voltage versus load resistance plots are practically linear on the logarithmic scale. This indicates an exponential relationship between the FEG output voltage amplitude and the resistance of the load.The slopes of the voltage versus the resistance plots vary with PZT 52/48 element thickness. The voltage generated by longitudinally shocked ferroelectrics across resistive loads depends on both the resistance of the load and the geometrical size and, correspondingly, the electrical parameters of the ferroelectric element.

- The experimentally obtained generated energy versus load resistance plots pass through a maximum at a certain resistance. There is an optimum load resistance for each size of ferroelectric element at which the FEGs provide maximum output power and energy in the active load. It can be explained by an optimal balance between the generated current and the leakage current in the shock-compressed portion of the ferroelectric element. The longitudinal FEG can be considered as an effective current source until the load resistance begins to approach the shunt resistance of the shock-compressed portion of the ferroelectric element.

Bibliography

1. L.L. Altgilbers, J. Baird, B. Freeman, C.S. Lynch and S.I. Shkuratov, *Explosive Pulsed Power* (Imperial College Press, London, 2010).

2. S.I. Shkuratov, M. Kristiansen, J. Dickens, A. Neuber, L.L. Altgilbers, P.T. Tracy and Ya. Tkach, Experimental study of compact explosive driven shock wave ferroelectric generators, *Proceedings of 13th International Pulsed Power Conference*, eds. R. Reinovsky and M.A. Newton, Las Vegas, Nevada, USA (2001) IEEE Catalog Number: 01CH37251, ISBN: 0-7803-7120-8, Vol. II, pp. 959–962.

3. Ya. Tkach, S.I. Shkuratov, J. Dickens, M. Kristiansen, L.L. Altgilbers and P.T. Tracy, Explosive driven ferroelectric generators, *Proceedings of 13th International Pulsed Power Conference*, eds. R. Reinovsky and M.A. Newton, Las Vegas, Nevada, USA (2001) IEEE Catalog Number: 01CH37251, ISBN: 0-7803-7120-8, Vol. II, pp. 986–989.

4. Ya. Tkach, S.I. Shkuratov, J. Dickens, M. Kristiansen, L.L. Altgilbers and P.T. Tracy, Parametric and experimental investigation of EDFEG, *Proceedings of 13th International Pulsed Power Conference*, eds. R. Reinovsky and M.A. Newton, Las Vegas, Nevada, USA (2001) IEEE Catalog Number: 01CH37251, ISBN: 0-7803-7120-8, Vol. II, pp. 990–993.

5. Y. Tkach, S.I. Shkuratov, E.F. Talantsev, J.C. Dickens, M. Kristiansen, L.L. Altgilbers and P.T. Tracy, Theoretical treatment of explosive-driven ferroelectric generators, *IEEE Trans. Plasma Sci.* **30**(5) (2002) pp. 1665–1673.

6. S.I. Shkuratov, E.F. Talantsev, J. Baird, H. Temkin, L.L. Altgilbers and A.H. Stults, Longitudinal shock wave depolarization of $Pb(Zr_{0.52}Ti_{0.48})O_3$ polycrystalline ferroelectrics and their utilization in explosive pulsed power, *AIP Conf. Proc. CP845, Shock Compression of Condensed Matter — 2005*, eds. M.D. Furnish, M. Elert, T.P. Russell, and C.T. White (American Institute of Physics, 2006), pp. 1169–1172.

7. S.I. Shkuratov, E.F. Talantsev, J. Baird, H. Temkin, Y. Tkach, L.L. Altgilbers and A.H. Stults, The depolarization of a $Pb(Zr_{0.52}Ti_{0.48})O_3$ polycrystalline piezoelectric energy-carrying element of compact pulsed power generator by a longitudinal shock wave, *Proceedings of the 16th IEEE International Pulsed Power Conference*, eds. E. Schamiloglu and F. Peterkin, Albuquerque, NM (2007), pp. 529–532.

Chapter 15

Theoretical Treatment of Explosive Ferroelectric Generators

15.1 Introduction

In this chapter, detailed descriptions of theoretical models [1–3] of longitudinal and transverse shock wave ferroelectric generators are presented. These models were used to determine the optimal operating parameters of the FEGs. The results of the parametric investigations obtained with the model of longitudinal FEG were further verified through experiments [4–6]. A comparison of the experimental and calculated results showed them to be in good agreement.

15.2 Factors Influencing the Operation of Explosive Ferroelectric Generators

The basic factors, which influence the energy output of the ferroelectric generator, include shock wave characteristics, characteristics of the ferroelectric material after shock compression, load resistance and inductance, bulk electrical breakdown resulting from the factors inherent to the shock compression of ferroelectrics (e.g. free charge carriers injection, microfractures), and geometrical dimensions of the ferroelectric element. Unfortunately, a consistent theoretical treatment that addresses the process by which the polarized domains in the ferroelectric materials reorient during the action of a shock

371

wave is still a challenging problem. Consequently, the previously mentioned parameters can be subdivided into three groups.

The first group of parameters includes those that can be relatively easily accounted for through experiments and empirical models. This group includes load resistance and the geometrical dimensions of the ferroelectric elements.

The second group includes those parameters that can be accounted for indirectly by adjusting their values in numerical models by fitting the calculated results to the experimental results. These parameters include the permittivity and conductivity of shocked ferroelectric materials. This group also includes those parameters whose possible maximum value can only be measured experimentally such as the stress-induced depolarization, which can be measured by depolarizing the ferroelectric elements at different stress and shock wave geometries. It should be noted here that the product of the density of stress-induced charge released by the ferroelectric and the area of ferroelectric element electrodes are the main factors which determine the amount of electric charge released to the active load during the depolarization process.

The third group includes those parameters that can only be accounted for through experimental measurements. These parameters include the electrical breakdown strength of the shocked ferroelectric material and the shock wave characteristics. For example, a shock wave, propagating through ceramics, has very complex characteristics, and generally speaking, this wave represents the superposition of a number of elastic and inelastic acoustic and shock waves. This complex wave strongly depends on the initial pressure generated by either the flyer plate or the direct action of the HE. It is currently believed that pressures that generate only elastic waves are the most beneficial for the normal operation of the FEG, since, when chaotically disorienting the domain structure, they do not cause mass injection of free charge carriers and related effects that could cause bulk electrical breakdown and an increase in the bulk leakage conductance in the shocked portion of the ferroelectric element.

15.3 Theoretical Model Description

Two simulation models have been developed [1–3]. The first is for the case where the shock wave propagates parallel or anti-parallel to the polarization vector (longitudinal FEG) and the second is for the case where the shock wave propagates perpendicular to the polarization vector (transverse FEG).

15.3.1 *Theoretical model of longitudinal FEG*

As the shock wave passes through the polarized ferroelectric element, its volume is divided into two zones (see Figures 15.1 and 15.2) differing by such parameters as bulk polarization, permittivity, and conductance. These zones are referred to as the "compressed" and "uncompressed" zones, where the compressed zone is that through which the shock wave has already passed and the uncompressed zone is that through which the shock wave has not passed.

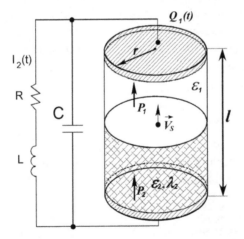

Fig. 15.1. Diagram illustrating the propagation of a shock wave through a longitudinally shock-compressed ferroelectric element connected to the reactive load and the field distributions in the element; Q_1 is charge released on the electrodes, P_1, ε_1 denotes bulk polarization and permittivity of the uncompressed zone of the ferroelectric element; $P_2, \varepsilon_2, \lambda_2$ denotes bulk polarization, permittivity and conductance of the compressed zone of the ferroelectric element; V_s is the velocity vector of the shock wave front moving in the ferroelectric material.

The compressed portion of the ferroelectric element is hatched. The uncompressed portion of the element is clear. These two zones are different, since their parameters, such as the bulk polarization, P, permittivity, ε, and conductance, λ, are different. The charge $Q_1(t)$ is that released at the contact plates of the ferroelectric element. The bulk polarization and permittivity in the uncompressed portion are denoted by P_1 and ε_1, respectively. The bulk polarization, permittivity, and conductance in the compressed portion of the element are denoted by P_2, ε_2, and λ_2, respectively. The velocity of the shock wave front moving in the ferroelectric element is denoted by V_S.

Taking these zones into account, the equivalent circuit diagrams for the longitudinal and transverse FEGs, connected to the reactive load with parameters R, L and C_L are presented in Figures 15.1 and 15.2, respectively.

In building the model, the following assumptions were made:

- Bulk compression of the ferroelectric material, which is equal to the ratio of the change of the ferroelectric element volume during shock compression to the initial volume of the element, was not taken into account, since experimental results indicate that this value does not exceed 0.05.
- Shock polarization inherently bears an inertial characteristic; that is, domain rearrangement is a kinetic process.
- The ferroelectric material is a linear dielectric in both the compressed and uncompressed zones. The dependence of permittivity on the electric field strength was not taken into account in these studies.

Taking these assumptions into consideration, the general set of ordinary differential equations (ODEs) and initial conditions for the FEG connected to a reactive load are:

$$\dot{Q}_1 + \dot{Q}_2 = I_0(t) - I_2 - I_{leak}(Q_1); \quad Q_1|_{t=0} = 0;$$

$$L\dot{I}_2 = Q_2 C_L^{-1} - RI_2; \quad I_2|_{t=0} = 0; \qquad (15.1)$$

$$\dot{Q}_1 C^{-1}(t) - \dot{Q}_2 C_L^{-1}(t) = \dot{C}(t) C^{-2}(t) Q_1; \quad Q_2|_{t=0} = 0;$$

where R in (Ω), L in (H), and C_L in (F) are the resistance, inductance, and capacitance of the load, respectively; $I_2(t)$ is current through the resistive part of the load.

Consider the case of a longitudinal FEG. The diagram illustrating the propagation of a shock wave through a longitudinally shock-compressed ferroelectric element connected to a reactive load and the field distributions in the element is shown in Figure 15.1.

In this design (Figure 15.1), the ferroelectric element polarization vector P_1 is parallel or anti-parallel to the shock wave velocity vector V_S. The capacitance of the ferroelectric element, $C(t)$, in the longitudinal FEG case is described by using the standard model for a layered parallel plate capacitor, where one layer is the compressed zone and the other layer is the uncompressed zone. For this type of capacitor, it is a requirement that the distance between the electrodes of the ferroelectric element, l, must be less than the radius of the electrodes, r.

When calculating the capacitance for geometries other than parallel plate capacitors, errors are introduced. However, taking into account the specific geometry of the element and the distribution of the electric fields, this does not pose a problem when solving the set of ODEs in Eq. (15.1). Therefore, the capacitance of the ferroelectric element shown in Figure 15.1 becomes:

$$C(t) = \begin{cases} \varepsilon_0 \varepsilon_1 \left[l + \left(\varepsilon_1 \varepsilon_2^{-1}(p) - 1 \right) V_S t \right]^{-1} & \text{for } l \geq V_S t; \\ \varepsilon_0 \varepsilon_2(p) S l^{-1} & \text{for } l < V_S t; \end{cases} \qquad (15.2)$$

where $S(m^2)$ is the area of the electrodes of the ferroelectric element, ε_1 is the relative permittivity of the uncompressed zone, $\varepsilon_2(p)$ is the relative permittivity of the compressed zone, ε_0 is the permittivity of free space, l is the total length of the ferroelectric element, r is the radius of the element, and V_S is the velocity of the shock wave in the ferroelectric material, which is approximately equal to the velocity of sound in the material. The permittivity of the shock-compressed ferroelectric material is a complex function of the pressure in the shock wave, p.

The total electric charge released on the electrodes of the ferro-electric element during the time it takes for the shock wave to travel through the element is $Q_{tot} = \sigma(p)S$. Since the free charge surface density, $\sigma(p)$, released on the electrodes is equal to the difference in the bulk polarization in the compressed and uncompressed regions of the material, $\sigma = P_1 - P_2$, the amount of charge released at any given moment in time, $Q_0(t)$, is proportional to the depolarized volume of the element and is described by the expression:

$$Q_0(t) = \sigma(p)V_S t l^{-1}. \tag{15.3}$$

Taking the derivative with respect to time yields the depolarization current:

$$I_0(t) = \theta(t)\frac{d\sigma(p)V_S t l^{-1}}{dt}; \tag{15.4}$$

where $\theta(t)$ takes into account the "switching on" of the polarization current at the origin, which corresponds to the moment at which the shock wave enters the ferroelectric element, and the "switching off" at the moment the shock wave exits the element $(t_f = lV_S^{-1})$.

Shock depolarization is assumed to be an inertial process due to the substantial time required for the dipole moments to rotate in solid dielectrics and the domain wall motion. Thus, the relaxation time is the period from the time the shock enters the element to the time at which the charge is first released from the end plates. Ferroelectrics require this amount of time to change polarization.

According to [7, 8], when a shock wave enters or leaves a ferroelectric element, the current increases or decreases, respectively, linearly with respect to the finite time, τ_{rel}. This linear behavior cannot be accounted for in the expressions for the capacitance of the element or the resistance of the load.

Therefore, it seems reasonable to introduce a similar increase or decrease of the current in the simulation by means of a hyperbolic tangent function, $\theta(t)$, which behaves linearly at the transient points where the shock wave enters and where it leaves the element. Since the rise time and the fall time of the current are assumed to be the same [7], they are chosen to equal τ_{rel}. Taking these assumptions into account, $\theta(t)$ is represented by the semi-empirical expression:

$$\theta(t) = \frac{1}{4}\left(1 - \tanh\left[4t_{rel}^{-1}(p)(t - lV_S^{-1}) - 2\right]\right)$$
$$\left(1 + \tanh\left[2t_{rel}^{-1}(p)t - 2\right]\right); \tag{15.5}$$

which is convenient for numerical calculations, since it is a smooth function. Since the calculation of $\tau_{rel}(p)$ using kinetic theory is relatively complex, experimentally measured values of $0.05 - 2.0\,\mu s$ are used. These values decrease significantly as the pressure in the shock wave increases.

The electrical conductance of ferroelectric ceramics sharply increases under shock compression and part of the charge released at the end plates (ferroelectric element electrodes) leaks through the compressed region of the element forming a leakage current, $I_{leak}(p)$. In the case of a longitudinally driving force, the leakage takes place over the entire surface of the element and, consequently, does not depend on time. The leakage current can be found from the field strength in the shock-compressed region of the element by using Ohm's law:

$$j_{leak}(Q_1) = \lambda_2(p)E_2;$$
$$I_{leak}(Q_1) = j_{leak}(Q_1)S = \lambda_2(p)Q_1\left[\varepsilon_2\varepsilon_0\right]^{-1} \tag{15.6}$$

where λ_2 $(\Omega \cdot \text{m})$ is the specific conductance of the shock-compressed ferroelectric material. The conductance increases as the pressure in the shock wave increases, due to the increase in the number of free charge carriers because of electron tunneling, ionization, and other phenomena.

The set of ODEs in Eq. (15.1) describes the operation of a longitudinally shock-compressed ferroelectric element before the onset of bulk breakdown in the compressed region of the element. Apparently, breakdown can occur in both the compressed and uncompressed regions but probably, it starts first in the compressed region because the electric strength of the ferroelectric ceramics can be less than that in the uncompressed region. This is related to the formation of local breakdown areas in the compressed zone due to the impact of the shock wave on grain boundaries, defects, dislocations, and air-filled cavities generated during the baking process. Thus, Eq. (15.1) is restricted to the time domain prior to the start of bulk breakdown in the compressed zone:

$$E_{br} > Q_1(t) \left[S\varepsilon_2(p)\varepsilon_0 \right]^{-1}; \tag{15.7}$$

where $E_{br} [kV/m]$ is the breakdown field of the ferroelectric material compressed by the shock wave.

Since the set of ODEs in Eq. (15.1) is stiff and cannot be efficiently solved with the required precision, they must be normalized by introducing the following normalized variables:

$$\tau = t/\tilde{t}; q = Q/\tilde{Q}; \quad \text{and} \quad i = I\tilde{t}/\tilde{Q}; \tag{15.8}$$

where $\tilde{t} = lV_S^{-1}$ and $\tilde{Q} = \sigma(p)S$. After making the appropriate substitutions, Eq. (15.1) becomes:

$$\frac{dq_1}{d\tau} = \frac{1}{C(\tau) + C_L} \left[\frac{\dot{C}(\tau)}{C(\tau)} C_L q_1 + C(\tau) (i_0(\tau) - i_2 - i_{leak}(q_1)) \right];$$

$$q_1|_{\tau=0} = 0;$$

$$\frac{dq_2}{d\tau} = \frac{1}{C(\tau) + C_L} \left[-\frac{\dot{C}(\tau)}{C(\tau)} C_L q_1 + C_L (i_0(\tau) - i_2 - i_{leak}(q_1)) \right];$$

$$q_2|_{\tau=0} = 0$$

$$\frac{di_2}{d\tau} = \tilde{t}^2 \frac{q_2}{LC_L} - \tilde{t}R i_2; \quad i_2|_{\tau=0} = 0; \tag{15.9}$$

where:

$$i_0(\tau) = \frac{1}{4} \left[1 - \tanh \left(4\tilde{t} t_{rel}^{-1}(p)(\tau - 1) \right) - 2 \right]$$

$$\left[1 + \tanh \left(4\tilde{t} t_{rel}^{-1}(p)\tau - 2 \right) \right];$$

$$C(\tau) = \begin{cases} \varepsilon_1 \varepsilon_0 \left[1 + (\varepsilon_1 \varepsilon_2^{-1}(p) - 1)/\tau \right]^{-1} & \text{for } \tau \leq 1 \\ \varepsilon_0 \varepsilon_2(p)Sl^{-1} & \text{for } \tau > 1; \end{cases}$$

$$\dot{C}(\tau) = \begin{cases} \varepsilon_1\varepsilon_2(p)\,\varepsilon_0 Sl^{-1}\,(\varepsilon_2(p) - \varepsilon_1) \\ \quad [\varepsilon_2(p) + (\varepsilon_1 - \varepsilon_2(p))\,\tau]^{-2}; \quad \tau \leq 1 \qquad (15.10) \\ 0; \quad \tau > 1 \end{cases}$$

$$I_{leak}(q_1) = (\varepsilon_2(p)\varepsilon_0)^{-1}\,\lambda_2(p)\tilde{t}q_1.$$

The normalized bulk breakdown condition for the longitudinal FEG is:

$$E_{br} > \sigma(p)q_1(\tau)\,[\varepsilon_2(p)\varepsilon_0]^{-1}.$$

This set of equations was solved numerically using the Gear method with a relative error of less than 10^{-5} for all variables.

15.3.2 *Theoretical model of transverse FEG*

Consider the operation of the transverse FEG, shown in Figure 15.2. In this design of the FEG, the shock wave velocity vector, V_S, is perpendicular to the bulk polarization vector, P_1.

The set of ODEs that describes the operation of the transverse FEG is the same as that which describes the operation of the longitudinal FEG (Eq. (15.1)). The capacitance of the ferroelectric element is, however, defined differently in that now the two capacitors are assumed to be connected in parallel:

$$C(t) = \begin{cases} \varepsilon_0 h d^{-1}\,[\varepsilon_1 l + (\varepsilon_2(p) - \varepsilon_1)\,V_s t]; & l \geq V_s t \\ \varepsilon_0\varepsilon_2(p)\,h l d^{-1}; & l < V_s t \end{cases} \qquad (15.11)$$

where all notations are the same as in Eq. (15.2). The geometrical dimensions of the ferroelectric element, h, l, d are shown in Figure 15.2. The depolarization current in this case is found by using a method similar to that used for the longitudinally shock-compressed ferroelectric element:

$$I_0(t) = \frac{1}{4}\left(1 - \tanh\left[4t_{rel}^{-1}(p)\left(t - lV_s^{-1}\right) - 2\right]\right)$$

$$\times\left(1 + \tanh\left[4t_{rel}^{-1}(p)\,t - 2\right]\right)\frac{d\left(\sigma(p)\,hV_s t\right)}{dt}; \qquad (15.12)$$

where all notations are the same as those in Eqs. (15.4) and (15.5).

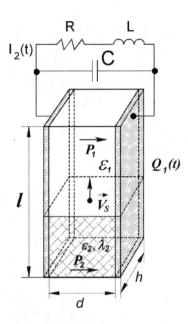

Fig. 15.2. Diagram illustrating the propagation of a shock wave through a transversely shock-compressed ferroelectric element connected to the reactive load and the field distributions in the element; Q_1 is the charge released at the electrodes, $\boldsymbol{P_1}, \varepsilon_1$ denotes bulk polarization and permittivity of the uncompressed zone of the ferroelectric element; $\boldsymbol{P_2}, \varepsilon_2, \lambda_2$ denotes bulk polarization, permittivity and conductance of the compressed zone of the ferroelectric element; $\boldsymbol{V_S}$ is the velocity vector of the shock wave front moving in the ferroelectric material.

Unlike the longitudinal FEG case, the leakage current through the shock-compressed portion of the ferroelectric element depends explicitly on time, because the surface through which leakage takes place increases as the shock wave propagates through the element:

$$
I_{leak}(Q_1, t) = \begin{cases} [\varepsilon_0(\varepsilon_1 l + (\varepsilon_2(p) - \varepsilon_1)V_s t)]^{-1} \lambda_2(p) Q_1 V_s t; \\ \qquad l \geq V_s t \\ [\varepsilon_2(p)\varepsilon_0 l]^{-1} \lambda_2(p) Q_1; \; l < V_s t \end{cases} \tag{15.13}
$$

The condition for bulk breakdown initiation changes as well, since in the case of the transversely shocked ferroelectric element, the field strength inside the ferroelectric material depends on the total capacitance of the element and acquires the following form:

$$
E_{br} > Q_1(t) C^{-1}(t) d^{-1} \tag{15.14}
$$

where E_{br} [kV/m] is the breakdown field of shock-compressed ferroelectric material.

The procedure for normalizing the set of ODEs for the transversely shock-compressed ferroelectric element is similar to that described above for the longitudinal FEG case and employs the following normalized variables $\tau = t/\tilde{t}$; $q = Q/\tilde{Q}$; $i = I\tilde{t}/\tilde{Q}$; where $\tilde{t} = lV_s^{-1}$; $\tilde{Q} = \sigma(p)hl$. The resulting normalized set of ODEs that describe the operation of the transverse FEG is identical to Eq. (15.9), where the following expressions for the FEG parameters are used:

$$i_0(\tau) = 0.25 \left[1 - \tanh\left(4\tilde{t}t_{rel}^{-1}(p)(\tau - 1) - 2\right)\right]$$
$$\left[1 + \tanh\left(4\tilde{t}t_{rel}^{-1}(p)\tau - 2\right)\right];$$

$$C(\tau) = \begin{cases} \varepsilon_0 h d^{-1}\left[\varepsilon_1 l + (\varepsilon_2(p) - \varepsilon_1)l\tau\right]; & \tau \leq 1 \\ \varepsilon_0 \varepsilon_2(p) h l d^{-1}; & \tau > 1. \end{cases}$$

$$\dot{C}(\tau) = \begin{cases} \varepsilon_0 h d^{-1}(\varepsilon_2(p) - \varepsilon_1)l; & \tau \leq 1 \\ 0; & \tau > 1. \end{cases} \tag{15.15}$$

$$i_{leak}(q_1, \tau) = \begin{cases} (\varepsilon_0(\varepsilon_1 + (\varepsilon_2 - \varepsilon_1)\tau))^{-1}\lambda_2(p)\tau\tilde{t}q_1; & \tau \leq 1; \\ (\varepsilon_0\varepsilon_2)^{-1}\lambda_2(p)\tilde{t}q_1; & \tau > 1. \end{cases}$$

The normalized bulk breakdown condition for the transverse FEG takes on the following form:

$$E_{br} > \sigma(p)hlq_1(\tau)C^{-1}(\tau)d^{-1}.$$

The set of equations for the transverse FEG was also solved numerically using the Gear method with a relative error of less than 10^{-5} for all variables.

15.4 Simulation Results

The developed model was employed for the calculation of longitudinal FEG outputs and their comparison to the results of experiments described in Chapter 14 of this book. The calculations were conducted through the substitution of known parameters of the FEG.

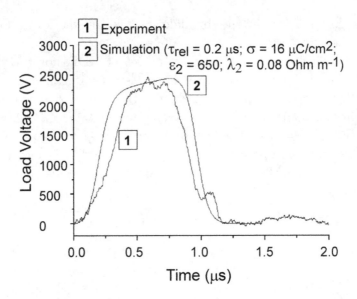

Fig. 15.3. Comparison of the experimental (curve 1) and simulated (curve 2) waveforms of output voltages generated by longitudinal FEG across a 40 Ω load. The PZT 52/48 element has diameter 25 mm and thickness 2.5 mm.

Since some of the parameters, such as τ_{rel}, σ, ε_2 and λ_2 are not known at shock pressures, they were corrected based on earlier experiments and used in calculations for later experiments.

Figure 15.3 presents the experimental and calculated outputs of the longitudinal FEG operating with a 40 Ω load. The FEG contained the PZT 52/48 element with diameter of 25 mm and thickness of 2.5 mm. The parameters of the model: $\tau_{rel} = 0.2\,\mu s$, $\sigma = 0.16\,C/m^2$, $\varepsilon_2 = 650$, $\lambda_2 = 0.08\,\Omega \cdot m^{-1}$. The calculated waveform is in good agreement with the experimental waveform.

One of the most easily adjusted parameters, which can facilitate maximizing the output of the FEG, is the load resistance. A parametric study of this parameter was conducted by using both the experimental results (Chapter 14 of this book) and a phenomenological computer code for the simulation of the FEG operation described above.

Experimental and calculated parametric plots of the amplitude of current generated by longitudinally shock-compressed PZT 52/48

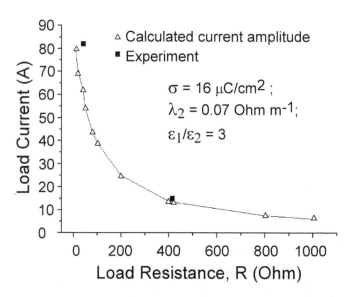

Fig. 15.4. Experimentally obtained (squares) and calculated (triangles) amplitudes of current generated by longitudinally shock-compressed PZT 52/48 elements (diameter 25 mm and thickness 2.5 mm) in the resistive load as a function of load resistance.

elements in the resistive loads are presented in Figure 15.4. These plots were obtained for PZT disk elements with a thickness of 2.5 mm and a diameter of 25 mm. As can be seen, the current in the active load decreases as the load resistance increases. The reason for this is the increase in the losses in the shock-compressed portion of the ferroelectric element due to bulk leaks stimulated by the shock wave.

Results of the calculation of the energy density of a longitudinally shock-compressed PZT element (thickness of 2.5 mm and diameter of 25 mm) are shown in Figure 15.5. The plot for the specific energy density indicates that this value passes through a maximum at a certain resistance, which indicates an optimal balance between the generated current and the leakage current. The similar behavior of the energy delivered to the resistive loads was observed in the experiments with longitudinally shocked PZT 52/48 elements of different sizes (see Figure 14.10 in Chapter 14 of this book).

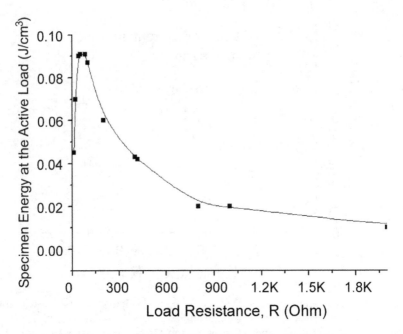

Fig. 15.5.   Calculated specific energy density of longitudinally shock-compressed PZT 52/48 elements (diameter 25 mm and thickness 2.5 mm) operating with resistive loads as a function of load resistance.

Both the theoretical and experimental results shown in Figures 15.4 and 15.5 suggest that the FEG can be effectively treated as a current source until the load resistance begins to approach the shunt resistance of the shock-compressed portion of the ferroelectric element.

The second important factor is the geometric size and shape of the ferroelectric elements. Plots of the calculated amplitude of the current and the energy density in the load as a function of PZT 52/48 element thickness for a fixed load resistance ($R = 10\,\Omega$) are presented in Figure 15.6.

As can be seen (Figure 15.6), short elements are optimal from the standpoint of generating higher currents and energy densities in the load. These results are physically sound since the surface charge density depends only on the bulk density of polarization dipoles and not the volume of the PZT element.

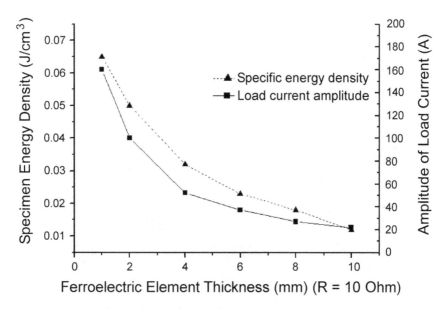

Fig. 15.6. The calculated current amplitude in an active load and the specific energy released to the load as a function of the PZT 52/48 cylindrical element thickness for load resistance $R = 10\Omega$ (PZT 52/48 element diameter is 25 mm).

The simulations were conducted for the transverse FEGs with PZT elements. Figure 15.7 shows the calculated waveforms of voltage generated by transversely shock-compressed ferroelectric elements having different permittivities of the compressed zone ferroelectric material across a 2 kΩ load. The calculation results indicate that the permittivity of the compressed zone of the ferroelectric element does not have a significant effect on the amplitude and waveform of the high-voltage pulse generated under transverse shock compression.

Figure 15.8 shows the calculated waveforms of voltage generated by transversely shock-compressed ferroelectric elements having different conductivity of compressed zone ferroelectric material across a 2 kΩ load.

In accordance with the results of calculations, the conductivity of the compressed zone has a significant effect on the amplitude and waveform of the high-voltage pulse produced by transversely shock-compressed ferroelectrics. An increase in conductivity from 0.03 to

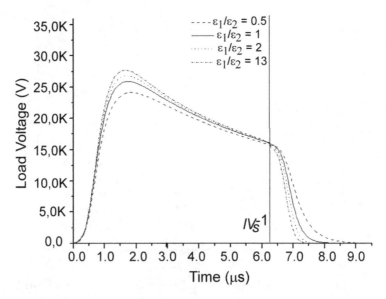

Fig. 15.7. Simulated waveforms of voltage generated by transverse FEG across a 2 kΩ load for different permittivities of shock-compressed ferroelectric material. The relative permittivity of uncompressed ferroelectric material was $\varepsilon_1 = 1300$. The conductance of compressed ferroelectric materials was $\lambda_2 = 0.03\Omega \cdot m^{-1}$. Solid vertical lines signify the moment that the shock wave exits the ferroelectric element.

$0.08\,\Omega \cdot m^{-1}$ results in a decrease of the voltage amplitude from 36 to 27 kV (Figure 15.8).

15.5 Summary

- The shock compression of ferroelectric elements connected to resistive loads has been investigated numerically. By comparing the experimental results obtained with those calculated, it has been established that the developed theoretical model provides reliable agreement with the measured voltage across the active load over the entire length of the pulse. The computer code can be used to calculate the pulse length, shape, amplitude, and energy produced by the FEG.
- In accordance with the calculation results, a decrease in the amplitude of current generated by longitudinally shock-compressed

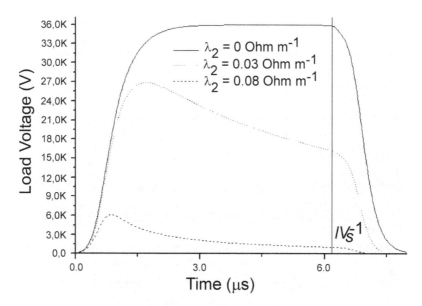

Fig. 15.8. Simulated output voltages of the transverse FEG, connected to a 2 kΩ load, with different conductivity of the compressed ferroelectric ceramic material. The relative permittivity of uncompressed ferroelectric material was $\varepsilon_1 = 1300$ and $\varepsilon_1/\varepsilon_2 = 2$. Solid vertical lines signify the moment that the shock wave exits the ferroelectric element.

ferroelectrics with an increase in the load resistance is caused by the losses in the shock-compressed portion of the ferroelectric element due to bulk leaks stimulated by the shock wave.

- The results of the calculation of the energy density of a longitudinally shock-compressed ferroelectric element show that the energy density passes through a maximum at a certain load resistance, which indicates an optimal balance between the generated current and the leakage current. The calculation results suggest that the FEG can be effectively treated as a current source until the load resistance begins to approach the shunt resistance of the shock-compressed portion of the ferroelectric element.

- In accordance with the calculation results, in the longitudinal shock depolarization mode, thin ferroelectric elements are optimal from the standpoint of generating higher currents and, correspondingly, higher energy densities in the load. These results are physically sound since the surface charge density depends only on the bulk density of polarization dipoles and not the volume of the PZT element.

- The calculation results obtained for ferroelectrics subjected to transverse shock compression indicate that the permittivity of the compressed zone of the ferroelectric element does not have a significant effect on the amplitude and waveform of the high-voltage pulse generated under transverse shock loading.

- In accordance with the calculation results, the conductivity of the compressed zone has a significant effect on the amplitude and waveform of the high-voltage pulse produced by transversely shock-compressed ferroelectrics.

- In order to extend the numerical treatment of the operation of explosive ferroelectric generators, the dependence of permittivity of ferroelectric materials on the electric field strength, microscopic domain kinetics under shock conditions and electrical breakdown within the elements must be taken into account.

Bibliography

1. Y. Tkach, S.I. Shkuratov, J. Dickens, M. Kristiansen, L.L. Altgilbers and P.T. Tracy, Explosive driven ferroelectric generators, *Proceedings of 13th International Pulsed Power Conference*, eds. R. Reinovsky and M.A. Newton, Las Vegas, Nevada, USA (2001) IEEE Catalog Number: 01CH37251, ISBN: 0-7803-7120-8, Vol. II, pp. 986–989.

2. Y. Tkach, S.I. Shkuratov, J. Dickens, M. Kristiansen, L.L. Altgilbers and P.T. Tracy, Parametric and experimental investigation of EDFEG, *Proceedings of 13th International Pulsed Power Conference*, eds. R. Reinovsky and M.A. Newton, Las Vegas, Nevada, USA (2001) IEEE Catalog Number: 01CH37251, ISBN: 0-7803-7120-8, Vol. II, pp. 990–993.

3. Y. Tkach, S.I. Shkuratov, E.F. Talantsev, J.C. Dickens, M. Kristiansen, L.L. Altgilbers and P.T. Tracy, Theoretical treatment of explosive-driven ferroelectric generators, *IEEE Trans. Plasma Sci.* **30**(5) (2002) pp. 1665–1673.

4. S.I. Shkuratov, M. Kristiansen, J. Dickens, A. Neuber, L.L. Altgilbers, P.T. Tracy and Y. Tkach, Experimental study of compact explosive driven shock wave ferroelectric generators, *Proceedings of 13th International Pulsed Power Conference*, eds. R. Reinovsky and M.A. Newton, Las Vegas, Nevada, USA (2001) IEEE Catalog Number: 01CH37251, ISBN: 0-7803-7120-8, Vol. II, pp. 959–962.

5. S.I. Shkuratov, E.F. Talantsev, J. Baird, H. Temkin, L.L. Altgilbers and A.H. Stults, Longitudinal shock wave depolarization of $Pb(Zr_{0.52}Ti_{0.48})O_3$ polycrystalline ferroelectrics and their utilization in explosive pulsed power, *AIP Conf. Proc. CP845, Shock Compression of Condensed Matter — 2005*, eds. M. D. Furnish, M. Elert, T. P. Russell, and C. T. White (American Institute of Physics, 2006), pp. 1169–1172.

6. S.I. Shkuratov, E.F. Talantsev, J. Baird, H. Temkin, Y. Tkach, L.L. Altgilbers, and A.H. Stults, The depolarization of a $Pb(Zr_{0.52}Ti_{0.48})O_3$ polycrystalline piezoelectric energy-carrying element of compact pulsed power generator by a longitudinal shock wave, *Proceedings of the 16th IEEE International Pulsed Power Conference*, eds. E. Schamiloglu and F. Peterkin, Albuquerque, NM (2007), pp. 529–532.

7. E.Z. Novitsky, V.D. Sadunov and G.Ia. Karpenko, Behavior of ferroelectrics in shock waves, *Physics of Combustion and Explosion* **14**(4) (1978) pp. 115–121

8. V.N. Mineev and A.G. Ivanov, EMF formed during shock compression of matter, *Sov. Physics Uspekhi* **119**(1) (1976) pp. 75–156

Chapter 16

Shock-Compressed Ferroelectrics Combined with Power-Conditioning Stage

16.1 Introduction

It has been shown in previous chapters that the dynamics of stress-induced charge released by ferroelectrics under longitudinal shock compression are an order of magnitude higher than those under transverse shock. The electric charge released by longitudinally shock-compressed ferroelectrics can be utilized for charging a capacitive load or powering a pulse-forming network.

The results of experimental investigations of longitudinal shock wave FEG combined with a capacitive-type power-conditioning stage are presented in this chapter. An explosive pulsed power technology has been successfully combined with a conventional pulsed power technology to produce an autonomous compact high-voltage nanosecond pulsed power system [1, 2]. An explosively driven longitudinal shock wave FEG was used for powering the spiral vector inversion generator (VIG). The FEG-VIG systems were capable of producing nanosecond pulses with amplitudes exceeding hundred of kilovolts.

In the course of the development of the FEG-VIG system, the operation of longitudinally shock-compressed ferroelectrics charging capacitive loads has been investigated experimentally and numerically [3–5]. The experimental results indicate that longitudinal shock wave FEGs are capable of generating pulsed powers exceeding 0.3 MW and hundreds of amperes of charging current and can charge

391

capacitor banks with capacitances ranging from 2 to 36 nF to a high voltage. In order to make the FEG-capacitor bank system operate in a predictable and transparent manner, a methodology for the numerical simulation of this system was developed [4, 5]. The results of these investigations are also presented in this chapter.

16.2   Vector Inversion Generator

The vector inversion generator was invented in the 1960s [6–8]. The VIG is a pulse generator, which, as a single unit, can store an electric charge at certain voltage and discharge it as a pulse having a peak value higher than the stored voltage. A spiral VIG is a variant of the well-known Blumlein pulser [9, 10], which is capable of doubling the voltage to which the energy storage lines are charged. The Blumlein pulser is at best a voltage doubler and is commonly used to drive loads typical of that associated with microwave producers. By contrast, a VIG is composed of two strip transmission-lines wound into a spiral that typically forms a hollow cylinder with output voltage determined by the number of turns in the VIG — "essentially a wound-up Blumlein."

Schematic diagrams of a spiral VIG are shown in Figure 16.1. The two conducting strips are insulated from one another with a solid-dielectric film, with an additional film of dielectric material included on the outside of the strip-line to provide turn-to-turn insulation. The net result is two parallel plate transmission lines, spirally wound, sharing a common conductor. The capacitance of the VIG can be calculated as follows:

$$C_{VIG} = \frac{2\varepsilon_r\varepsilon_0 A}{d} \qquad (16.1)$$

where C_{VIG} is the capacitance of the vector inversion generator (i.e. rolled foil capacitor), ε_r is the relative permittivity of the VIG insulating material, ε_0 is the permittivity of free space, A is the area of the transmission line, and d is the distance between the electrodes of the transmission line.

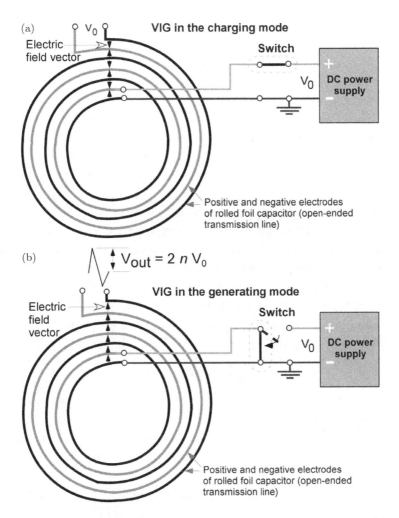

Fig. 16.1. Schematic diagrams of the vector inversion generator in (a) the charging mode and (b) the generating mode.

The solid dielectric insulation is wider than the conductor width to provide a margin that inhibits edge-to-edge breakdown at high transient fields. In the highest voltage versions, techniques such as edge grading, potting, oils, or high-pressure gasses such as sulfur hexafluoride can be used to provide further edge-to-edge insulation.

During operation, the strip-line is DC charged to voltage V_0, usually between $1\,\text{kV}$ and $50\,\text{kV}$, with electrostatic energy stored in the capacitance C_{VIG}. A plotting of the field vectors, turn-to-turn, shows that the array will either be electrically neutral or at the charge voltage, depending on the geometry.

When a pulse is desired, the input switch ("Switch" in Figure 16.1) is triggered. This trigger action starts an electromagnetic wave traveling up one of the strip lines, but to a first approximation, due to the single-turn inductance, not up the second line. When the wave has propagated up the active line to the open end and reflected back to the input switch, the electric field vectors in the active line have been reversed, aligning the potential vectors radially (Figure 16.1(b)). The net result is a transient high voltage between the lines at "Switch" and the outermost turn on the spiral. The voltage waveform is a ramp function with rise time equal to two electrical wave transit times, determined by the dielectric properties of the material in the lines, along the length of the spiral. Assuming no losses, the peak voltage across the generator is:

$$V_{out} = 2n\,V_0, \tag{16.2}$$

where V_{out} is the output voltage, V_0 is the charging voltage, and n is the number of turns in the roll.

A rise time of the generated pulse is equal to double the electrical length of the transmission line which appears at the contacts of the VIG.

The output capacitance of the VIG can be determined as follows:

$$C_{out} = C_{VIG}(2n)^{-2} \tag{16.3}$$

where C_{out} is the output capacitance of the VIG, C_{VIG} is the static capacitance of the VIG (Eq. (16.1)), and n is the number of turns in the roll.

For VIG total energy, W_{VIG}, the calculation is:

$$W_{VIG} = C_{VIG}V_0^2 \tag{16.4}$$

The advantages of this system (Figure 16.1) are its simplicity and the short nanosecond rise-time of the pulse it produces.

Losses and geometry could reduce the output voltage amplitude to a fraction of the ideal amplitude (Eq. (16.2)). Losses have at least three components. As the electromagnetic wave travels up the transmission lines, it is subject to attenuation due to the dielectric properties of the media in which it is traveling. A second loss factor is the inductance of the input switch. For efficient switch performance, its impedance should be much less than the characteristic impedance of the strip line. A third loss factor is due to the inductance of a single turn of the spiral since it determines the rate at which the second strip line can discharge. In general, it is desirable to increase the discharge time for the second line as much as possible. For some geometries, the efficiency of the system can be as high as 95% and are usually associated with large diameters or small total line length.

The output of the VIG can be connected directly to a load with the pulse waveform being a ramp function of voltage or an output switch can be utilized to connect the VIG to the load at an optimal point for maximum voltage. In that case, the waveform would be somewhat adjustable from a simple $C_{VIG}R_{load}$ waveform to a more "flat top" wave shape more typical of that needed by most microwave producers. In addition, passive elements such as transfer capacitors, non-linear elements, and pulse shaping lines can be added to produce almost any wave shape that a microwave producer would need. Voltages as high as one megavolt can be achieved with a pulse repetition rate determined by the basic energy store, the switch technology employed, and the desired life of the device.

Vector inversion generator technology effectively eliminated the need for multi-switching technology and discrete capacitive storage elements in devices designed to produce high-voltage and high-power pulses. The distributed nature of the energy store allowed for flexible design and proved to be volumetric and mass efficient — greater than 10 MW/kg for individual pulses. Devices with output voltages in excess of 1 million volts, for 20 kilovolts charge, were manufactured and tested with single pulse energy storage greater than 1 kilojoule [7, 8]. The VIG provides a precisely controllable method for producing high-voltage pulses with a predetermined rise time in a single dynamic step, requiring only one active component.

It was demonstrated [7, 8] that the VIG-based systems are capable of generating high-power microwaves of a wide range of frequencies, pulse shapes, and total energy per pulse.

16.3 Pulse Charging Capacitor Banks by Longitudinal Shock Wave FEGs

The operation of the VIG is critical to the parameters of a charging power supply. The longitudinal shock wave ferroelectric generators were taken into consideration as prime power sources for autonomous combined FEG-VIG systems because longitudinal FEGs provide high dynamics of stress-induced charge that can be utilized for charging a spiral VIG.

To determine an optimal design of ferroelectric generators for a combined FEG-VIG system, the operation of longitudinally shock-compressed ferroelectrics with capacitive loads has been investigated experimentally and numerically [3–5]. The results of these investigations are presented in this and the following sections.

Schematics of the longitudinal shock wave FEG and the measuring circuit used in the investigations are shown in Figure 16.2. The FEG was placed in the blast chamber. The output terminals of the FEG were connected to the capacitor bank placed in the oil bath

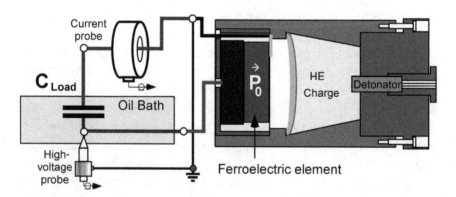

Fig. 16.2. Schematics of longitudinal shock wave FEG with a cylindrical ferroelectric element and the measuring circuit used in the FEG-capacitive load experiments. P_0 is the remanent polarization vector.

outside the blast chamber. The voltage across the capacitive load was monitored with a Tektronix P6015A high-voltage probe. The current in the load circuit was monitored with a Pearson Electronics current probe (model 411).

16.3.1 *FEG-capacitor bank system: Oscillatory mode*

The operation of longitudinal FEG-capacitive load systems was investigated with PZT 52/48 ferroelectric elements in a few sizes and in a wide range of load capacitances. The first series of experiments was conducted with PZT 52/48 ceramic disks having a thickness of 0.65 mm and a diameter of 26 mm (disk volume 0.35 cm^3). The capacitance of these PZT 52/48 elements was found to be 7.0 ± 0.1 nF from standard low electric field measurements. The load capacitance was chosen to be in the range from 9 to 36 nF.

Figure 16.3 shows a typical waveform of voltage generated by a PZT 52/48 element across the capacitor bank of 18 nF. It was not a single pulse but a series of oscillations. Figure 16.3 also shows voltage generated by a PZT 52/48 element with identical size in the open circuit mode. It was a single pulse with amplitude of 3.53 kV.

In the charging mode experiments, it was expected that the waveform of voltage generated by an FEG across the capacitor bank would be a single pulse, similar to that shown in the open circuit mode. However, the charging mode voltage waveform was not a single pulse, but a series of oscillations with a frequency of about 1.0 MHz. Three more experiments were performed with identical FEG-capacitor bank systems. The results of these experiments were very similar to those obtained in the first one (Figure 16.3).

The amplitude of the first wave of oscillations generated across the 18 nF capacitor bank was $V_{CLoad}(9.1\,\mu s) = 2.16$ kV, FWHM was 0.54 μs, and rise time was 0.34 μs. The energy delivered into an 18 nF capacitor bank was $W_{CLoad} = 42$ mJ. The amplitude of the first wave of oscillations averaged from the three experiments was 2.1 ± 0.2 kV.

Figure 16.4 shows waveforms of charging current, power and electric charge transferred from the longitudinally shock-compressed PZT 52/48 element into the 18 nF capacitor bank. The amplitude of

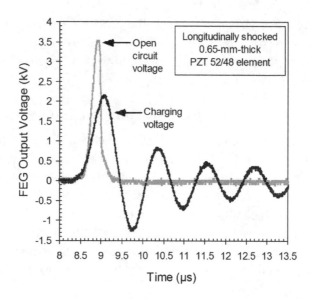

Fig. 16.3. Typical waveform of charging voltage generated by a longitudinally shock-compressed PZT 52/48 element (thickness 0.65 mm and diameter 26 mm) across the 18 nF capacitor bank and typical waveform of voltage generated by an identical PZT 52/48 element in the open circuit mode.

the first current pulse was 140 A with FWHM 0.3 μs. The amplitude of the second current pulse was higher than the first one, -180 A.

The experimental results indicate that the electric charge transferred into the capacitive load, 50 μC, was 32% of the initial charge stored in the ferroelectric element (161 μC). The power delivered in the load can be determined as: $P_{CLoad}(t) = I_{CLoad}(t) \cdot V_{CLoad}(t)$, where $I_{CLoad}(t)$ is the instantaneous value of the load current and $V_{CLoad}(t)$ is the instantaneous value of the FEG output voltage. The peak output power delivered in the capacitive load was $P_{CLoad}(8.9\,\mu\text{s}) = 0.24$ MW.

The next series of experiments was performed by using half as much load capacitance, i.e., 9 nF. The output voltage oscillated as it did in the experiments with the 18 nF capacitor bank (see Figure 16.3). The frequency of these oscillations was slightly higher; i.e., $\sim$1.2 MHz, than that obtained with the 18 nF capacitor bank. The average amplitude of the first wave of oscillations produced by

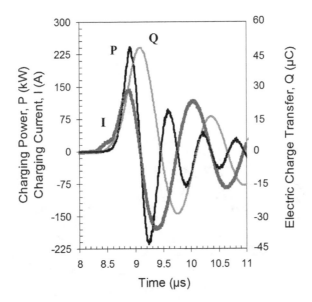

Fig. 16.4. Waveforms of charging current (I), charging power (P) and electric charge (Q) transferred from a longitudinally shock-compressed PZT 52/48 element (thickness 0.65 mm and diameter 26 mm) into the 18 nF capacitor bank (see corresponding voltage waveform in Figure 16.3).

the FEG across the 9 nF capacitor bank was 2.4 kV. The average peak energy delivered into the 9 nF capacitor bank reached a value of 26 mJ.

An immediate conclusion from these experiments; i.e., FEGs charging 9 and 18 nF capacitor banks, is that there is an increase in the energy transferred from the FEG to the capacitor bank when the capacitance of the bank is higher. This conclusion was confirmed in the third series of FEG-capacitor bank experiments, which are described below.

The third series of experiments was performed with a 36 nF capacitor bank. Figure 16.5 shows a typical voltage waveform produced by an FEG across this capacitor bank. Note that the results of these experiments were different from those obtained with the 18 nF and 9 nF capacitor banks.

The FEG produced a series of oscillations, but the amplitude of the first pulse was significantly higher than that of the next pulse

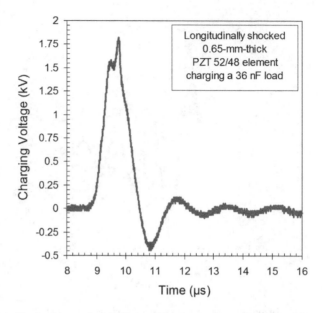

Fig. 16.5. Typical waveform of the charging voltage generated by a longitudinally shock-compressed PZT 52/48 element (thickness 0.65 mm and diameter 26 mm) across the 36 nF capacitor bank.

in the train and the oscillations damped out quickly. The amplitude of the first voltage pulse was $V_{CLoad}(9.8\,\mu s) = 1.82\,\text{kV}$, FWHM was $0.85\,\mu s$, and rise time was $0.93\,\mu s$. The energy delivered to the 36 nF capacitor bank was $W_{CLoad} = 60\,\text{mJ}$. The energy density of the PZT 52/48 element was $171\,\text{mJ/cm}^3$.

Figure 16.6 shows waveforms of the charging current, charging power and charge transferred from the FEG into the 36 nF capacitor bank. The total charge transferred into the load was $73\,\mu C$, which is 55% of the initial remanent polarization FEG charge. The peak charging current was $I_{CLoad}(9.34\,\mu s) = 197\,\text{A}$ and the charging power was $P_{CLoad}(9.42\,\mu s) = 0.29\,\text{MW}$.

Table 16.1 summarizes the results of the FEG-capacitor bank experiments for all three load capacitances considered. The experimental results indicate that it is fundamentally possible to pulse charge a capacitor bank with capacitance ranging from 9 to 36 nF to a high voltage by longitudinally shock-compressed PZT 52/48

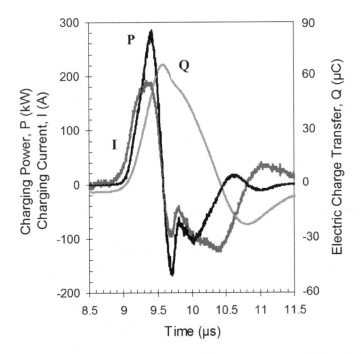

Fig. 16.6. Waveforms of charging current (I), charging power (P) and electric charge (Q) transferred from a longitudinally shock-compressed PZT 52/48 element (thickness 0.65 mm and diameter 26 mm) into the 36 nF capacitor bank (see corresponding voltage waveform in Figure 16.5).

Table 16.1. Experimental results obtained with longitudinally shock-compressed PZT 52/48 disk elements (thickness 0.65 mm and diameter 26 mm) operating in the charging mode.

Load (nF)	Load voltage (kV)	Charging current (A)	Charging power (MW)	Charge transfer (μC)	Operation mode
9	2.4 ± 0.3	111 ± 17	0.18 ± 0.01	31 ± 3	oscillatory
18	2.1 ± 0.2	149 ± 17	0.25 ± 0.01	52 ± 3	oscillatory
36	1.7 ± 0.1	204 ± 16	0.30 ± 0.01	74 ± 3	single pulse

ceramics. A miniature PZT 52/48 element with volume 0.35 cm^3 was capable of producing few hundreds of amperes of charging current and hundreds of kilowatts of charging power.

16.3.2 *Theoretical description of the FEG-capacitor bank system*

As mentioned above, the oscillatory behavior of FEG-capacitor bank systems was not expected and, at this time, the cause of these oscillations was not completely understood. In order to understand the physical nature of these oscillations, a theoretical model that describes the basic physical mechanisms that cause these oscillations of the FEG-capacitor bank system was developed in [4, 5]. The innovation in this approach was to use the voltage produced by the FEG in the open circuit mode as the basic parameter for determining the voltage and current delivered to the capacitive loads. It was demonstrated that the predicted FEG output current and voltage amplitudes and rise times are in good agreement with the experimental data obtained for a variety of loads and generator types. A schematic diagram illustrating the longitudinal shock wave depolarization of a ferroelectric element is shown in Figure 16.7.

The equivalent circuit diagram employed in the simulation of the longitudinal FEG-capacitor bank system is shown in Figure 16.8. The capacitance of the ferroelectric element is represented by C_{FEG}. The compressed zone of the ferroelectric element can have a higher

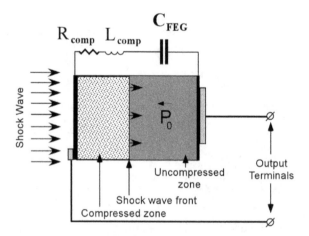

Fig. 16.7. Schematic diagram illustrating the longitudinal shock wave depolarization of a PZT element. P_0 is the remanent polarization vector.

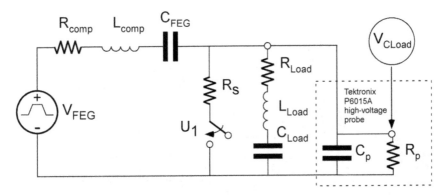

Fig. 16.8. The equivalent circuit diagram employed in the numerical simulation of a FEG-capacitor bank system.

electrical conductivity than the uncompressed zone. The inductance and resistance of the compressed zone are represented by L_{comp} and R_{comp}, respectively (Figure 16.8). The inductance, resistance, and capacitance of the load (capacitor bank) and the connecting cables are represented in the equivalent circuit diagram as L_{Load}, R_{Load}, and C_{Load}, respectively. The capacitance and resistance of the Tektronix P6015A high-voltage probe are represented in the circuit as $C_p = 3\,\text{pF}$ and $R_p = 100\,\text{M}\Omega$, respectively. Internal electrical breakdown in the PZT element is simulated by switch U_1, which has a resistance of R_S. It closes when the voltage across the capacitor bank reaches its maximum value.

The results of the simulation of an FEG-charged 18 nF capacitor bank are presented in Figure 16.9. The voltage across the bank oscillates, as it did in the experiment (Figure 16.3). The system parameter values used in the simulation were: $C_{FEG} = 7\,\text{nF}$, $L_{comp} = 5\,\mu\text{H}$, $R_{comp} = 0.2\,\Omega$, $C_{Load,} = 18\,\text{nF}$, $L_{Load} = 2\,\mu\text{H}$, $R_{Load} = 2\,\Omega$, and $R_S = 0.3\,\Omega$.

The results of the simulation for the 36 nF capacitor bank are shown in Figure 16.10. The parameters of the system; i.e., C_{FEG}, L_{comp}, R_{comp}, L_{comp}, and R_{Load} were equal to those used for the case with the 18 nF capacitor bank, except for C_{Load} being 36 nF and R_S being 4.3 Ω. The output voltage of the FEG was 20% lower than in the case of the 18 nF bank (Figure 16.9).

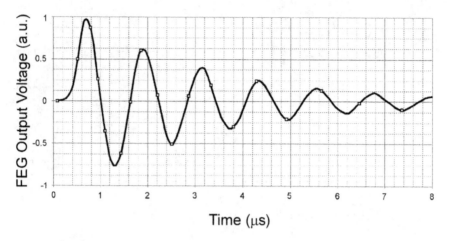

Fig. 16.9. Results of the numerical simulation of the operation of an FEG-capacitor bank system. The PZT 52/48 element had the thickness 0.65 mm and diameter 26 mm. The load capacitor was 18 nF.

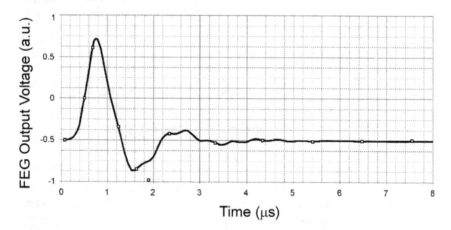

Fig. 16.10. Results of the numerical simulation of the operation of an FEG-capacitor bank system. The PZT 52/48 element had the thickness 0.65 mm and diameter 26 mm. The load capacitor was 36 nF.

These results are similar to those obtained in the experiments (Figures 16.2 and 16.5, Table 16.1). The waveform of the calculated charging voltage across the 36 nF capacitor bank (Figure 16.10) is practically a single pulse, as it was in the experiment (Figure 16.5).

Based on the results of these numerical simulations, it can be concluded that a key parameter responsible for the oscillatory mode of operation of the FEG-capacitor bank system is the resistance of the shock-compressed PZT element after electric breakdown that was represented in the equivalent circuit by the resistance of the switch, R_S. This means that electrical conductivity and the intensity of internal electrical breakdown in the shock-compressed portion of the PZT module have a significant effect on the processes that occur in the FEG-capacitor bank system.

In the experiments described above, increasing the capacitance of the capacitor bank led to a decrease in the voltage produced by the FEG across the capacitor bank and, correspondingly, a decrease in the voltage applied across the shock-compressed zone of the ceramic disk. Decreasing the voltage across the capacitor bank below a certain threshold level results in lower impedances in the conductive channels formed in the ceramics due to electrical breakdown, and it correspondingly causes the aperiodic behavior of the signals in the FEG-capacitor bank system.

16.3.3 *Electric charge and energy transfer from longitudinally shocked PZT 52/48 elements into capacitive loads*

The amplitude of voltage generated by longitudinally shock-compressed PZT 52/48 elements with thickness 0.65 mm in the charging mode (see previous sections) was not high enough to power the VIG. It is shown in previous chapters that the thickness of the ferroelectric element is a key parameter affecting the amplitude of the voltage generated under shock compression. The next series of the charging mode experiments were conducted with longitudinal FEGs containing PZT 52/48 elements with two thicknesses, 2.1 and 5.1 mm. It should be noted that PZT 52/48 elements of these types always generate single pulse charging voltages in a full range of load capacitances. The oscillatory mode had not been observed for load capacitances lying in the range from 2.25 to 36 nF.

Figure 16.11 shows the results of a typical experiment in which a 4.5 nF capacitor bank was pulse charged by a longitudinally

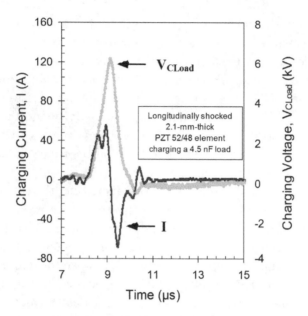

Fig. 16.11. Typical waveforms of the charging current (I) and the charging voltage (V_{CLoad}) generated by a longitudinally shock-compressed PZT 52/48 element (thickness 2.1 mm and diameter 27 mm) across the 4.5 nF capacitor bank.

shock-compressed 2.1-mm-thick PZT 52/48 element. The charging voltage amplitude was $V_{CLoad}(9.05\,\mu s) = 6.16\,\text{kV}$ with an FWHM of $0.7\,\mu s$ and a rise time of $0.87\,\mu s$. The slope of the $V_{CLoad}(t)$ curve at the moment that depolarization began was $7.1\,\text{kV}/\mu s$. It is almost two times less than the same quantity measured when the FEG was operated in the open circuit mode (see Chapter 8 of this book).

The rapid decrease in the right edge of the voltage pulse $V_{CLoad}(t)$ in Figure 16.11 is direct evidence of electrical breakdown in the shock compressed zone of the ferroelectric element. The electrical breakdown field, $E_{break} = 2.9\,\text{kV/mm}$, was lower than in the open circuit mode. The peak energy delivered to the capacitor bank was 85 mJ. Figure 16.11 also shows the waveforms of the charging current, $I(t)$. The charging current amplitude was $I_{CLoad}(8.97\,\mu s) = 55.5\,\text{A}$. The electric charge transferred from the PZT element into the capacitor bank was $32\,\mu C$, which is 22% of the initial charge.

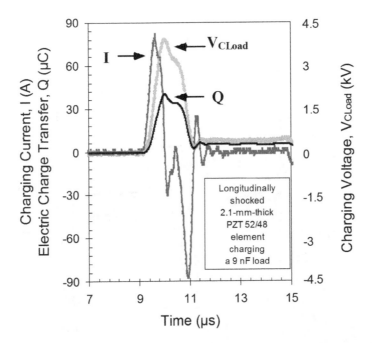

Fig. 16.12. Typical waveforms of the charging current (I), the charging voltage (V_{CLoad}) and the electric charge (Q) transferred from a longitudinally shock-compressed PZT 52/48 element (thickness 2.1 mm and diameter 27 mm) into the 9 nF capacitor bank.

Figure 16.12 shows the results of waveforms of charging current, charging voltage and electric charge transferred from a longitudinally shocked 2.1-mm-thick PZT 52/48 element into the 9 nF capacitor bank. The charging voltage amplitude was $V_{CLoad}(10.0\,\mu s) = 3.95\,\text{kV}$ and its rise time was $0.8\,\mu s$. The charging current amplitude was $I_{CLoad}(9.65\,\mu s) = 82\,\text{A}$. The electric charge transferred from the PZT element into the capacitor bank was $41\,\mu C$. The energy generated in the load was 72 mJ.

An increase of load capacitance from 4.5 to 9 nF resulted in a decrease in the load energy and, at the same time, in an increase in the electric charge transferred from a longitudinally shocked 2.1-mm-thick PZT 52/48 element into the load. Similar results were observed with 0.65-mm-thick PZT elements (see previous sections).

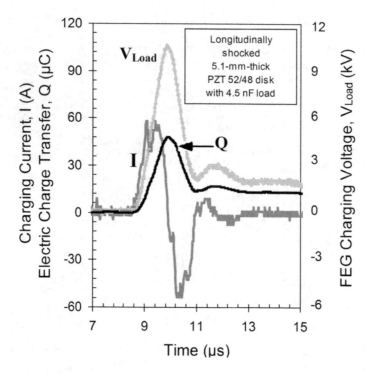

Fig. 16.13.   Typical waveforms of the charging current (I), the charging voltage (V_{CLoad}) and the electric charge (Q) transferred from a longitudinally shock-compressed PZT 52/48 element (thickness 5.1 mm and diameter 25 mm) into the 4.5 nF capacitor bank.

Figure 16.13 shows the results of waveforms of charging current, charging voltage and electric charge transferred from a longitudinally shock-compressed 5.1-mm-thick PZT 52/48 element into the 4.5 nF capacitor bank. The charging voltage amplitude was $V_{CLoad}(9.84\,\mu s) = 10.4\,kV$. The charging current amplitude was 58 A. The electric charge transferred from the PZT element into the capacitor bank was 46 μC, which is 30% of the initial charge stored in the ferroelectric element.

16.4   FEG-VIG System

It is shown in the sections above that miniature longitudinally shock-compressed PZT 52/48 ferroelectric elements with volumes of

0.35 to 2.5 cm^3 generate hundreds of amperes of charging current and hundreds of kilowatts of charging power and they are capable of charging a capacitor bank with capacitance ranging from 2.25 to 36 nF to high voltages during hundreds of nanoseconds. The energy generated by PZT 52/48 elements under longitudinal shock compression can be used for powering the VIG.

16.4.1 *Experimental*

The FEG-VIG experiments were performed at the experimental test stand to study explosive pulsed power and microwave sources [1, 2]. The main guideline for the design and development of this experimental test stand was to use commercial probes for monitoring the pulsed power signals. Out of several possibilities, the design was chosen that is shown in Figure 16.14. It contained the explosive ferroelectric generator and a diagnostic/test station.

The explosive FEG was placed inside the blast chamber near a stainless-steel side port. The diagnostic/test station, containing the VIG, probes, oscilloscopes, and other diagnostic and experimental equipment, was placed near the side port, but outside of the blast chamber. Some of the FEG output cables were connected to the diagnostic/test station through air-sealed connectors in the port. The other FEG output cables were connected to the diagnostic/test station directly.

In order to avoid mechanical strains being transmitted through the generator's output cables to the pulse measuring and recording systems during generator firing, the output cables were fixed in the port cover using specially developed cylindrical clamps. During the explosive operation of the generator, the cables were cut off at their FEG connections instead of at the measuring system connections. Since mechanical strains are not transferred to the diagnostic/test station through the cables, there was no mechanical effect from the explosive detonation on the results of the electrical measurements. Positioning the sensitive equipment outside the blast chamber in this manner protected the equipment from the explosive environment within the chamber, thereby preventing test related damage. More details on the experimental setup can be found elsewhere [1, 2].

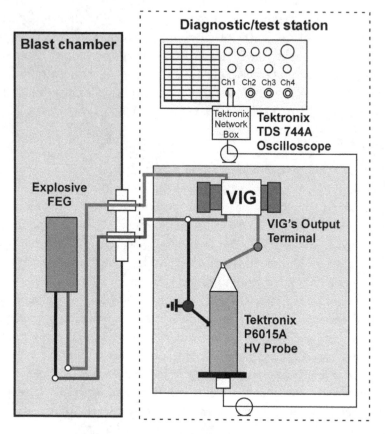

Fig. 16.14.   Schematic of the experimental setup used for investigations of FEG-VIG systems.

16.4.2   *Ultrahigh-voltage generation by FEG-VIG system*

A schematic diagram illustrating the operation of the FEG-VIG system is shown in Figure 16.15. The positive output terminal of the FEG was connected to the positive electrode of the transmission line. The negative electrode of the transmission line was connected to the negative output terminal of the FEG and grounded. The opposite end of the transmission line was open with the positive electrode connected to a Tektronix P6015A high-voltage probe (resistance of

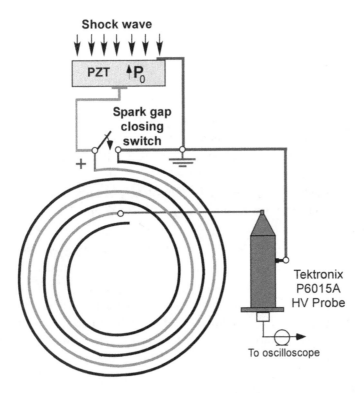

Fig. 16.15. Schematic illustrating the operation of the FEG-VIG system.

100 MΩ, rise-time of 4 ns). The pulsed signals were recorded with a Tektronix TDS744A oscilloscope (bandwidth of 500 MHz, 2 GS/s).

The operation of the FEG-VIG system was as follows. The explosive FEG produced a microsecond pulse that impulse charged the VIG (i.e. the rolled foil capacitor). When the charge voltage exceeds the VIG spark gap holdoff threshold, the VIG erects in a time equal to two wave transit times (a few nanoseconds) through the device, producing a transient voltage that is several times greater than the breakdown voltage of the VIG spark gap switch.

The VIG chosen for the first experiments was an eight-turn unit made with 0.1-mm-thick capacitor grade Teflon as the dielectric that had a width of 50.8 mm and 0.05 mm copper shims as the capacitor conducting plates. These VIGs were wound on ferrimagnetic mandrels (ferrite 2535) with a width of 25.4 mm. The VIG had a "rectangular cross section," which did not affect its efficiency. The voltage efficiency (measured by voltage multiplication) of the devices was in the 80–90% range. The calculated capacitance of these devices was 8.9 nF.

The first series of experiments was performed with FEGs containing PZT 52/48 ceramic elements of thickness 2.1 mm and diameter 27 mm. A typical waveform of voltage generated by a longitudinally shock-compressed PZT 52/48 element of this type in the open circuit mode is shown in Figure 8.2 in Chapter 8 of this book. The open circuit voltage amplitude was $V_{OC} = 6.88$ kV and its rise time was 0.8 μs.

A typical waveform of the voltage generated by a longitudinally shocked PZT 52/48 element of this type across the 9 nF capacitor bank is shown in Figure 16.12 above. The charging voltage amplitude was $V_{CLoad} = 3.95$ kV and its rise time was 0.8 μs.

The design and implementation of the VIG spark gap were mostly a matter of trial and error. A standard paper punch was used to make a repeatable hole in dielectric films, which could be stacked to lengthen the gap. In this way, the switch inductance was kept at a minimum and the breakdown voltage could be somewhat controlled. To get some idea of the impulse behavior of the gap, we developed a simple test fixture to allow an impulse to be applied to the switch ensemble. A preliminary characterization of the VIG spark gap was done in the laboratory in real time. For the first series of experiments the gap was tuned using a high-voltage DC power supply and was set to break down at approximately 3 kV.

A typical waveform of the voltage produced by an FEG-VIG system is shown in Figure 16.16. The voltage pulse amplitude was $V_{VIG} = 28.8$ kV, its FWHM was 20 ns, and its rise time was 6 ns. The actual VIG charge voltage at the gap trigger point can be calculated from the spiral efficiency and the output voltage. The unit triggered at about 2.6 kV. Even at voltages of only 28.8 kV, the effects of corona in the VIG were appreciable and reduced the efficiency of the system to approximately 75%. Correspondingly, the actual amplitude of the voltage pulse generated by this FEG-VIG system was about 36 kV.

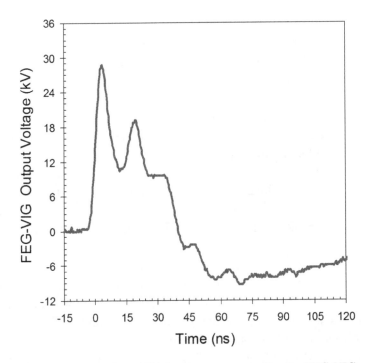

Fig. 16.16. Typical waveform of high voltage generated by the FEG-VIG system. The longitudinal shock wave FEG contained a PZT 52/48 element with thickness 2.1 mm and diameter 27 mm. The capacitance of the eight-turn VIG was 8.9 nF.

In the next series of experiments, a five-turn VIG was used. The device was prepared similarly to the eight-turn unit (see description above), but was impregnated with oil to eliminate corona effects and was capable of producing output voltages in excess of 100 kV. The calculated capacitance of the VIG was 5.6 nF.

In these experiments, longitudinal FEGs containing PZT 52/48 ceramic disks of thickness 5.1 mm and diameter 25 mm were used. A typical waveform of voltage produced by longitudinally shock-compressed PZT 52/48 elements of this type in the open circuit mode is shown in Figure 8.2 in Chapter 8 of this book. The open circuit voltage amplitude was $V_{OC} = 16.7$ kV and its rise time was $1.0\,\mu$s.

A typical waveform of voltage generated by a longitudinally shocked PZT 52/48 element of this type across the 4.5 nF capacitor bank is shown in Figure 16.13 above. The charging voltage amplitude was $V_{CLoad} = 10.4$ kV and its rise time was $0.9\,\mu s$.

The breakdown voltage of the VIG spark gap was once again tuned to trigger at a set voltage using a high-voltage DC power supply to ensure that the unit would discharge at the appropriate time. The spark gap spacing was set to DC discharge at approximately 6 kV, expecting that it would probably discharge at a higher voltage than that under impulse conditions.

The experimental setup for this series of experiments was similar to that in the previous series (see Figures 16.13 and 16.14). The only difference was that a custom voltage divider was used with a coefficient of 5.02 connected to the Tektronix P6015A high-voltage probe.

A typical waveform of voltage produced by an FEG-VIG system is shown in Figure 16.17. The voltage pulse amplitude was $V_{VIG} = 102.3$ kV, the FWHM was 6.5 ns, and the rise time was 5.2 ns. The rise time of the pulse approached the resolution limit of the Tektronix high-voltage probe. Probe resolution, as well as stray capacitive effects, may distort the rise time of the FEG-VIG system and introduce considerable rise time error. The actual rise time for this system was on the order of 3 ns. The unit triggered at approximately 11.9 kV, which was almost twice the level at which the spark gap switch was set to break under DC conditions.

The waveform of high voltage produced by the FEG-VIG system is not a single pulse but kind of a sine wave. This nanosecond pulse can be used for powering a subnanosecond pulse-forming stage or can be applied directly to the transmitting antenna to radiate high-power microwaves.

In conclusion, the succesful operation of an autonomous compact completely explosive high-voltage nanosecond pulsed power system using an explosive FEG as the primary power source with a VIG as a power-conditioning stage has been demonstrated. Adding a VIG stage increases the voltage output of the FEG by a multiplication factor depending on the VIG's parameters, while simultaneously compressing the pulse width into the range of a few nanoseconds. This combination produces an extremely high-power, single-shot pulser.

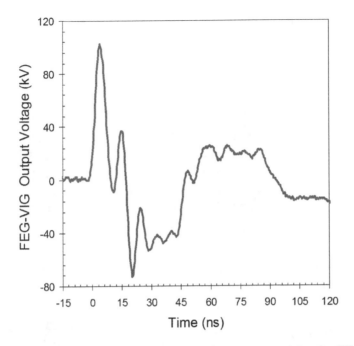

Fig. 16.17. Typical waveform of the high voltage generated by the FEG-VIG system. The longitudinal shock wave FEG contained a PZT 52/48 disk of thickness 5.1 mm and diameter 25 mm. The capacitance of the five-turn VIG was 4.9 nF.

16.4.3 *Theoretical description of FEG-VIG system*

A theoretical model that describes the operation of FEG-VIG systems has been developed. The model of the VIG was created by conceptualizing a VIG as Blumlein lines connected in series. A circuit diagram of this system is shown in Figure 16.18. R_1, L_1 and C_4 are the parasitic resistance, the inductance, and the capacitance of the output circuit, respectively. L_2, L_3 and L_4 are parasitic inductances of the switch.

The parameters of each transmission line (capacitance, inductance, and resistance) correspond to the parameters of a single turn of the VIG. Each Blumlein line is formed by the current turn and the previous (or the next) one. All three Blumlein lines are connected in series.

The results of the simulation are shown in Figure 16.19. The input voltage was 10 kV, the output voltage was 48 kV, and the shape of

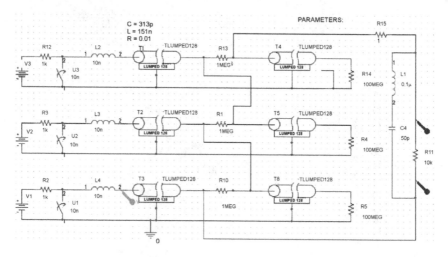

Fig. 16.18. Circuit diagram of the three-turn VIG represented as three Blumlein lines connected in series.

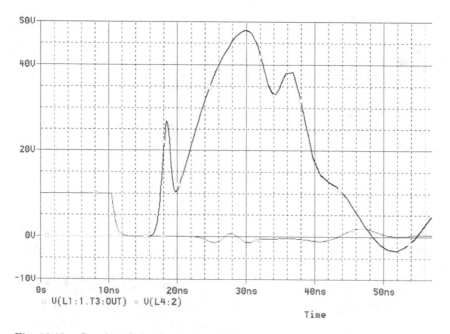

Fig. 16.19. Results of the simulation of the operation of the FEG-VIG system. The three-turn VIG was represented as three Blumlein lines connected in series (see circuit diagram in Figure 16.18).

the output pulse was the same as that of the pulse we obtained in the experiments. The multiplication coefficient is very close to the theoretical prediction, and this model is stable with different loads.

16.5 Summary

- Longitudinally shock-compressed ferroelectrics operating with a capacitive type power-conditioning stage have been investigated experimentally. An explosive pulsed power technology has been combined with a conventional pulsed power technology to produce an extremely high-power autonomous compact high-voltage nanosecond pulser.

- The explosive-driven longitudinal shock wave FEG was successfully used for powering a spiral vector inversion generator. The FEG-VIG systems were capable of producing nanosecond pulses with amplitudes exceeding hundred of kilovolts.

- Adding a VIG stage increased the voltage output of the FEG by a multiplication factor depending on the VIG's parameters, while simultaneously compressing the pulse width into the range of a few nanoseconds.

- A theoretical model of the FEG-VIG system is discussed. The VIG was represented by Blumlein lines connected in series. The results of numerical simulations obtained with this model are in good agreement with the experimental results.

- The operation of longitudinally shock-compressed PZT 52/48 ferroelectrics in the charging mode has been investigated experimentally and numerically. The experimental results indicate that longitudinally shocked PZT 52/48 ceramics can charge capacitor banks with capacitances ranging from 2 to 36 nF to high voltages during hundreds of nanoseconds. Miniature PZT 52/48 ferroelectric elements with volumes of 0.35 to 2.5 cm^3 are capable of generating hundreds of amperes of charging current and hundreds of kilowatts of charging power.

- In order to make the longitudinal FEG-capacitor bank system operate in a predictable and transparent manner, a methodology for numerical simulation of this system was developed. The innovation in this approach was to use the voltage produced by the FEG in the open circuit mode as the basic parameter for determining the voltage and current delivered by the FEG to the capacitive loads. The predicted FEG output current and voltage amplitudes and rise times are in good agreement with the experimental data obtained for a variety of loads and generator types.

- The experimental results indicate that longitudinally shock-compressed PZT 52/48 ferroelectrics connected to a capacitive load can operate either in the single-pulse mode or oscillatory mode with the frequency of the oscillations ranged from 1 to 1.5 MHz.

- The developed theoretical model of the longitudinal FEG-capacitive load system described the basic physical mechanisms that cause the oscillations of the charging voltage and current. Based on the results of these numerical simulations, it can be concluded that a key parameter responsible for the oscillatory mode is the resistance of the shock-compressed portion of the PZT element after electric breakdown. The electrical conductivity and the intensity of internal electrical breakdown in the PZT element have a significant effect on the processes that occur in the longitudinal FEG-capacitive load system.

Bibliography

1. S.I. Shkuratov, E.F. Talantsev, J. Baird, M.F. Rose, Z. Shotts, L.L. Altgilbers and A.H. Stults, Completely explosive ultracompact high-voltage nanosecond pulse-generating system, *Rev. Sci. Instrum.* **77**(4) (2006) p. 043904.

2. S.I. Shkuratov, J. Baird, E.F. Talantsev, M.F. Rose, Z. Shotts, L.L. Altgilbers, A.H. Stults and S.V. Kolossenok, Completely explosive ultracompact high-voltage pulse generating system, *Proceedings of the 16th IEEE International Pulsed Power Conference*, eds. E. Schamiloglu and F. Peterkin, Albuquerque, NM (2007), pp. 445–448.

3. S.I. Shkuratov, J. Baird, E.F. Talantsev, Y. Tkach, L.L. Altgilbers, A.H. Stults and S.V. Kolossenok, Pulsed charging of capacitor bank by compact explosive-driven high-voltage primary power source based on longitudinal shock wave depolarization of ferroelectric ceramics, *Proceedings of the 16th IEEE International Pulsed Power Conference*, eds. E. Schamiloglu and F. Peterkin, Albuquerque, NM (2007), pp. 537–540.

4. S.I. Shkuratov, E.F. Talantsev, J. Baird, A.V. Ponomarev, L.L. Altgilbers and A.H. Stults, Operation of the longitudinal shock wave ferroelectric generator charging a capacitor bank: Experiments and digital model, *Proceedings of the 16ᵗʰ IEEE International Pulsed Power Conference*, eds. E. Schamiloglu and F. Peterkin, Albuquerque, NM (2007), pp. 1146–1150.

5. S.I. Shkuratov, J. Baird, E.F. Talantsev, A.V. Ponomarev, L.L. Altgilbers and A.H. Stults, High-voltage charging of a capacitor bank, *IEEE Trans. Plasma Sci.* **36**(1) (2008) pp. 44–51.

6. R.A. Fitch and R.T.S. Howell, Pulsed Generator, U.S. Patent number 3,289,015, granted Nov. 29, 1966.

7. J.W. Rice, R.J. Gripshover, M.F. Rose and R.C. Van Wagoner, Spiral Line Oscillator, U.S. Patent number 4,217,468, granted Aug. 12, 1980.

8. S.A. Merryman, F. Rose and Z. Shotts, Characterization and application of vector inversion generators, *14th IEEE International Pulsed Power Conference, 2003*, Los Alamitos, CA, Digest of Technical Papers, Vol. 1 (2003) pp. 249–252.

9. G.A. Mesyats, *Pulsed Power* (Kluwer/Plenum, New York, 2005).

10. H. Bluhm, *Pulsed Power Systems: Principles and Applications* (Springer, New York, 2006).

Chapter 17

Case Studies

17.1 Introduction

It was demonstrated in the previous chapter that explosive FEG technology can be successfully combined with conventional pulsed power technology to produce an extremely high-power, single-shot pulser. The FEG combined with the vector inversion generator is capable of producing ultrahigh-voltage nanosecond sine waves that can be used to direct-drive the transmitting antenna in order to power a microwave source or a pulse-forming network. In this chapter, two more examples of how explosive pulsed power devices have been used will be briefly discussed. The first is a high-voltage generator based on longitudinal compression of ferroelectric materials developed by Diehl BGT Defence GmbH and Company in Germany. Diehl approached the development of this generator in a very systematic way [1–3]. The second is the results of a study in which FEG was used to direct-drive the transmitting antenna [4, 5].

17.2 Case Study 1: Ferroelectric High-Voltage Generator

Staines *et al.* [1–3] developed a prototype of an explosive- (propellant-) driven high-voltage generator using longitudinally compressed PZT-5A ferroelectric elements that had a diameter of 15 mm and a length of 20 mm, and a pyrotechnic gas generator that drives a piston. The diameter of the generator is 65 mm and its length is 275 mm, not including the electrical connector for the gas generator.

A small pyrotechnic gas generator, a modified DM82 cartridge, which has been produced in large quantities by Diehl for many years, produces at one end of the generator a pressure pulse that increases linearly over a 10 ms period to 200 bars in a small volume of gas in front of the piston. During this time, the piston transmits a force of up to 40 kN to the ferroelectric elements, thus longitudinally compressing them. The force is contained by a robust fiberglass housing.

The generator can contain up to six ferroelectric elements split into two groups or modules. The limitation on the voltage that can be generated by this generator is not the ferroelectric elements, but rather the voltage that can be sustained both inside the generator and across the exterior of its housing without electrical breakdown. The generator can be filled with either oil or SF_6 gas to inhibit breakdown, but outside the generator the breakdown voltage is limited by the atmosphere. When the electric field on the housing reaches 3 MV/m, a discharge can form, which can short-circuit the load. Since the charging current provided by the ferroelectric elements is small, on the order of 1 mA, even a highly resistive path will be sufficient to discharge the load. If the exterior of the source housing is wet or if the air humidity is greater than 60%, the problem will be even more serious. This is especially critical for small sources, where voltages of 100 kV and up can produce electric fields of more than 1 MV/m.

The simplest technique for pulse charging is to arrange two groups or modules containing multiple ferroelectric elements so that their polarizations are in opposite directions. A metal connecting rod in the center of the generator between the ferroelectric modules will be charged to two times the voltage on each ferroelectric element. A voltage of 200 kV or higher can be achievable in this manner.

In order to demonstrate that higher voltages could be obtained through the simple series cascading of ferroelectric elements, a test configuration for a generator with multiple ferroelectric elements, shown in Figure 17.1, was developed.

For the tests, it was necessary that the voltage be measured during ferroelectric element compression. Therefore, the generator

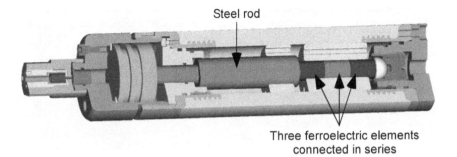

Fig. 17.1. Diehl high-voltage generator with three ferroelectric elements connected in series.

was placed into a bucket containing transformer oil to prevent breakdown during the measurements. This exterior insulation is only necessary for test purposes. The complete ferroelectric generator was explicitly designed to avoid the need for any exterior insulation.

The maximum voltage measured with the three-element module was 240 kV using a modified DM82 gas generator with 310 mg of K525 (propellant) powder (Figure 17.2). The voltage obtained with the three-element module was 2.4 times the voltage from a single element.

A final point for demonstration was that this generator with three elements could be used to charge a capacitive load. The capacitance of the module with three elements in series should be about 42 pF. To confirm this, a short length of high-voltage coaxial cable was connected across the output of the generator. The capacitance of the open-circuit cable was 80 pF. The maximum voltage with the capacitive load was about 80 kV.

With no load, the measured output voltage was about 200 kV. Ideally, the open-circuit voltage should be about three times the voltage that would be obtained with an 80 pF load. The measured result agrees reasonably well with this prediction, with the voltage across the capacitive load being slightly higher than the expected value of 66 kV. This discrepancy could be accounted for by variations in the pressure from shot to shot. This test showed that the ferroelectric

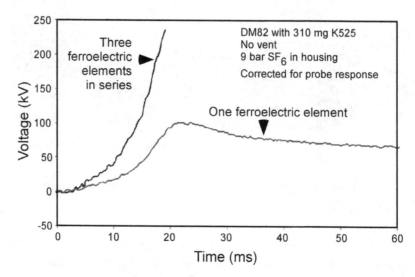

Fig. 17.2. Output voltage from a Diehl high-voltage generator with three ferroelectric elements connected in series.

generator was producing both high voltage and significant amounts of energy, with about 250 mJ stored in the coaxial cable.

The ferroelectric generator could then be used to charge a high-energy antenna, which, in turn, produces an RF pulse. Alternatively, there may be sufficient energy to drive a microwave source. Such a ferroelectric generator could also be used as a general purpose single-shot charging system for general pulsed power applications. The advantages of this type of system are its compact size, direct high-voltage output, and virtually unlimited maintenance-free storage time.

17.3   Case Study 2: FEG-Driven Antenna

The Naval Research Laboratory [3, 4] conducted a series of tests in which they used an FEG to drive a simple dipole antenna through a simple pulse-forming network. They conducted three test shots using the same dipole antenna and pulse-forming network and using FEGs that had identical or similar physical configurations. A similar receiving dipole antenna was placed approximately 3 m from the

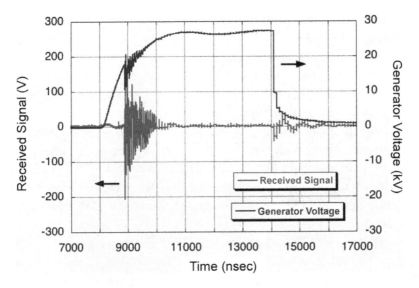

Fig. 17.3. Voltage generated by an FEG and radiated signal produced when this FEG was used to drive a dipole antenna. The signal was detected by a receiving antenna at 3 m.

transmitting antenna. The received waveforms were recorded along with the voltage pulse (Figure 17.3) delivered by the FEG to the pulse-forming network.

The FEG contained a $12.7 \times 12.7 \times 50.8 \, \text{mm}^3$ PZT 52/48 (EC-64) element. The FEG used 25 g of RDX high explosives. This generator produced peak voltages that were as high as 40 kV into an open circuit load (Figure 17.3). The pulse-forming network was constructed by using approximately 280 mm of RG-58 coaxial cable. The cable length may be used to control the frequency spectrum of the output signal. Since the cable was a 1/4 wavelength impedance matching circuit, the expected radiated frequency was 21.67 MHz.

A spark gap was placed between the inner conductor of the coaxial cable and the antenna. The antenna was V-shaped and each leg had a length of 1 m. The resonant frequency of the antenna was estimated to be about 75 MHz. The receiving antenna was similarly constructed.

Each shot produced what appeared to be two RF bursts that coincided with rapid changes in the source voltage (Figure 17.3).

The first RF burst produced a pulse with a peak-to-peak voltage amplitude of $\sim 30\,\mathrm{V}$ that lasted for $1\,\mu\mathrm{s}$. The second burst produced a pulse with an estimated peak-to-peak voltage amplitude of 50 V that also lasted for $1\,\mu\mathrm{s}$. The true peak voltage of the second burst could not be determined accurately, since the second burst saturated the input of the oscilloscope.

Using the peak voltages picked up by the receiving antenna, the peak power density at the receiving antenna was calculated to be $1.64\ \mathrm{W/cm^2}$ and the effective radiated power at the source antenna was calculated to be 2 MW, assuming that the receiving antenna had a gain near unity. The FEG generated about 2.4 MW of electrical power.

A Fast Fourier Transformation (FFT) (Figure 17.4) revealed that the spectrum of the RF bursts was concentrated between 18 and 26 MHz, with the largest signal being at 21.4 MHz, which is in good agreement with the predicted value of 21.67 MHz.

Utilizing two independent field diagnostic techniques, it was demonstrated in other experiments that an FEG-dipole antenna setup similar to the one described above exhibited similar behavior with a radiated frequency of around 100 MHz [3]. One field probe was an electric field sensor that transmitted the detected signal back to the diagnostics screen room over a fiber optic cable, and the other

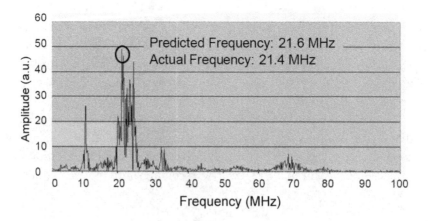

Fig. 17.4.   FFT of the signal generated by the FEG-driven dipole antenna.

sensor was a D-dot probe matched with a wideband balun. Both probes recorded a maximum peak-to-peak electric field of around 850 V/m at a distance of 5 m, giving a field-range product of about 4.2 kV.

17.4 Summary

- In this chapter, we have looked at two case studies. The first test case was a high-voltage generator based on the longitudinal compression of ferroelectric materials. This explosive-driven ferroelectric generator could then be used to direct-drive the transmitting antenna or to power a microwave source. Such a ferroelectric generator could also be used as a single-shot charging system for general pulsed power applications. The advantages of this type of system are its compact size, direct high-voltage output, and virtually unlimited maintenance-free storage time.

- The second case reported on was the results of FEG-driven dipole antenna tests. The configuration of the FEG-antenna system can be different from that described above. Using a sinuous antenna can change the radiated frequency and simultaneously increase the number of radiated pulses [4].

- There are of course other examples that could have been discussed. It is recommended that the reader refer to the proceedings of the various Megagauss and Pulsed Power Conferences that have been conducted since the 1960s.

- In this book, we have looked at different types of explosive ferroelectric generators, their capabilities and limitations, and how they can be used in practical systems. The operation of these generators is affected by the power conditioning circuit and the nature of the load to be driven. Therefore, the selection of an explosive FEG design, the size of the ferroelectric element, and the positioning of the element within the FEG will require a complete system analysis. That is, one needs to start with the load and work back through the power conditioning system to determine those parameters the generator should have, which will, in turn, determine which type of generator is the best one for this application.

- Further work in this area continues to improve the performance of explosive ferroelectric generators by leading to the development of new high-energy density ferroelectric materials, new generator designs and novel ways of explosively driving generators.

Bibliography

1. G. Staines, H. Hofman, J. Dommer, L.L. Altgilbers and Ya. Tkach, Compact piezo-based high voltage generator. Part I: Quasi-static measurements, *Journal of Electromagnetic Phenomenon* **3**(11) (2003) pp. 373–383.
2. G. Staines, H. Hofman, J. Dommer, L.L. Altgilbers and Ya. Tkach, Compact piezo-based high voltage generator. Part II: prototype generator, *Journal of Electromagnetic Phenomenon* **4**(6) (2004) pp. 477–489.
3. J. Bohl, J. Dommer, T. Ehlen, F. Sonnemann and G. Staines, High Voltage Generator, Especially for Using as a Noise Frequency Generator, U.S. Patent No. 6,969,944, 29 Nov. 2005.
4. M.S. Rader, C. Sullivan and T.D. Andreadis, Experimental observation of RF radiation generated by an explosively driven voltage generator, *Naval Research Laboratory Report* NRL/FR/5745-05-10,122 (2005).
5. L.L. Altgilbers, A.H. Stults, M. Kristiansen, A. Neuber, J. Dickens. A. Young, T. Holt, M. Elsayed, R. Curry, K. O'Connor, J. Baird, S.I. Shkuratov, B. Freeman, D. Hemmert, F. Rose, Z. Shotts, Z. Roberts, W. Hackenberger, E. Alberta, M. Rader and A. Dougherty, Recent advances in explosive pulsed power, *Journal of Directed Energy* **3** (2009) pp. 149–191.

Index

CPSIA information can be obtained
at www.ICGtesting.com
Printed in the USA
BVHW012145020819
554489BV00007B/1/P